CMOS RF and mm-Wave Transceivers and Synthesizers

CMOS RF and mm-Wave Transceivers and Synthesizers

Bharatha Kumar Thangarasu
Nagarajan Mahalingam
Kaixue Ma
Kiat Seng Yeo

JENNY STANFORD
PUBLISHING

Published by

Jenny Stanford Publishing Pte. Ltd.
Level 34, Centennial Tower
3 Temasek Avenue
Singapore 039190

Email: editorial@jennystanford.com
Web: www.jennystanford.com

British Library Cataloguing-in-Publication Data
A catalogue record for this book is available from the British Library.

CMOS RF and mm-Wave Transceivers and Synthesizers

ISBN 978-981-5129-80-9 (Hardcover)
ISBN 978-1-003-67356-9 (eBook)

Chips make computers powerful, but AI will reinvent computers.

—Kiat Seng Yeo

The top 5 countries leading in artificial intelligence (AI) have the letters "A" and "I" in their names. The Total score in the table below is based on Stanford's Report Global Vibrancy Ranking, assessing the AI outlook in countries across various essential criteria such as research and development, economic activity, and infrastructure, with data collected up to 2023.

Top 5 countries leading in AI

Rank	**Country**	**Total score**
1	US (United States)	70.06
2	China	40.17
3	UK (Great Britain + Northern Ireland)	27.21
4	India	25.54
5	UAE (United Arab Emirates)	22.72

Source: *The Economic Times* | News, Online, Feb 10, 2025.

ASEAN could be an AI regional hub because many of its member states, Brunei Darussalam, Cambodia, Indonesia, Malaysia, Singapore, Thailand and Vietnam, contain the letters "A" and "I" in their names. In fact, Singapore was ranked 10th with a Total Score of 18.15 in the Stanford's Report Global Vibrancy Ranking.

Everyone must know AI because it is not just a technology, but also a tool and a global currency that can be traded in the future. It will be everywhere on Earth from *A* to *Z* (Afghanistan, Algeria, Argentina, Austria, Australia, Azerbaijan, Brazil, Estonia, Finland, Haiti, Iran, Iraq, Ireland, Israel, Italy, Kuwait, Lithuania, Mali, Pakistan, Russia, Saudi Arabia, Tajikistan, Zambia, Zimbabwe, etc.).

—Kiat Seng Yeo

About the Authors

Bharatha Kumar Thangarasu holds a BE in electronics and communication engineering from Rashtreeya Vidyalaya College of Engineering, India (2002), MSc from the German Institute of Science and Technology (2010), and PhD in electrical engineering from Nanyang Technological University (NTU), Singapore (2015). From January 2010 to August 2015, he was with VIRTUS, IC Design Centre for Excellence, NTU, as a research associate, where he worked on SiGe HBT and CMOS-based reconfigurable amplifiers for microwave and millimeter-wave RF integrated circuit design. He was also a Research Fellow II at the Singapore University of Technology and Design (SUTD), Singapore. Since 2023, he is an associate professor at Tianjin University, China. Dr Thangarasu's research interests include RF and millimeter-wave reconfigurable integrated circuit design. He has authored or co-authored over 40 journals and conference papers. He was a recipient of the DAAD scholarship during his MSc study.

Nagarajan Mahalingam holds a BE in electronics and communication engineering from Bharathidasan University, India (2001), MS in electrical engineering from the University of Texas at Arlington, USA (2005), and PhD from NTU (2016). In 2006, he joined Advanced RFIC (S) Private Limited as an IC design engineer and worked on frequency synthesizers for portable wireless and data converter applications. He joined the Circuits & Systems Division, School of Electrical and Electronic Engineering, NTU, in 2008 as a research associate and focused on low-power designs for wireless and biomedical applications. He was an integral member of the 60 GHz team at NTU, which developed the VIRTUS chipset. He was also a Research Fellow II at SUTD. Since 2023, he is an associate professor at Tianjin

University. Dr Mahalingam's current research interests include radio and millimeter-wave integrated circuit design with a focus on oscillators and frequency synthesizers.

Kaixue Ma received his BE and ME degrees from Northwestern Polytechnical University, China, and PhD from NTU. From 1997 to 2002, he was with China Academy of Space Technology (Xi'an), where he was group leader of the millimeter-wave group for space-borne microwave and mm-Wave components and sub-systems for satellite payload and VSAT ground station. From 2005 to 2007, he was with MEDs Technologies as an R&D Manager, where he provided design services and product development. From 2007 to 2010, he was with ST Electronics as an R&D Manager. From 2010 to 2013, he was a senior research fellow and team leader of RFIC/millimeter-wave group researching on 60 GHz design at NTU. From 2013 to 2018, he was a full professor at the University of Electronic Science and Technology of China. Since 2018, he is the dean of the School of Microelectronics, Tianjin University, China, and director of the Tianjin Key Laboratory of Imaging and Sensing Microelectronics Technology, Tianjin. He has authored 2 books and more than 400 science citation index/engineering index (SCI/EI)-indexed articles in the fields of microwave, millimeter-wave, and THz integrated circuits and systems related to satellite communications, software radio, and mobile communications, including over 150 IEEE journal articles, 30 International Microwave Symposium) articles, and 3 International Solid State Circuits Conference articles. Dr Ma was a recipient of the Chinese National Science Fund for Distinguished Young Scholars in 2016. He received the Emerging Technology Awards from the Singapore Information and Communication Technology Association, 10 technique awards, including the Best Paper Award. He was the Chairperson of the Microwave and Circuit System Committee for the Young Scientists of the Chinese Institute of Electronics from 2016 to 2018, and the IEEE MTT-S R10 Coordinator for China and Singapore.

Kiat Seng YEO (M'00–SM'09–F'16) received his BEng (EE) in 1993, and PhD (EE) in 1996 both from Nanyang Technological University (NTU), Singapore. Currently, he is Advisor (Global Partnerships), President's Office, at Singapore University of Technology and Design (SUTD) and Distinguished Professor at Tianjin University, Tianjin, China. He was Chairman of the University Research Board, Chairman of Graduate Studies, Member of the Academic Council, Associate Provost for Research, Founding Associate Provost for Graduate Studies and Founding Associate Provost for International Relations at SUTD. He has 36 years of experience in industry, academia, and consultancy. Before joining SUTD, he was Full Professor at NTU and spent 13 years in management positions as Associate Chair (Research), Head of Circuits and Systems and Sub-Dean (Students Affairs). Professor Yeo was also a Fellow of the Renaissance Engineering Programme (REP) and served as Senator and Advisory Board Member at NTU. He has made many outstanding contributions to advance Singapore's education and research ambitions over the course of his career. As the Founding Director of VIRTUS, a S$52 million IC Design Centre of Excellence jointly set up by NTU and Singapore Economic Development Board (EDB), he contributed extensively to the economic development of integrated circuit design in Singapore by leading multidisciplinary research, with a focus on industry collaboration. Since 1996, he has been providing consultancy services to statutory boards, local SMEs, and multinational corporations in the areas of electronics, semiconductor, and IC design. He has secured over S$70M of research funding and published 14 books, 7 book chapters, 650+ top-tier refereed journal and conference papers and holds 55 patents of which 27 are US patents. He is also the principal author of *Integrated Circuit Design Research Ranking for Worldwide Universities 2008 and World University Research Rankings (WURR)* 2020.

Dr Yeo is a world-renowned authority on low-power RF/mm-wave IC design and a recognized expert in CMOS technology. He is one of the 7 Academicians elected to both the Singapore

Academy of Engineering (SAEng) and the Singapore National Academy of Science (SNAS). He is Fellow of the Canadian Academy of Engineering, ASEAN Academy of Engineering and Technology, International Artificial Intelligence of Industry Alliance (AIIA), Asia-Pacific Artificial Intelligence Association (AAIA) and IEEE for his contributions to low-power integrated circuit design. He is the World's AI Top Scientist by AIIA, as well as Top Scholar by ScholarGPS, and was recognized among the top 2% of scientists worldwide by Stanford University from 2020 to 2024. He also holds/held key positions in many international conferences as Advisor, General Chair, Co-Chair and Technical Chair. In 2009, he was conferred the Public Administration Medal (Bronze) on National Day by the President of the Republic of Singapore and the Nanyang Alumni Achievement Award by NTU for his outstanding contributions to the university and society. In 2020, he was conferred the Long Service Medal on National Day by the President of the Republic of Singapore.

Education is **A**chieving, not **B**elieving.
Research is **C**onceiving, not **D**eceiving.
Innovation is **E**nabling, not **F**earing.

—Kiat Seng Yeo

The Semiconductor Evolution: From 4 C's to 4 I's

The semiconductor industry market will grow from US$ 650 billion today to over US$ 1 trillion by 2030. Traditionally, the growth of the semiconductor industry is dominated by the **4 C's: Computers, Communications, Consumers, and Cars**. But its growth will now come from the **4 I's: Intelligence, Integration, Innovation, and Interdisciplinarity**. The combination of these I's and C's will result in more ICs. Therefore, in addition to new energy, drones, and smart cars, ICs will also be widely used in the human body, robots, future communications, biomedical, aerospace, and many other fields. In the foreseeable future, there will be more ICs in the human body than in any electronic products.

—Kiat Seng Yeo

Contents

Most technological inventions make our life better, but they do not necessarily make our brain better. For example, the invention of the calculator makes us forget about basic mathematics. The invention of computers does not automatically enhance our learning. The invention of artificial intelligence may stop us from thinking more creatively. In fact, smart devices are gradually stifling our imagination and creativity. Therefore, we must find and discover new ways of teaching and learning.

—Kiat Seng Yeo

Preface

With a wide range of new applications from 5G/6G communications, mobile computing, artificial intelligence, and advanced Internet of Things (IoT) to big data, cloud, and edge computing, power consumption has become a critical concern in chip design. Although the energy required for data and information processing can still be reduced to less than one-thousandth of what is required today without violating physical laws, big challenges (power, performance, efficiency, sensitivity, noise, linearity, selectivity, etc.) lie ahead for chip designers when they choose to develop integrated circuits (ICs) using silicon technology for low power and high data rate applications.

Silicon technology suffers from undesirable energy dissipation in terms of capacitive and magnetic losses due to its lossy substrate. High resistive loss of aluminum and copper metal lines at GHz frequencies also degrades the performance of on-chip inductors and transformers. Additionally, RF/mm-Wave circuits and systems fabricated using silicon are more sensitive to changes in the process parameters and/or technologies. Nonetheless, silicon remains the most suitable material satisfying the demands of a rapidly growing semiconductor market through low fabrication cost and ease of achieving system-on-chip or system-in-package integration with RF front-end and digital/analog baseband co-existing on the same chip or in a common package, respectively.

While long being neglected, low-power RF/mm-Wave design has vaulted to the forefront of attention in recent years due to the demand for ultra-low power (ULP) transceivers to achieve sustainability. Designing genuinely ubiquitous transceivers for these new applications requires innovations in both system architecture and circuit implementation. Recently, there has been a lot of research interest in efficient wireless communication. The conveyance of information is possible with the help of a transmitter and a receiver combinedly termed a transceiver. There is an immense demand for large amounts of data to be transferred instantly across multiple devices. To support this demand, it passive devices are used for frequency selectivity, impedance transformation, matching networks, resonant tanks, etc. It is therefore desirable to model and characterize them to optimize their quality factors to achieve smaller area and lower power consumption in RF circuits.

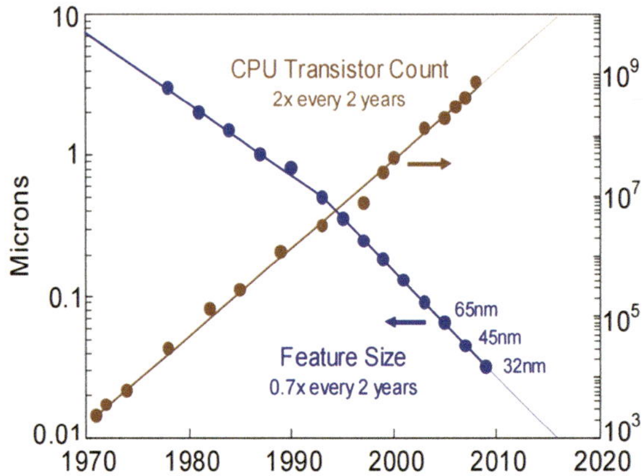

Figure 1.1 Moore's Law: Semiconductor technology scaling trends [1].

This chapter presents in-depth description of the model and characterization of active and passive components in RF and mmWave IC design. The optimized model and characterization of these passive devices ensures consistent performance and accuracy of the RFICs.

1.2 RF Transistor Models

1.2.1 MOS Transistors

CMOS transistor or MOSFET is one of the basic types of field effect transistor consisting of a minimum of four terminals namely the gate, drain, source and bulk as shown in Fig. 1.2. The operating principle of the MOSFET is based on modulating the resistance of the conduction channel by applying a signal voltage across the gate-source terminal [2]. For the usage of MOSFETs in the IC industry, simulation programs with integrated circuit emphasis (SPICE) models are developed to predict the circuit behavior and the Berkeley Short-Channel IGFET Model (BSIM) family of device model is the industry standard. The evolution of BSIM compact device model is shown in Fig. 1.3 [3]. The BSIM 1 to BSIM 5 are planar MOSFET models and only BSIM 4 is still under maintenance [4].

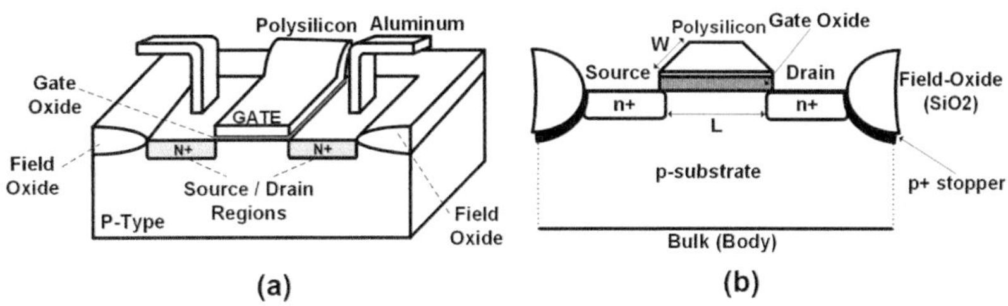

Figure 1.2 MOS transistor (a) cross section of NMOS, and (b) schematic view.

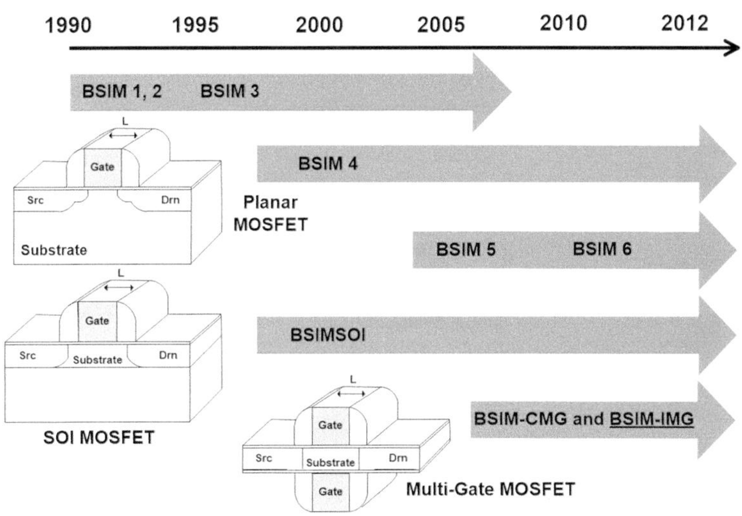

Figure 1.3 Evolution of BSIM compact model.

1.2.2 BSIM Models

Analog and RF circuits are sensitive to noise and mismatch. Therefore, accurate models are important for designers to capture the non-linearity of the real devices in circuit design phase. Without the real-device effects captured in the model, the inaccuracy can be large [5]. BSIM3 was a famous industry standard followed until the release of BSIM4 in 2000. BSIM3 is a DC scalable model with a strong correlation between the model parameters and actual devices [5], [6]. Though BSIM3 was accurate for low-frequency analog and digital circuits, but for high-frequency RF circuits, additional parasitic components were added to the core model [7, 8]. In BSIM3v3, the common approach is by adding sub-circuit components to model the gate resistance R_g and substrate resistance R_{sub} while in BSIM4, the R_g effect was included in the core model [9]. BSIM3 and BSIM4 models are threshold voltage-based models and other models based on surface potential and charge scaling are reported in [10–12]. The typical device model, which consists of a core model and real-device effects, is shown in Fig. 1.4 [13].

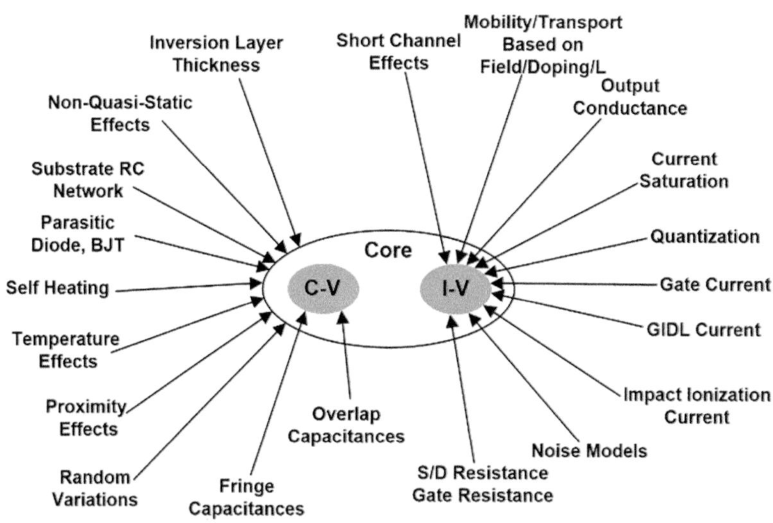

Figure 1.4 Compact models are composed of core and real-device models. The core models must be symmetric and accurate for an ideal long-channel device.

1.2.3 MOSFET Scaling

Over the years since its introduction in the IC Industry, MOSFET dimensions have been scaled continuously resulting in a doubling of transistor density and improved performance with lower supply voltages. As shown in the cross-sectional view of the MOSFET in Fig. 1.2, the critical parameters that have been scaled in the MOSFET are the gate length and the gate oxide thickness affecting both the lateral and vertical dimensions, respectively. The lateral dimensions are scaled by a factor of $1/\sqrt{2}$ each generation and the oxide thickness have been reduced from 100 nm to 1 nm. The scaling of lateral dimensions improves the chip density due to an increase in the transistor count. The reduction in oxide thickness ensures reduced hot electron effects as the thin gate oxide device has smaller mobility and threshold voltage degradation due to the shift of the damaged interface region towards the drain contacts [14]. Also, reducing the oxide thickness increases the gate leakage and increases the power consumption [15]. With advancement in MOSFET scaling below 100 nm technology nodes, digital circuits have followed the scaling while RF circuits are still designed in 28 nm and 40 nm technology nodes. The critical challenges in RF design in lower technology nodes are lower intrinsic gain of transistors, proximity of back end-of-line (BEOL) stack to the lossy substrate and stringent metal density rules that can cause considerable degradation of passive components quality factor [16].

1.2.4 MOS Parameters

The scaling of CMOS technology has resulted in improved performance of MOS transistors with increased speed and low power consumption. Typically, the models are extracted for lower frequencies but for higher frequencies parasitic resistances and capacitance play an important role and need to be modeled accurately.

For most RF circuits, there are different choices of RF transistor layout. The multi-finger layout with minimum gate length is mostly chosen due to higher speed and drain current for a given transistor width. In the multi-finger layout, the unit width (W_f) of the RF transistor is fixed, and the transistor finger number (N_f) is optimized. The layout and the equivalent small-signal circuit of the multi-finger RF layout with unit finger width are shown in Fig. 1.5. Following are the important figures of merit (FOM) for the RF transistor:

(a) Cut-off- frequency f_T and maximum oscillation frequency f_{max}.

The cut-off frequency of the MOS transistor is defined as the frequency where the current gain is equal to unity and is given by

$$f_T = \frac{g_m}{2\pi(C_{gs} + C_{gd})}, \tag{1.1}$$

where g_m is transconductance and C_{gs} and C_{gd} are the gate-source and gate-drain capacitance, respectively. The maximum oscillation frequency f_{max} is defined as the frequency where power gain is equal to unity and is given by

$$f_{max} = \frac{f_T}{2\sqrt{g_{ds}\cdot(R_g + R_s) + 2\pi f_T R_g C_{gd}}}, \tag{1.2a}$$

$$\approx \sqrt{\frac{f_T}{8\pi R_g C_{gd}}}, \tag{1.2b}$$

where g_{ds} is the output conductance and R_g and R_s are the gate and source resistance, respectively. Equations (1.1) and (1.2) indicate that f_T and f_{max} are strong functions of transconductance g_m and parasitic resistance and capacitance which need to be modeled carefully.

(b) Minimum noise figure NF_{min}

The noise performance of the front-end RF circuits namely the low noise amplifiers (LNA) and voltage-controlled oscillators (VCO) are critical for the overall transceiver noise performance. Hence, the choice of transistor with minimum noise figure in the design of these circuits is important. In high frequencies, thermal noise is the main source of noise and is dominated by the transistor channel noise. The NF_{min} is typically given by

$$F_{\min} = 1 + 2\frac{f}{f_T}\sqrt{\left(\gamma g_m \cdot g_m \cdot (R_g + R_s)\right)} \tag{1.3a}$$

$$R_n = \frac{\gamma g_m}{g_m} \tag{1.3b}$$

$$\mathrm{NF}_{\min} = 10\log F_{\min}\left[\mathrm{dB}\right] \tag{1.3c}$$

where R_g, R_s and R_n are the equivalent gate, source and noise resistances, respectively, and γg_m is the drain current excess noise factor. The theoretical value of γg_m is 2/3 in the saturation region for long-channel devices [17] and greater than 2/3 in short-channel devices due to channel length modulation [18].

(c) Intrinsic gain

The transistor intrinsic gain is defined as the ratio of transconductance (g_m) to output conductance (g_{ds}). In short-channel MOS devices, due to velocity saturation effects, the transconductance is approximated as in Eq. (1.4) and g_{ds} is approximated as shown in Eq. (1.5). From Eq. (1.6), it can be observed that, the intrinsic gain reduces with MOS scaling as the oxide thickness t_{ox} is not scaled in the same proportion with gate length L. The reduction of transistor intrinsic gain is one of the reasons high-frequency circuits are designed with higher technology nodes such as 28 nm or 40 nm or 65 nm in comparison to 5 nm or 7 nm or 14 nm for digital circuits.

$$g_m \propto W\frac{v_{\mathrm{sat}}}{t_{\mathrm{ox}}} \tag{1.4}$$

$$g_{\mathrm{ds}} \propto \frac{V_{\mathrm{ds}}}{L} \tag{1.5}$$

$$\frac{g_m}{g_{\mathrm{ds}}} \propto W\frac{v_{\mathrm{sat}}}{V_{\mathrm{ds}}}.\frac{L}{t_{\mathrm{ox}}} \tag{1.6}$$

1.2.5 MOS Model Extraction

Though CMOS downscaling has achieved a small form factor, the requirement of low power consumption in wireless and medical applications is one of the major challenges for RF designers. Therefore, it is important to understand the basic MOSFET model of RF circuit design. The RF models are typically extracted at high bias conditions and have captured the RF behavior accurately as presented in [19, 20]. Therefore, the extraction of RF models based on lower power consumption is dealt with in this section.

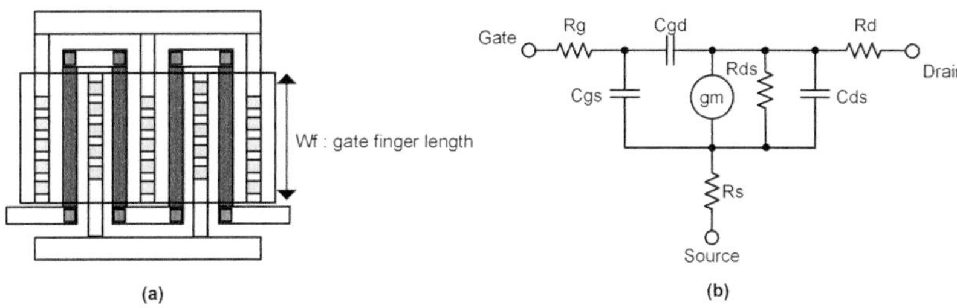

Figure 1.5 View of MOSFET (a) multi-finger layout, and (b) simplified small-signal equivalent circuit.

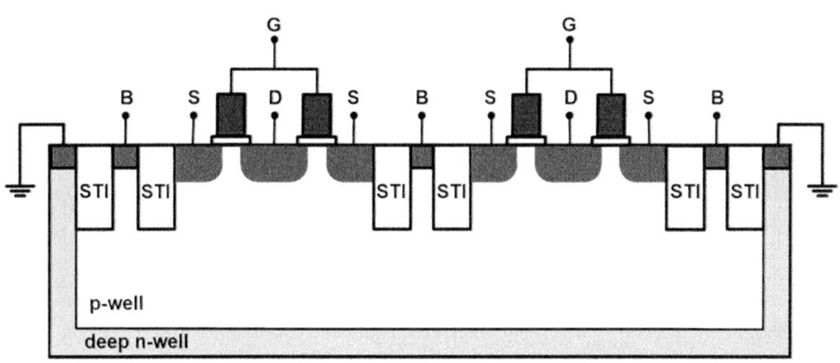

Figure 1.6 Cross section of NMOS device along length.

Typically, RF devices come with different geometry and the most common ones are the MOS transistors with multiple devices and fingers as shown in Fig. 1.6. The simple equivalent circuit is shown in Fig. 1.7. In the model extraction process, the bias dependency of the parameters cannot be ignored and different ways to extract the parameters are reported in [21–23]. Using

compact models like BSIM6 helps to mitigate the bias dependencies as most of the elements are already implemented in the model [24].

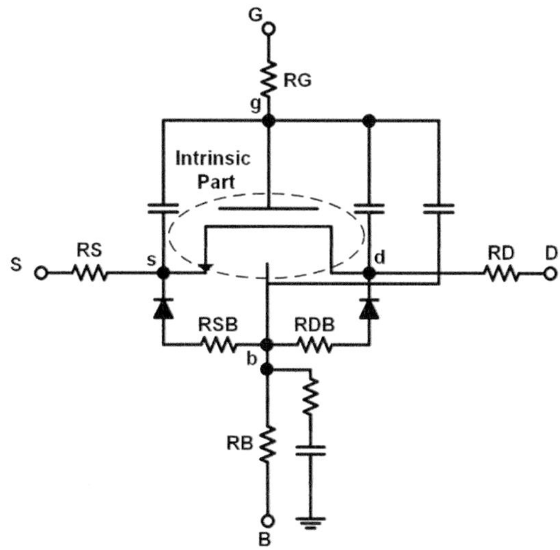

Figure 1.7 RF MOS equivalent circuit.

(a) DC parameter and FOM

The initial step in the extraction of any model is to extract the DC parameters. The fundamental drain current and transconductance for a 40 nm transistor for a low power consideration reported in [24] are shown in Fig. 1.8. In Fig. 1.8c, the transconductance efficiency is plotted vs. IC = I_{DS}/I_{spec}, a metric of inversion level of the transistor. Further, to the DC parameters, the extraction of parasitic capacitance for particular interest. In high frequencies, the capacitance dominates the RF behavior of the active transistor and a precise representation of capacitance in the weak and strong inversion region is important and shown in Fig. 1.8d [24]. In the next steps, the extractions of RF FOMs f_t and f_{max} and FOM_{RF} are of particular interest to the circuit designers and the results reported for the DUT are shown in Fig. 1.9 [24]. The FOM_{RF} is a normalized FOM defined as a ratio of current efficiency (G_m/I_{DS}) to the

RF gain (connected to f_t) [24, 25]. The FOM_{RF} can be used to find the best operating point for the circuit considering the trade-off between the current efficiency and f_t.

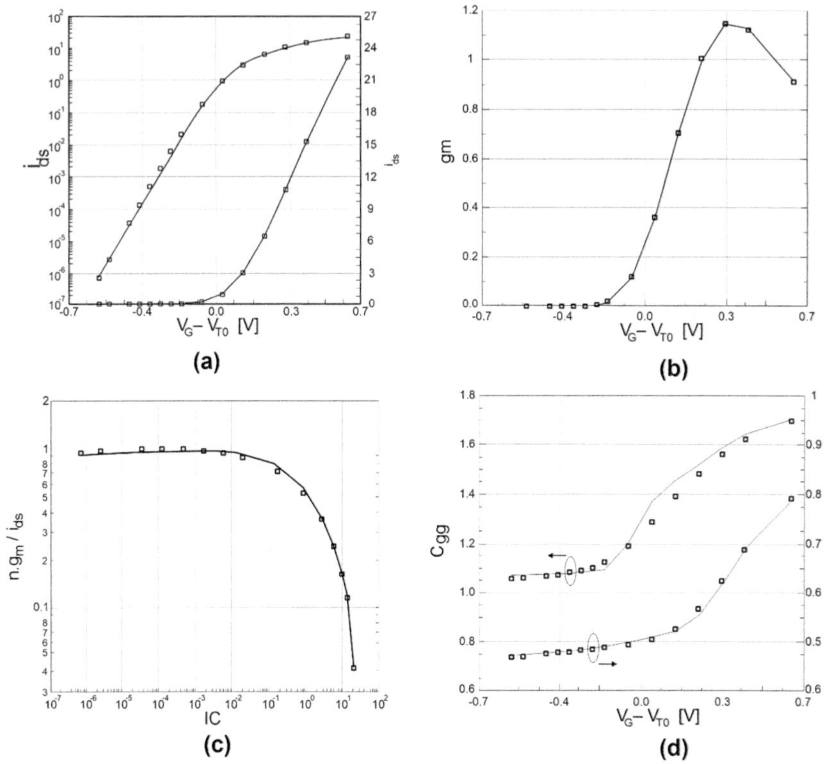

Figure 1.8 RF MOS parameters (a) I_{DS} vs. V_G-V_{T0}, (b) g_m vs. V_G-V_{T0}, (c) transconductance efficiency vs. IC, and (d) capacitance vs. V_G-V_{T0} [24].

(b) Noise models

The understanding and modeling of high-frequency noise in the MOSFETs is important in the design of RFIC. Though the downscaling of MOSFETs has improved their performance in terms of speed and power consumption with lower supply voltages, the noise performance has not scaled with the technology making high-frequency noise modeling essential [26, 27].

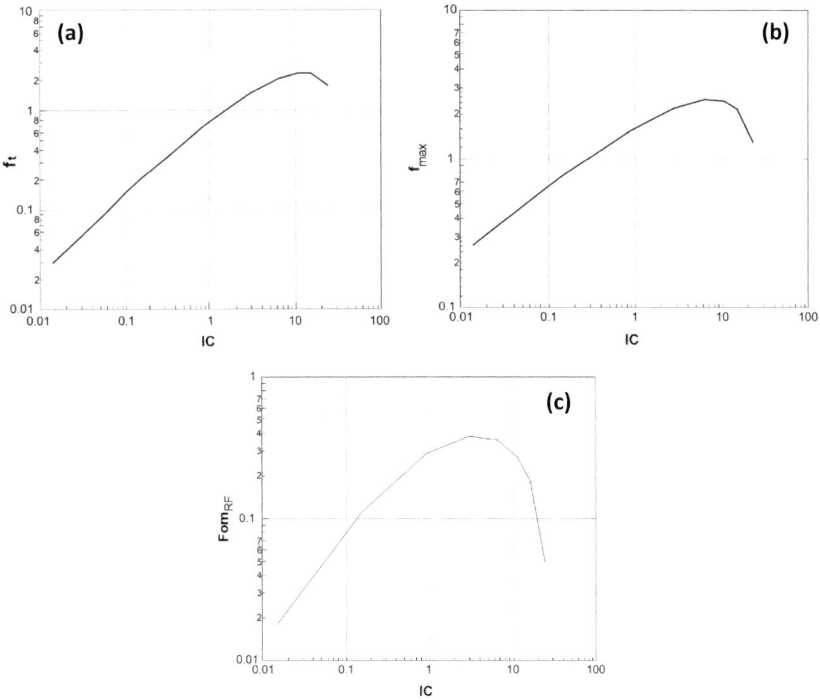

Figure 1.9 RF MOS parameters (a) f_t vs. *IC*, (b) $f_{\max}$ vs. *IC*, and (c) $\mathrm{FOM}_{\mathrm{RF}}$ vs. IC [24].

The high-frequency small-signal equivalent circuit of the MOSFET with accompanying noise sources is shown in Fig. 1.10 [28]. The most important noise sources in the MOSFET are thermal noise and channel noise. The thermal noise is due to the resistive elements including the gate resistance R_G, source and drain series resistance R_S and R_D, respectively and substrate resistances. The channel noise is a combination of drain current noise and induced gate noise due to the coupling of channel noise through the gate oxide at higher frequencies and is the largest contributor of MOSFET noise [29]. However, the high-frequency noise modeling of sub-nm MOSFETs is not straightforward as additional noise can be observed even after considering the non-linearities due to short-channel effects [30].

Based on the results from long-channel MOSFETs, the channel noise is identified as thermal noise generated by the resistive channel [17] with the general assumption being the channel is in thermal equilibrium and the generated noise sources are also thermal in nature. However, a MOSFET is a non-linear device, and the device current is not always linear with voltage. Therefore, to calculate the thermal noise, the channel needs to be divided into multiple small sections before integrating the results to calculate the overall noise. However, this is not practical in most cases and noise coefficient was introduced in [31]. When V_{DS} = 0 V, the MOS channel can be regarded as a linear resistor and the noise coefficient γ defined as the ratio of actual channel noise to the thermal noise of the channel resistance is given as [31]

$$\gamma = \frac{S_{\mathrm{id}}}{4kTg_{d0}} \tag{1.7}$$

where g_{d0} is the channel conductance when V_{DS} = 0 V. The noise coefficient γ = 1 in linear region and γ = 2/3 when the device is in saturation [17]. In short-channel sub-nm technology nodes, as channel length decreases, the noise coefficient γ increases and it is as around 2 for channel length 40 nm MOS transistor [30].

1.3 Advanced Topics in MOSFETs

With the advancement in technology and growing constraints for optimized system-on-chip (SoC) implementations, new technologies are being explored. In recent years, due to the aggressive scaling of silicon technology, the channel length has become narrower and non-linear effects such as short-channel effects, high leakage currents are observed. The transistor exhibits a high leakage current as the gate voltage cannot completely shut down the transistors [32–34]. Though thinner oxides and high-*k* dielectric materials have helped to alleviate the issue, the gate oxide thickness has reached a threshold to provide the necessary isolation between the transistor terminals [35–37]. Among the evolution of advanced CMOS technologies, fully depleted silicon-on-insulator (FDSOI) is replacing bulk CMOS to achieve higher performance with high integration capability. The FDSOI is a planar technology similar to bulk CMOS and makes it simpler to manufacture in comparison to 3D technology like Fin field effect transistor (FinFET). This section provides a brief overview of the active device in FDSOI technology.

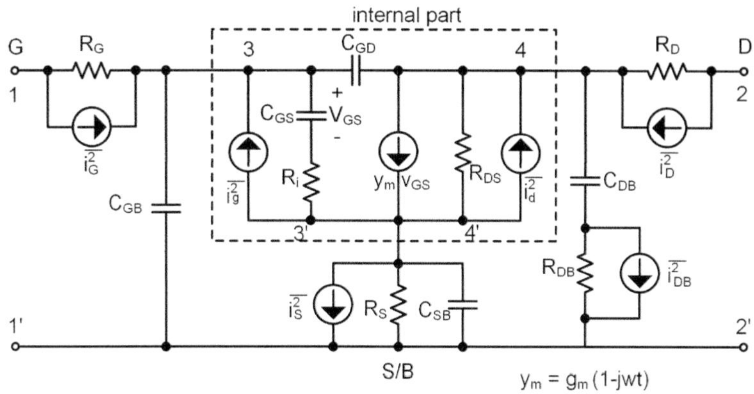

Figure 1.10 High-frequency RF noise model.

1.3.1 FDSOI MOS

The cross-sectional views of bulk CMOS and FDSOI are shown in Fig. 1.11 [38]. In comparison to the bulk CMOS technology, FDSOI consists of an ultra-thin silicon layer covering a very thin buried oxide (BOX). The BOX layer isolates the channel from the underlying substrate and thus reduces the leakage between the channel and the substrate by limiting the flow of electrons [39, 40]. In addition, the source and drain junction diodes are eliminated due to the presence of the BOX layer. In the FDSOI, lower threshold device (LVT) is processed with a flipped-well where NMOS is laid in N-Well and PMOS in P-Well as shown in Fig. 1.12.

The comparison of *I–V* characteristics of the bulk CMOS and FDSOI is presented in Fig. 1.13 [41]. Due to the isolated channel in FDSOI, a higher current efficiency is achieved. The other MOS performance parameters such as g_m/I_d, total gate capacitance, noise behavior, f_t and $f_{\max}$ are presented in Fig. 1.14. The superior performance of FDSOI especially in terms of f_T and $f_{\max}$ > 300 GHz makes it an ideal choice for the design of high-speed integrated circuits with lower power consumption and lower supply voltages.

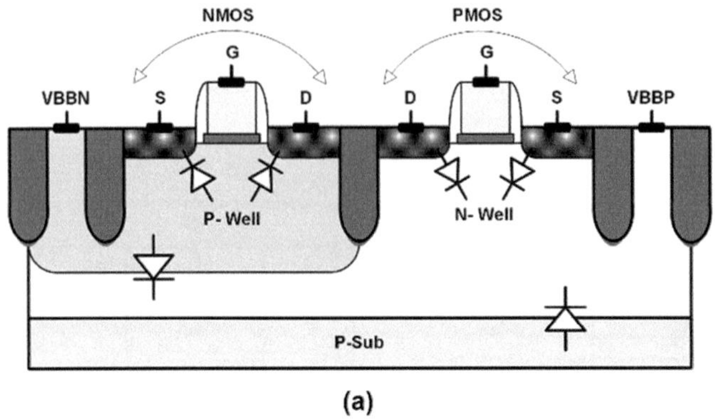

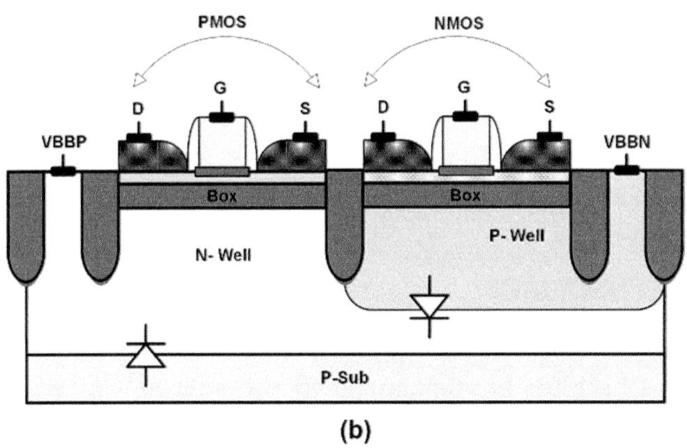

Figure 1.11 Cross section of regular CMOS transistor (a) bulk CMOS and (b) FDSOI CMOS.

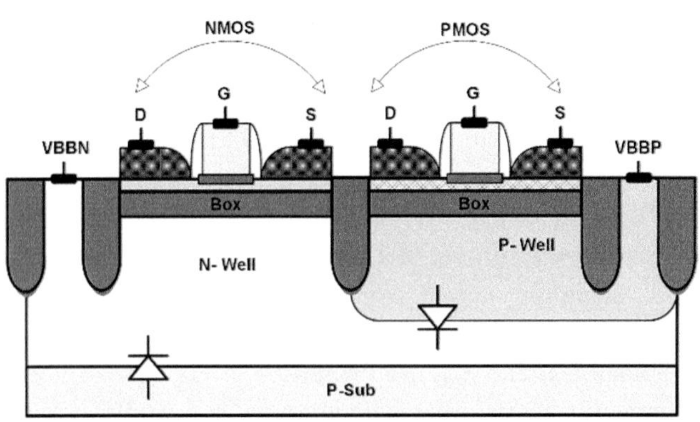

Figure 1.12 Cross section of FDSOI low-V_T CMOS transistor.

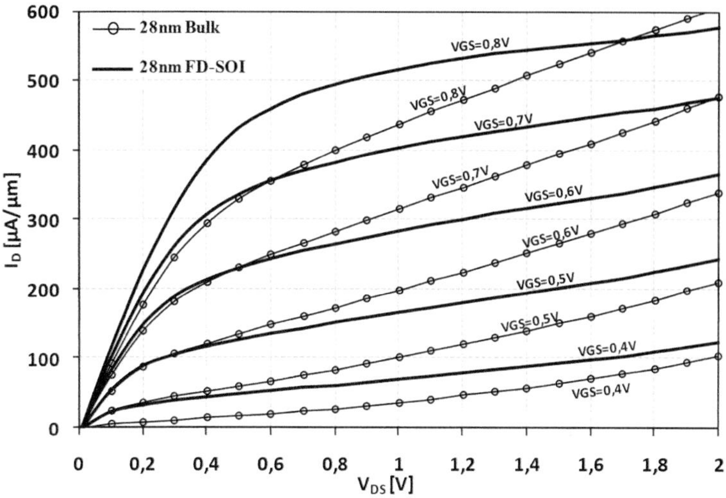

Figure 1.13 *I*–*V* characteristics of 28 nm Bulk CMOS and FDSOI CMOS [41].

One of the main advantages of the FDSOI in comparison to the bulk CMOS technology is the supply of an additional body-bias voltage. The variation of body-bias voltage to increase the V_{th} (increase V_{sb}) is called reverse body bias (RBB) which reduces leakage at the expense of transistor's performance. In the same manner, a decrease in V_{th} (reduce V_{sb}) is called forward body bias (FBB) and improves the transistor performance but the leakage is poor [42]. Bulk technologies offer very limited voltage range for RBB and FBB due the turn-on of the junction diodes between the drain/source and the substrate [43]. Also, variation of body bias is highly ineffective for lower technology nodes in bulk CMOS [44].

The body effect describes the variation of threshold voltage V_{th} due to the change in the bulk-source voltage V_{sb} when the bulk is not tied to the source and is given by [45]

$$V_{th} = V_{tho} + \gamma\sqrt{|2\varnothing_F + V_{SB}|} - \sqrt{|2\varnothing_F|} \tag{1.8}$$

where V_{tho} is the threshold voltage with zero source-to-body voltage, γ is the body effect coefficient given as γ = sqrt($2q\varepsilon_{si}N_{sub}$)/$C_{ox}$, ε_{si} is the dielectric constant of silicon, q is the electron charge, N_{sub} is the doping concentration of the substrate, C_{ox} is the oxide capacitance per unit area and $\Phi_F = (k_T/q)ln\ (N_{sub}/n_i)$ is the bulk potential, k is the Boltzmann constant, T is the temperature, n_i is the intrinsic carrier concentration and V_{sb} is the source body potential.

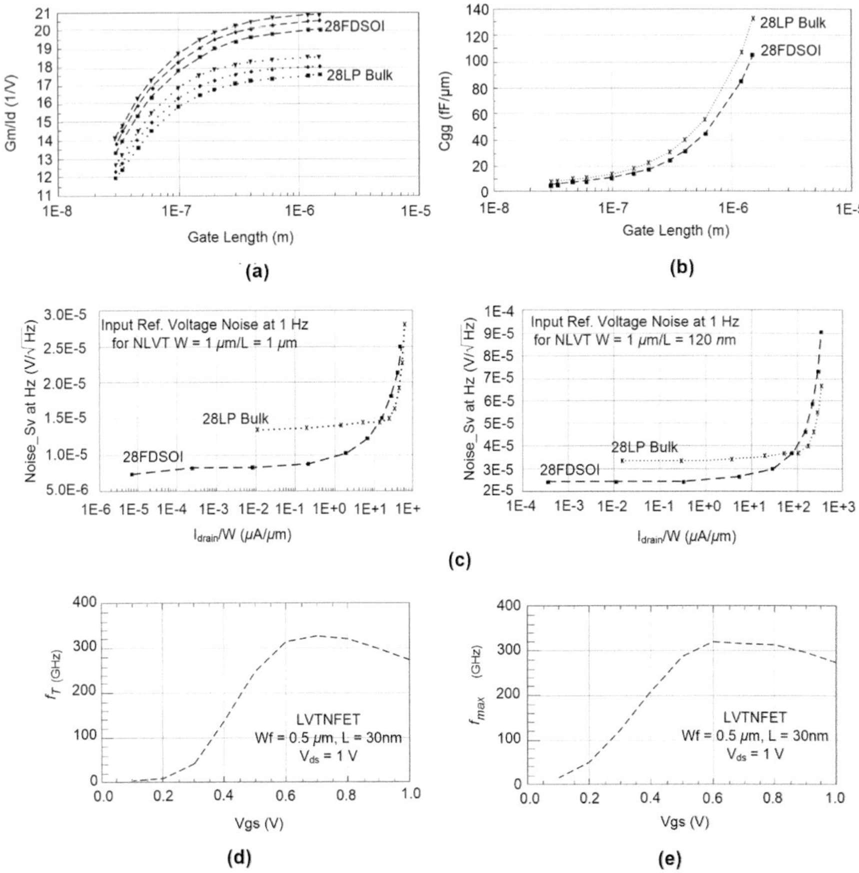

Figure 1.14 FDSOI MOS characteristics (a) current efficiency (g_m/I_d), (b) total gate capacitance, (c) noise behavior, (d) f_T, and (e) f_{max} [38].

In the FDSOI, since the BOX layer isolates the body terminal, V_T modulation through body biasing means dynamically controlling the gate and region below the oxide to control the channel behavior independently to optimize the transistor performance for low leakage or high-performance requirements. Due to the presence of

BOX layer isolation, the only limiting parasitic diodes are between the respective well region and the P-substrate. These diodes typically have a higher voltage of around 3 V as shown in Fig. 1.15 [38].

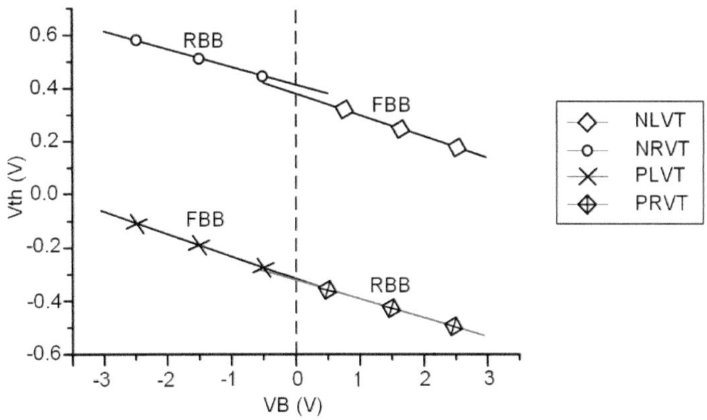

Figure 1.15 Threshold voltage variation vs. body-bias voltage in 28 nm FDSOI.

1.4 On-Chip Inductors and Transformers

Inductors and transformers are one of the basic passive elements in the integrated circuit design. Inductors and transformers store energy in the form of magnetic flux and is widely used for frequency selectivity, matching and impedance transformation. In silicon technologies, inductors and transformers can be realized on-chip due to smaller wavelength at RF and mmWave frequencies. Furthermore, incorporating lumped components on-chip reduces the external connections realizing a small package size.

1.4.1 Inductor Models

In silicon technologies, inductors and transformers are constructed based on the thick top metal layers in the same plane adjacent to each other or overlaid as stack metal layers. The inductors can be implemented in circular, square, rectangle and octagonal shapes. The realization, performance and optimizations of different inductor and transformer configurations are presented in [46–51]. The most popular implementations for the inductor and transformer are square and polygon [48]. The polygon shape can be circular, hexagon or octagon. Since the semiconductor foundries typically limit the angles in multiples of 45° for ease of fabrication, octagon is most popular polygon shape. Though the best performance is achieved with a circular topology [48, 52], while the square coils have larger inductance value for the same area due to the greater total length the square device presents. Irrespective of the geometry, the basic physical phenomena affecting the inductors and transformers are similar. The simple cross-sectional view of the inductor and its associated magnetic and electric field in silicon are shown in Fig. 1.16.

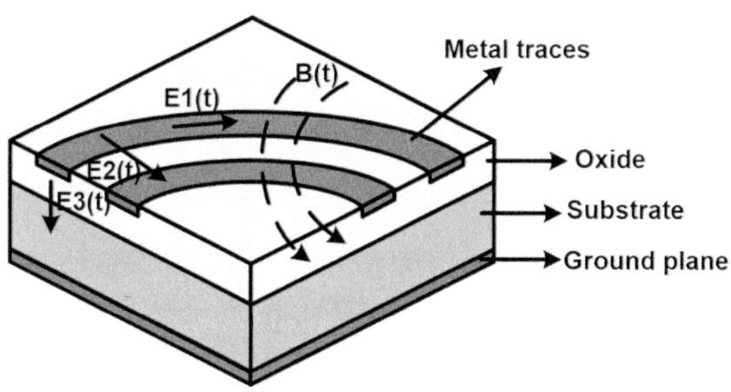

Figure 1.16 Cross section of inductor with magnetic and electrical field.

The operation of an inductor is based on the electromagnetic (EM) fields present in the monolithic inductor when the inductor metal layer is excited. There exists a time-varying magnetic field B(t), due to the current flow in the metal layers when they are excited resulting in inductance. However, due to the losses in the metal traces resulting from the non-ideal physical properties, there exist other EM fields that decrease the performance of the inductors.

In RFICs, inductors are used extensively, and it is important to model the inductor to predict the circuit performance during simulations. For lower frequencies, a simple RLC model or a lumped element pi-model consistent with the physical effects is sufficient to describe the inductance and its parasitic effects. The simple model of the integrated inductor is shown in Fig. 1.17.

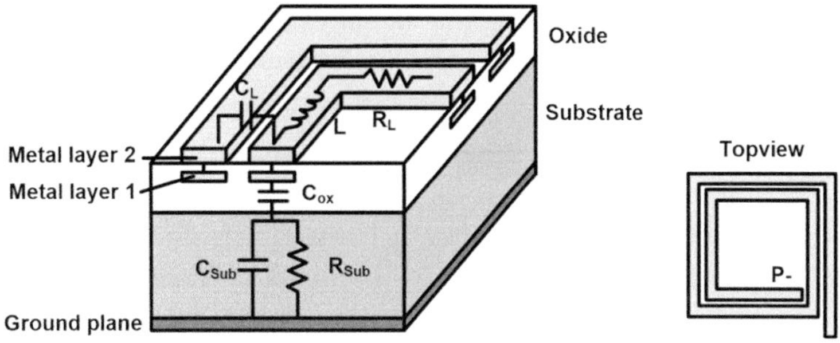

Figure 1.17 3D view inductor with associated lumped elements.

The model is described based on the lumped RLC components consisting of

- L: the inductance due to magnetic flux in the coil;
- R: the resistive losses in of the metal trace due to skin effect, current crowding and finite conductivity;
- C_L: The parasitic winding to winding capacitance due to the closely based metal traces;
- C_{ox}: The parasitic oxide coupling capacitance;
- C_{sub}: The parasitic capacitance coupling to the substrate; and
- R_{sub}: The resistive loss in the conductive substrate.

1.4.1.1 Loss mechanism in integrated inductors

One of the major limitations of the integrated inductors is the loss associated with the inductor coils which can classified as resistive loss due to metal layer, substrate and eddy current losses. The losses affect the quality of the inductor often leading to the reduction in the quality factor and its self-resonant frequency. Several works have focused on the improvement of inductor quality factor in [53–59]. The loss mechanism of integrated inductor is presented in detail in the following:

(a) Resistive loss due to metal trace

The resistive loss in the inductors is due to the finite conductivity of the metal traces. This loss is typically represented by a resistor in series with the inductor coil. In low frequencies, resistive loss is given as

$$R = \frac{\rho L}{WT}, \tag{1.9}$$

where R is the resistance, ρ is the resistivity of the metal trace, and L, W and T are the dimensions (length, width and thickness) of the metal trace. In high frequencies, the losses due to eddy currents also contribute to the resistive losses [60, 61].

(b) Eddy current losses

In the integrated inductors, there exists a time-varying magnetic field due to the time-varying current in the metal trace that generates eddy currents. According to Faraday–Lenz law, these eddy currents generate image eddy current in the highly conductive silicon substrate underneath the inductor. Due to the low resistivity of the silicon substrate, this image current can flow easily opposite to the current direction of the inductor. This image eddy current generates magnetic fields that oppose the applied inductor magnetic fields. Therefore, the effective magnetic field and the overall inductance value is reduced [60].

At higher frequencies, the current density at the center of the inductor decreases due to the opposing magnetic field and most of the current flows at the surface of the inductor resulting in a non-uniform current distribution [62]. This phenomenon is known as the skin effect and is directly proportional to the operating frequency. The measure of this current penetration in the metal trace is known as skin depth and is given by [63]

$$\delta = \frac{1}{\sqrt{\pi f \mu_o}} \tag{1.10}$$

where f is the operation frequency and μ_o is the permeability of the metal trace and σ is the conductivity. From Eq. (1.10), the skin depth δ decreases with increased frequency and this suggests there is a lesser low resistive path within the metal trace for current to flow easily [54] reducing the effective cross-sectional area of the metal trace and increasing the series resistive losses. The series resistance due to skin depth in terms of inductor dimensions is given as [60]

$$R = \frac{\rho L}{W\delta\left(1 - e^{-\frac{T}{\delta}}\right)}. \tag{1.11}$$

Therefore, to reduce the series resistance and improve the quality factor, the metal thickness of the inductor is increased.

(c) Substrate losses

Integrated inductors are built on the silicon substrate and the substrate losses arise due to the interaction of electric and magnetic fields of the inductor metal traces with the substrate. The substrate loss is illustrated in Fig. 1.18.

The capacitive substrate loss is due to the interaction between the electric field of the metal trace and the substrate. This electrical field generates conduction and displacement currents in the substrate while only the displacement current is observed in the oxide layer. In high frequencies, the displacement current in the substrate dominates the conduction currents [64] and this effect is modeled as a parallel combination of resistance and capacitance of the substrate *Rs* and $C_{S,}$ respectively in series with the oxide capacitance C_{ox} [65, 66].

In addition to the vertical electric fields, there exists lateral electric fields between the adjacent coils resulting in displacement currents. The lateral electric fields result in the winding-winding capacitance C_L and determine the self-resonant frequency (SRF) of the inductor. The SRF is defined as the frequency at which the inductance resonates with capacitance C_L. In integrated inductors, the SRF is typically chosen to be at least 2.5 to 3 times the operating frequency. To increase the SRF, the spacing between the coils is increased.

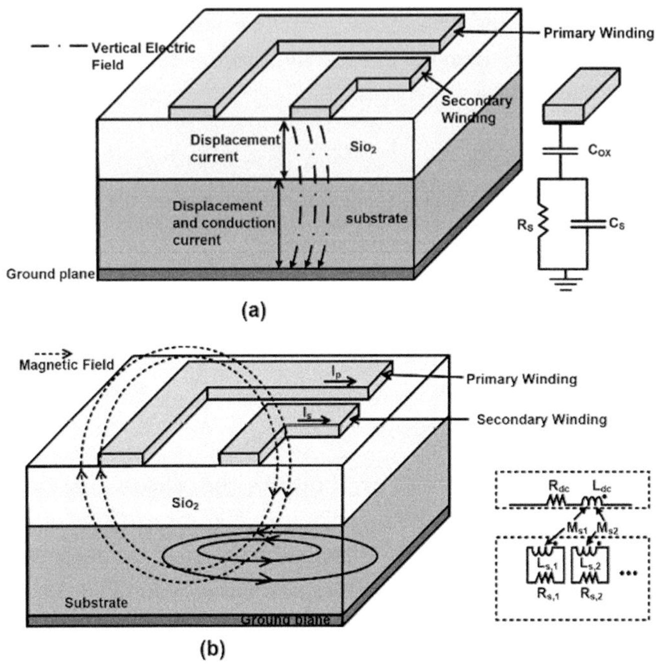

Figure 1.18 Illustration of substrate losses (a) electric field, and (b) magnetic field.

1.4.2 On-Chip Transformer Models

In RFIC design, transformers are indispensable as inductors. RF transformers are mainly used for impedance transformations and unlike inductors have separate metal winding called the primary and secondary. Though the number of secondary windings can be more than 1 [67], the following discussions are based on one primary and one secondary winding and are applicable to multiple secondary windings.

A simple transformer with two coupled inductors is shown in Fig. 1.19. In the transformer, the current through the primary winding generates a time-varying magnetic flux in the secondary winding inducing the voltage across its terminals. Also, the windings can be at different potentials as the two windings are isolated and there is no direct current flow from the primary to the secondary [68]. The terminal voltages and currents in the transformer are given as

$$\begin{bmatrix} V_1 \\ V_2 \end{bmatrix} = \begin{bmatrix} R_p + j\omega L_p & j\omega M \\ j\omega M & R_s + j\omega L_s \end{bmatrix} \begin{bmatrix} I_1 \\ I_2 \end{bmatrix}, \quad (1.12)$$

where L_p and L_S are the self-inductance of the primary and secondary, R_p and R_s are the resistive losses of the transformer primary and secondary coils and M is the mutual inductance. The magnetic coupling coefficient is given by

$$k = \frac{M}{\sqrt{L_p L_S}}. \quad (1.13)$$

In integrated transformers, the value of k is between 0 and 1 and depends on the transformer configuration as well as the physical dimensions such as width, metal spacing and other parameters. There are different variants in the transformers and the most popular topologies for monolithic implementation are the interleaved and stacked transformers.

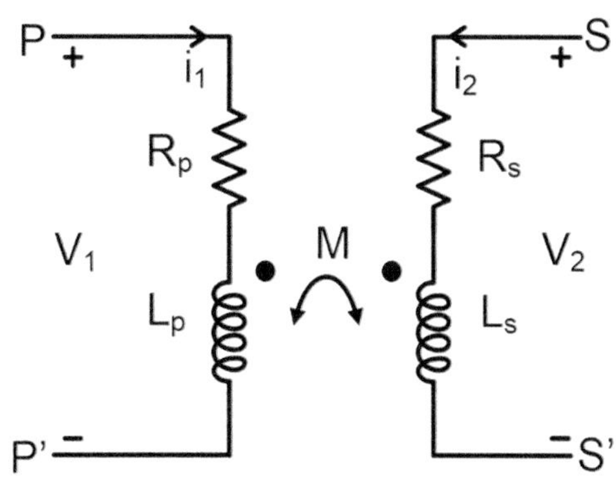

Figure 1.19 An ideal transformer.

The popular integrated transformer structures are shown in Fig. 1.20 [69]. The interleaved transformers can either be laid as truly interleaved or implemented in the same way as multi-turn inductors in the same metal layers laid adjacent to each other as shown in Figs. 1.20(a) and (b), respectively. Therefore, interleaved

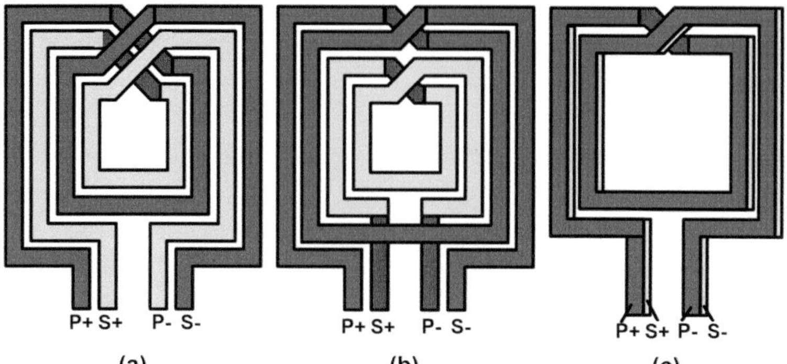

Figure 1.20 Differential transformer structures (a) interleaved, (b) tapered, and (c) stacked.

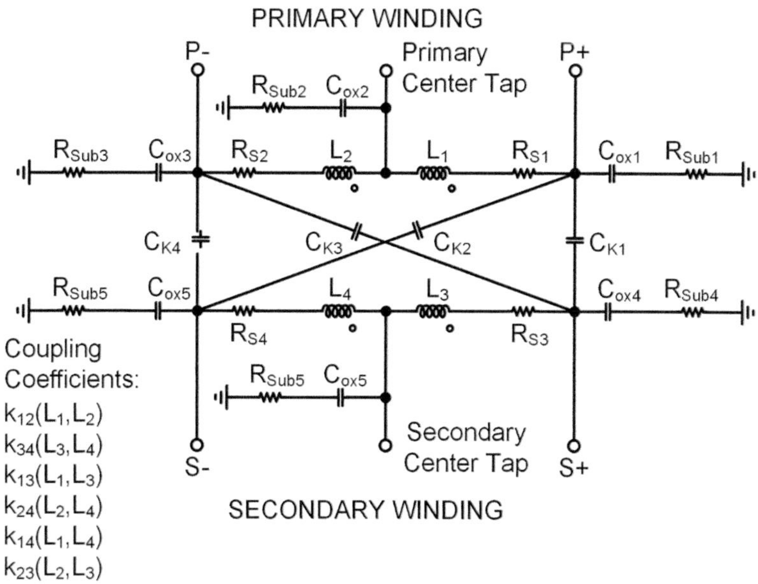

Figure 1.21 Lumped model of differential transformer.

transformers occupy more chip area in comparison to stacked transformers. Also, the mutual inductance is low in interleaved transformers as the edge coupling between the primary and secondary winding is reduced. The stack is shown in Fig. 1.20(c). The primary and the secondary windings are stacked are laid on top of each other. Due to the vertical nature of the coils, the vertical coupling of the magnetic field is higher than the edge coupling [70]. Also, due to the less metal spacing between the metal layers, the mutual inductance is higher in stacked transformers. However, the downside is the higher interwinding capacitance reducing the SRF of the transformer. Also, stacked transformers are asymmetric in nature since the top winding is typically implemented with the thick top metal layer and the bottom binding is implemented in lower metal layers close to the substrate increasing the losses. Like inductors, transformers are represented based on the lumped RLC model and the typical lumped model of a differential transformer is shown in Fig. 1.21 [71].

1.4.2.1 Practical considerations for transformer implementations

The practical implementation of an integrated transformer is shown in Fig. 1.22. In RFIC design, due to the use of differential circuits, symmetry is preferred in the transformer layout. The center point of the transformer is the electrical center and is often known as the center-tap. The comparison of a practical stacked and interleaved transformer is shown in Fig. 1.23. For the simulations, the transformer coils are constructed with the same metal layers and the same dimension and inductance of the primary coil is maintained for both the stacked and tapered transformers.

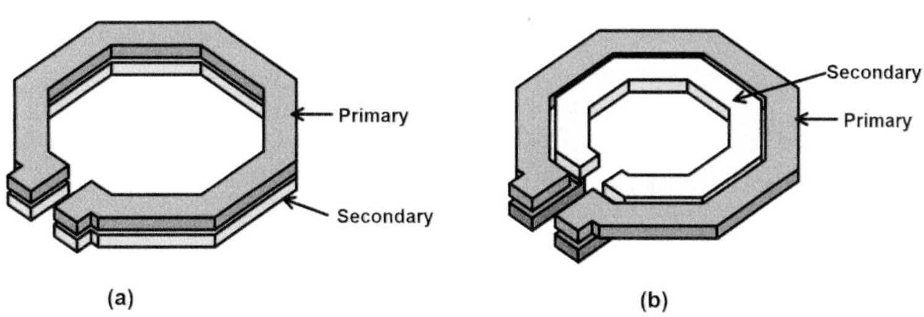

Figure 1.22 Differential transformer (a) stacked, and (b) tapered.

The inductance and quality factors of a stacked transformer are shown in Figs. 1.23(a) and 1.23(b), respectively. Due to the stacking of transformer coils, the primary and secondary have almost equal inductance values. However, the quality factor of the stacked secondary coil starts to decrease faster in comparison to the primary coil due to higher losses as it is closer to the substrate. In the case of tapered transformers, since and primary and secondary are placed adjacent to each other, the secondary coil is smaller and has a smaller inductance in comparison to the primary coil as shown in Fig. 1.23(c). The quality factor of tapered transformers as shown in Fig. 1.23(d) is low as they experience comparable losses and have a similar rate of decrease at higher frequencies. The comparison of the coupling coefficient of stacked and tapered transformer is shown in Fig. 1.23(e). Due to the stacking of transformer coils, the coupling coefficient is higher than tapered transformer.

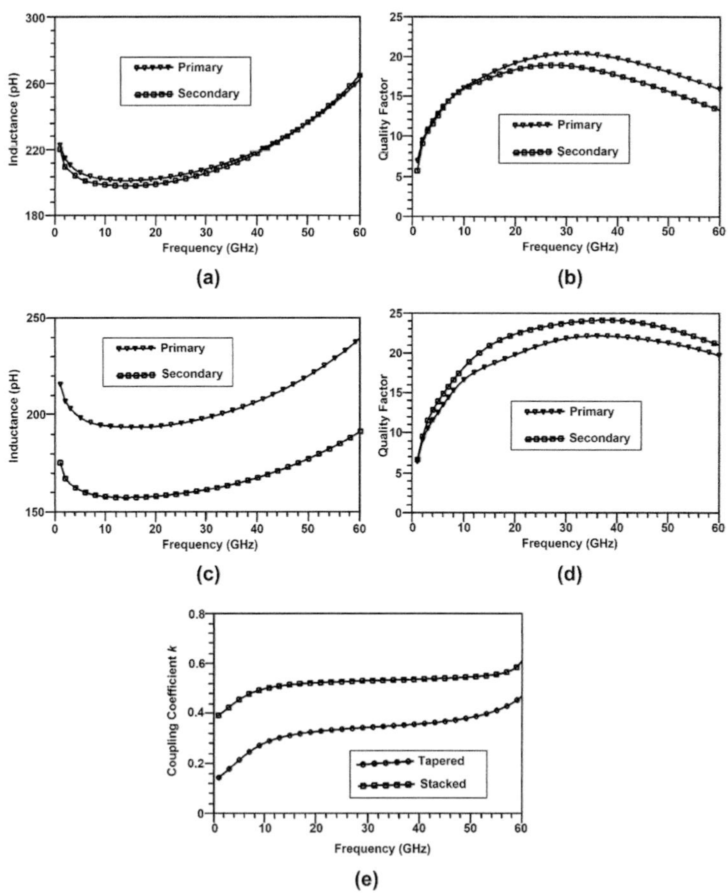

Figure 1.23 Transformer characteristics (a) inductance (stacked), (b) quality factor (stacked), (c) inductance (tapered), (d) quality factor (tapered), and (e) coupling coefficient.

Sometimes, it is desirable to have a multi-ratio transformer. The multi-ratio transformers can be laid as interleaved topology as shown in Fig. 1.24. In the interleaved structure, the primary and secondary windings are intertwined resulting in a much higher coupling coefficient, but the SRF is low due to higher interwinding capacitance. The simulation of a 1:2 interleaved transformer is shown in Fig. 1.25.

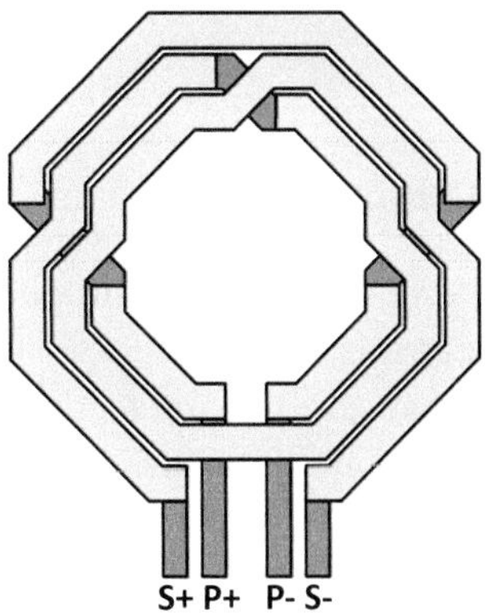

Figure 1.24 Interleaved 1:2 differential transformer.

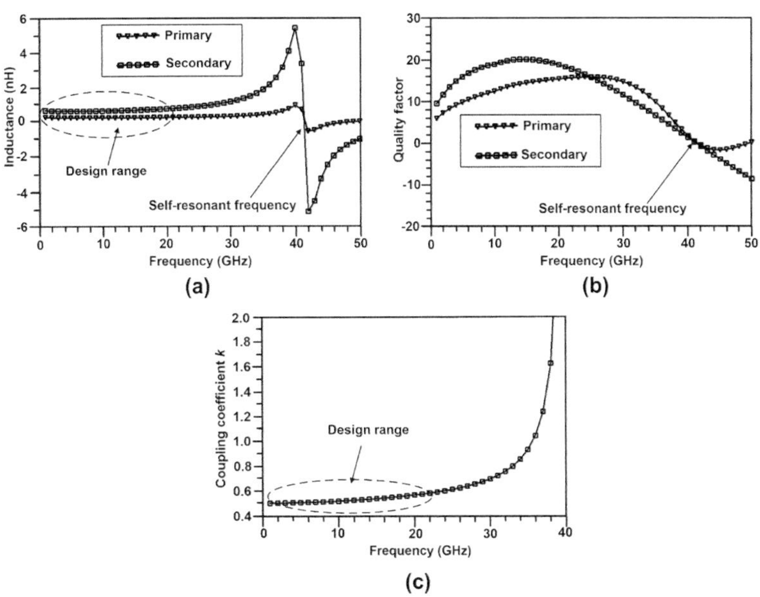

Figure 1.25 Interleaved 1:2 differential transformer (a) inductance, (b) quality factor, and (c) coupling coefficient.

As shown in Fig. 1.25, due to a higher turn ratio of the secondary, the secondary coil has a higher inductance, and the transformer has a low self-resonant frequency in comparison to the tapered transformer. Also, the secondary coil inductance rises sharply beyond 20 GHz resulting in a large variation in the inductance. Also, the secondary coil achieves a lower peak Q frequency in comparison to the primary. The coupling coefficient is shown in Fig. 1.25(c). Due to the interleaved topology, the transformer achieves a high coupling coefficient.

1.4.3 Ground Shielding

The quality factor is one of the important performance metrics of the inductors and transformers. From the previous discussions, it can be observed that the quality factor of the inductor and transformer is limited by the interaction of the electric field and

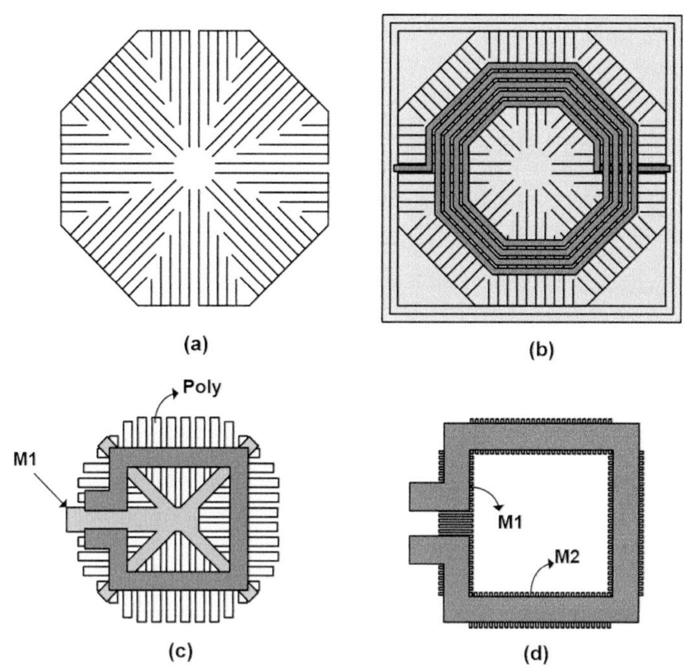

Figure 1.26 Inductor with PGS (a) PGS structure, (b) inductor with PGS [79], (c) inductor with PGS and M1, and (d) inductor with floating metal shield [78].

the substrate. Ground shielding reduces the substrate currents by preventing the electric fields of the metal traces from reaching the substrate. Though the best possible solution is to remove the area of the substrate underneath the inductor or transformer, this involves an additional semiconductor process for removing the substrate [72]. Alternatively, several works related to ground shields are reported in literature [72–79] and the popular one is the patterned ground shield (PGS) formed with polysilicon and metal as shown in Fig. 1.26 [78, 79]. One of the drawbacks of the PGS is the requirement of true ground potential (0 V) which is difficult to achieve in integrated circuits due to various noise sources and non-idealities presented in the integrated circuits [74]. Also, in advanced semiconductor nodes, there are strict metal density requirements that need to be satisfied. Floating metals are added as metal dummies beneath the inductors and transformers which in a way act as floating metal shields isolating the metal windings and the substrate thus reducing the substrate currents.

1.5 On-Chip RF Interconnects

Interconnects in integrated circuit design is used to electrically connect the different circuit blocks and provide interface from the bond-pads to the circuit blocks and vice-versa. In the RF and high-frequency circuit domain, the external connection to the RF signals is typically 50 Ohms. Interconnecting such signals with normal metal traces without consideration for any impedance matching results in impedance mismatch, resistive losses and delay due to metal traces used as interconnects and unwanted coupling with other circuits in the chip.

In integrated circuits, other than electrical connection, the interconnects can be used to synthesize filters [80], couplers [81], power dividers [82] and impedance matching. The interconnects are typically transmission lines (TL) and the two main types are microstrips and co-planar waveguides (CPW) as shown in Fig. 1.27. In the micro-strip, the field lines are terminated in the lossy substrate and a layer of metal can be used between the signal line and the substrate to avoid the losses. However, since the ground is now closer, the lines must be narrower to maintain the same impedance. Therefore, such transmission lines may not be suitable in applications where large current needs to be delivered. In comparison, in CPW, the ground lines are placed beside the signal lines on both sides. Therefore, the field lines are terminated at the ground lines resulting in low loss due to less penetration of electric field onto the substrate. In addition, in CPW, the signal line width and separation from the ground line can be adjusted to achieve the desired impedance. Also, the ground lines can be constructed with multiple metal layers and vias resulting in a ground-wall around the signal line. However, one of the major drawbacks of CPW is the large area required due to the additional ground lines adjacent to the signal line. The important considerations for the on-chip interconnects are the characteristic impedance, attenuation and phase velocity.

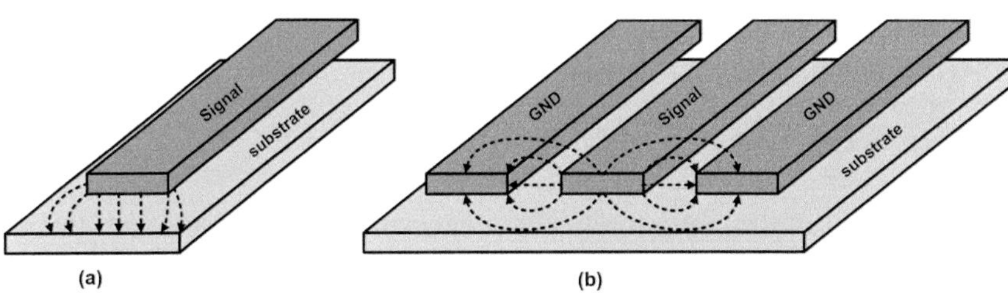

Figure 1.27 Transmission line topologies (a) micro-strip, (b) co-planar waveguide (CPW).

1.5.1 Transmission Line Characterization

The transmission line is characterized by different parameters and the methodology for the characterization of the transmission line can be explained based on the telegraph model as shown in Fig. 1.28 [83]. The series components R and L represent the resistive losses (Ω/m) and inductance (H/m), respectively, while the parallel components C and G represent the coupling capacitors (F/m) between the conductors and per unit length conduction losses in (S/m), respectively. The behavior of the transmission line can be described based on its characteristics impedance (Z_c)

and propagation constant (γ). For a lossless TL, these parameters are given as [84]

$$Z_c = \sqrt{\frac{R + j\omega L}{G + j\omega C}} \tag{1.14}$$

$$\gamma = \sqrt{(R + j\omega L)(G + j\omega C)} = \alpha + j\beta. \tag{1.15}$$

The propagation constant (γ) is the sum of the attenuation constant α (dB/m) and the phase constant β (rad/m). To characterize the transmission line, the S-parameter simulation is typically performed, and the transmission line characterization helps to determine the characteristic impedance Z_c and complex propagation constant γ given as [85]

$$Z_c = 50\sqrt{\frac{(1+S_{11})(1+S_{22}) - S_{12}S_{21}}{(1-S_{11})(1-S_{22}) - S_{12}S_{21}}} \tag{1.16}$$

$$\gamma = \frac{1}{l} a\cosh\left(\frac{(1+S_{11})(1-S_{22}) + S_{12}S_{21}}{S_{12} + S_{21}}\right), \tag{1.17}$$

where l is the length of the transmission line. The RLCG model parameters can be extracted as

$$R = \text{real}(Z_c \cdot \gamma), \qquad L = \frac{\text{im}(Z_c \cdot \gamma)}{\omega} \tag{1.18a}$$

$$G = \text{real}(\gamma / Z_c), \qquad C = \frac{\text{im}\left(\frac{\gamma}{Z_c}\right)}{\omega}, \tag{1.18b}$$

where ω is the operation frequency. The phase constant β represents the phase-shift per unit length and the quality factor of the transmission line is given by [85]

$$Q = \frac{\beta}{2\alpha}. \tag{1.19}$$

The CPW transmission line RLCG model depends on several design parameters. Though the formulas for the calculation of shunt parameters G and C are readily available [86, 87], the series parameters R and L are functions of CPW geometrical parameters such as conductor width, conductor thickness, ground conductor width, spacing, material properties like conductivity and frequency of operation. In IC design, the performance of CPW is limited by the technology. From Eq. (1.19), it can be observed that the quality factor of the CPW can be improved by increasing the phase constant β. Also, the RLCG model is typically correct for line length smaller than $\lambda/20$. Otherwise, it is necessary to divide the line into several stages to further improve the modeling accuracy.

1.6 Capacitors

In RFIC, capacitors are used extensively for filters, matching circuits, de-coupling, frequency selectivity, sampling circuits and many more applications. Capacitors are lumped components that store electrical energy, and the energy storage capacity of the capacitor depends on the design parameters. In the context of integrated circuits, capacitors are passive components that occupy a large area. Therefore, the important properties of integrated capacitors are capacitance density or low area (less cost) and process, voltage, and temperature (PVT) stability and leakage. The important parameters that determine the quality of the capacitors are linearity, quality factor (low loss) and self-resonant frequency. The quality factor of the capacitor is given by [88]

$$Q_c = \frac{1}{2\pi f R_s C}, \tag{1.20}$$

where R_S is the parasitic series resistance, C is the capacitance and f is the operation frequency. From Eq. (1.20), we can note that the quality factor of the capacitor decreases at higher frequencies. In RFICs, the most popular capacitor implementation topologies are based on metal-insulator-metal (MIM), metal-oxide-metal (MOM) and MOS capacitors as discussed below.

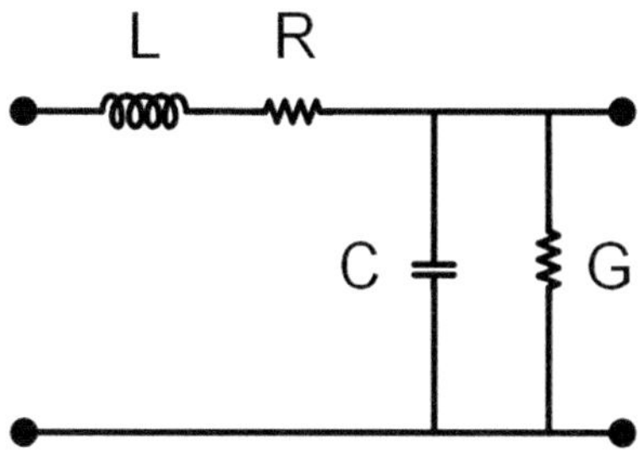

Figure 1.28 RLCG model of transmission line.

1.6.1 MIM Capacitor

The MIM capacitor is formed by two metal layers with an insulator in between as shown in Fig. 1.29. It is essentially a parallel-plate capacitor which has good linearity, temperature stability, high Q-factor and capacitance density [89]. However, the capacitance density of the MIM capacitors is dependent on the spacing between the metal layers and the permittivity of the insulator. The capacitance of the MIM capacitor based on the parallel-plate capacitor can be given as [90, 91]

$$C = \frac{\epsilon_0 A}{d}, \tag{1.21}$$

where ε_0 is the permittivity of free space, A and d are the area and separation between the metals, respectively. One of the drawbacks of the MIM capacitors is the parasitic capacitance formed due to the distance between the bottom plate of the MIM capacitor and the substrate and is highly process dependent.

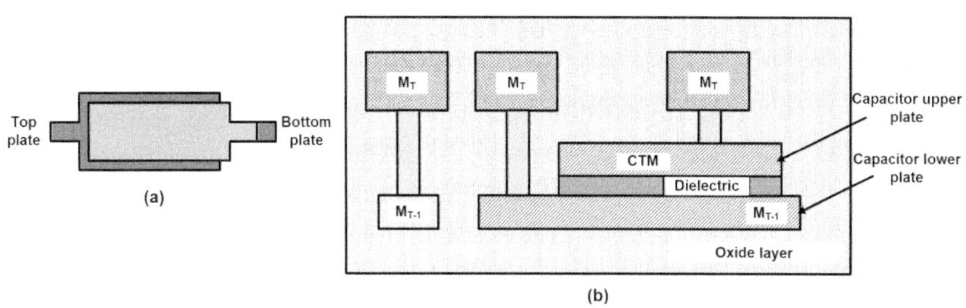

Figure 1.29 MIM capacitor (a) top view, and (b) structure.

1.6.2 MOM Capacitor

The MOM capacitor is a vertical capacitor in which the capacitor plates are non-planar and typically interdigitated with oxide as an insulating layer as shown in Fig. 1.30. In the advanced technology nodes, since there are multiple metal layers that are much closer, a high capacitor density can be easily achieved in MOM capacitor in comparison to a MIM capacitor. The capacitor density depends on the metal layers and the metal thickness. Since MOM are vertical capacitors, a large number of metal layers can be combined to increase the capacitor density per unit area [92]. The parasitic capacitance of the MOM capacitor is also formed between the bottom metal layer and the substrate with an oxide layer in between. One of the main advantages of the MOM cap in IC design is the requirement of no additional masks (reducing cost) during the IC fabrication process and thus enabling EM simulations during the IC design phase. The quality factor of the MOM capacitor is given by [91]

$$Q = \frac{Q_c Q_d}{Q_c + Q_d}, \tag{1.22}$$

where Q_c and Q_d are the quality factor due to the conductor losses and dielectric losses, respectively.

1.6.3 MOS Capacitors

The other type of integrated capacitor used in RFIC design is the MOS capacitor which is the most basic capacitor formed between the metal layer, gate oxide and the semiconductor channel between the source and drain terminals. The MOS capacitors typically exhibit higher metal densities as their oxide thickness is much thinner in comparison to MOM capacitors. However, their capacitance varies with the difference in the gate to source-drain voltage and is highly non-linear.

The most basic form of a MOS capacitor can be formed using the PMOS transistor with source and drain terminals connected as the bottom plate and bulk connected to a bias as shown in

Fig. 1.31. The depletion mode PMOS capacitors have also been reported in [93–95] with a deep N-well to reduce the parasitic capacitance. However, the parasitic capacitance formed is a function of the deep-N-well bias voltage [94]. The NMOS transistors in an N-well can also be used as a MOS capacitor as shown in Fig. 1.32 known as the accumulation-mode MOS capacitors. Since the NMOS is placed inside the N-well, the inversion is avoided, and the capacitance can be quite stable when the gate-source/drain voltage is greater than 0 V. Also, the parasitic capacitance is less in the accumulation-mode MOS varactor as there is only one parasitic capacitor between the N-well and the P-substrate.

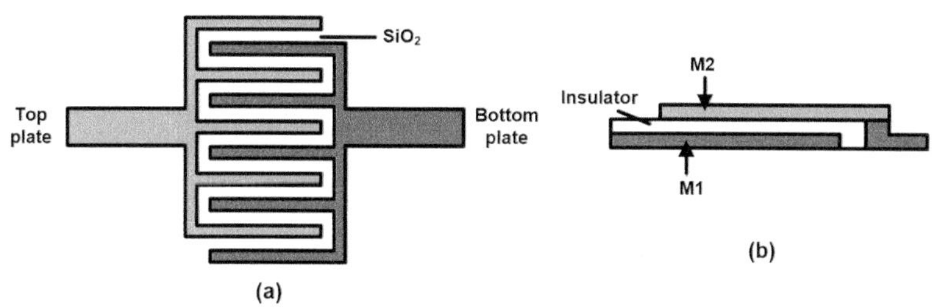

Figure 1.30 MOM capacitor (a) top view, and (b) side view.

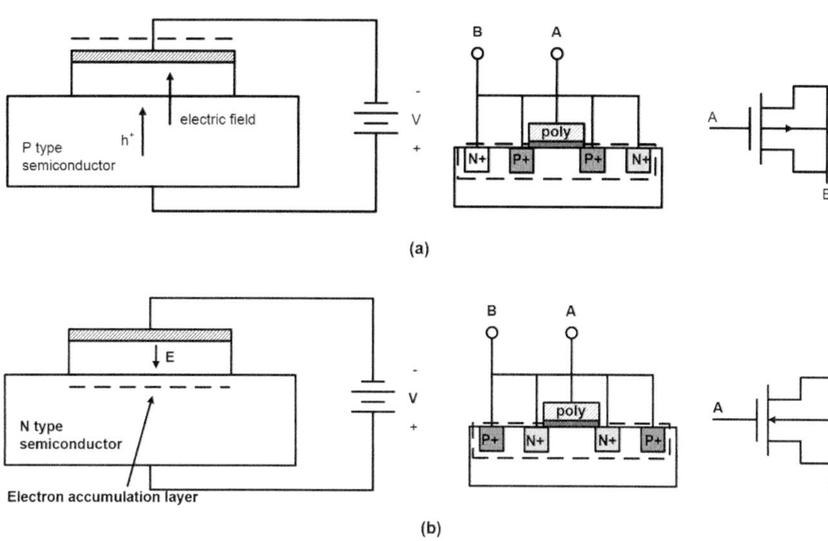

Figure 1.31 MOS capacitor (a) PMOS, and (b) NMOS.

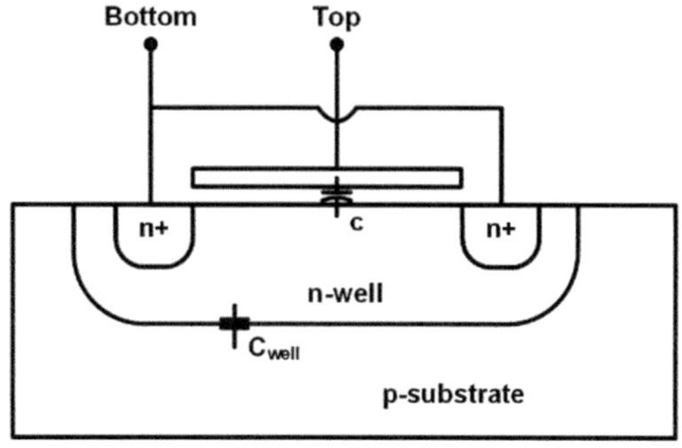

Figure 1.32 Accumulation-mode MOS capacitor.

1.6.4 Varactors

Varactors in integrated circuits are specifically variable capacitance used primarily in applications where capacitance tuning is required such as VCO for frequency tuning, phase shifts, amplifiers for tuning of LC tanks and so on. In CMOS technology, the varactors can either be p-n junction varactors implemented using the p+/n+ diffusions and N/P wells [96, 97] or MOS varactors. However, the p-n junction varactors are not area efficient [98, 99]. The MOS varactors are typically implemented either in P-substrate (MOS varactor) or in an isolated N-well also known as accumulation-mode MOS (A-MOS) varactors. A-MOS varactors are the preferred type of varactors as they tend to have a higher Q [100, 101] and will be discussed in the following.

The varactors implemented in the P-substrate have distinct capacitance characteristics based on the applied voltage at the gate terminal and operate in three different regions namely the strong inversion, depletion and accumulation as shown in Fig. 1.33. The overall capacitance is the series combination of varactors given by

$$\frac{1}{C_{\text{var}}} = \frac{1}{C_{\text{ox}}} + \frac{1}{C_d}. \tag{1.23}$$

When the negative voltage is applied to the gate, there are surplus holes at the semiconductor surface and the device is said to be in the accumulation region. With the increase in the gate voltage, a flat-band situation is reached and the semiconductor beneath the gate is neutral. Above the flat-band voltage, the holes are repelled, and a depletion region is formed. As the gate voltage is further increased and crosses a certain threshold, there is a surplus of electrons at the semiconductor surface and the device enters the inversion region. As shown in the *C*–*V* characteristics in Fig. 1.34, the capacitance is a strong function of the gate voltage resulting in a non-monotonic behavior [102] and limiting the use in practical circuits as devices are normally operated with the positive voltage.

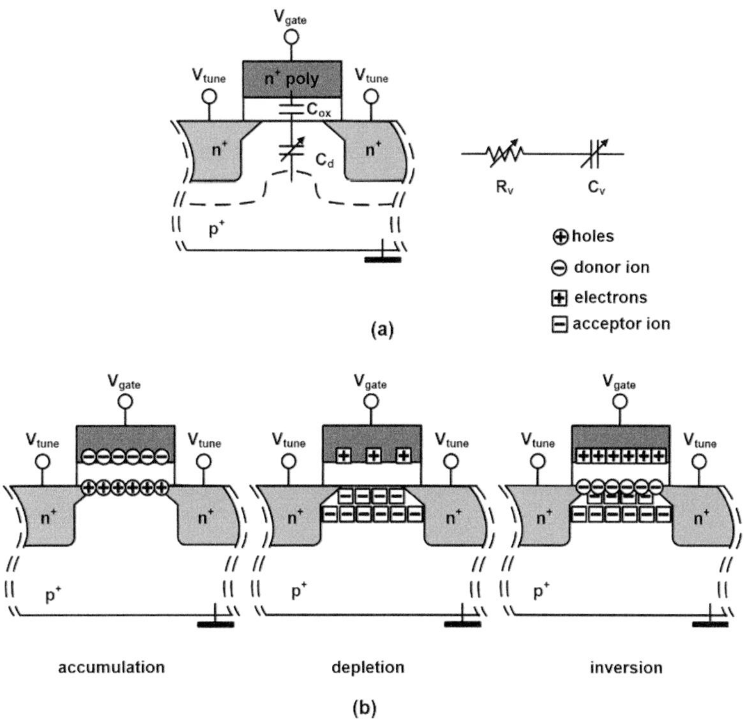

Figure 1.33 Conventional NMOS varactor (a) cross section, and (b) operation regions.

In RFIC design, the accumulation-mode MOS varactors are typically used as these devices operate in the depletion or accumulation modes with positive voltages. The cross section and the operation regions of A-MOS varactor are shown in shown in

Fig. 1.35. When the applied gate voltage is higher than the source-drain voltage, the electrons are attracted to the surface and an accumulation region is formed. As the source-drain voltage increases, the electrons are forced back from the surface of the N-well around the gate oxide creating a depletion layer. Since the NMOS is implemented in the N-well, in A-MOS varactors there is no inversion region as the material under the gate is also N-type [102]. The capacitance–voltage characteristics of the accumulation-mode MOS varactor are shown in Fig. 1.36.

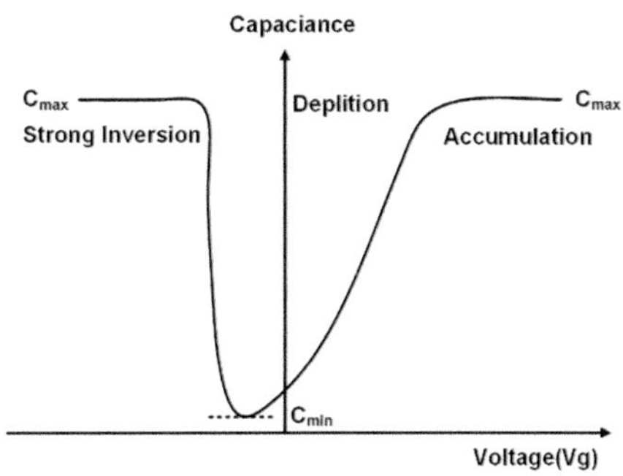

Figure 1.34 Capacitance–voltage characteristics of MOS varactor.

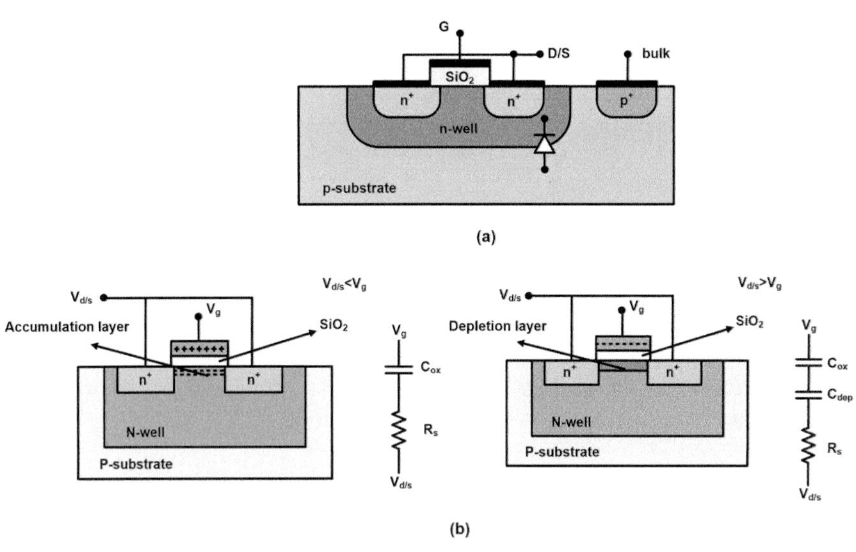

Figure 1.35 Accumulation-mode MOS varactor (a) cross section, and (b) operation regions.

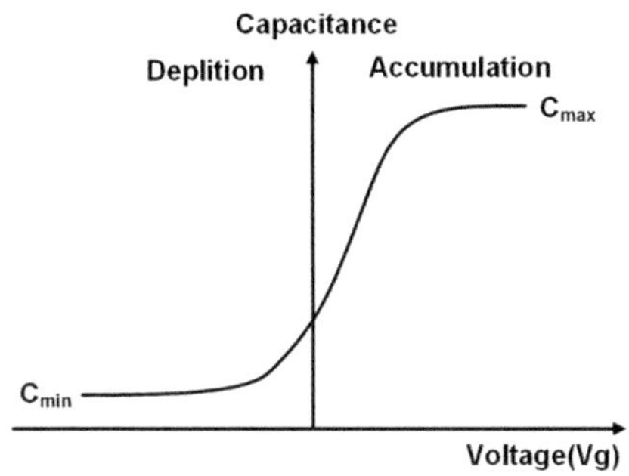

Figure 1.36 Capacitance–voltage characteristics of A-MOS varactor.

1.6.4.1 Parameters of MOS varactors

The important performance parameters of the varactors are capacitance ratio (C_{max}/C_{min}) and quality factor of the varactors. The main advantage of the MOS varactor is higher C_{max}/C_{min} in comparison to p-n junction varactors [100, 101].

One of the applications of the varactor is frequency tuning in the VCO. Since VCO typically has a large output swing, the voltage at the varactor terminals is also large. In such cases, the small-signal capacitance variation is no longer valid, and the C–V characteristics are determined based on time-varying capacitance due to large signal [103–105]. The quality factor of the varactor is given as

$$Q_{\text{var}} = \frac{1}{2\pi fRC}, \tag{1.24}$$

where f is the operation frequency, R and C are the series resistance and total capacitance, respectively. The Q of A-MOS varactor is dominated by the resistance of the gate and N-well contacts and can be given as [106, 107]

$$Q \cong \frac{12}{2\pi fC_{\text{ox}}(R_{nW}L^2 + R_{\text{poly}}W_f^2)}. \tag{1.25}$$

Typically, the resistance of the gate is much higher than the N-well resistance and the gate resistance of the varactor is dependent on the device layout. The gate resistance is given by

$$R_{\text{gate}} \propto N_s R_s, \tag{1.26}$$

where N_S and R_S are the number of squares and square resistance of the gate, respectively. Therefore, to reduce this resistance and improve the quality factor, multi-finger layout structures with multiple gate contacts are used. In multi-finger structures, the total width of the gate is given by

$$w_g = l_f N_F, \tag{1.27}$$

where l_f is the length of each finger and N_F is the number of fingers. Therefore, based on Eqs. (1.26) and (1.27), in sub-nm technologies, the quality factor of the A-MOS varactors can also be improved with a reduction of L [107, 108].

1.7 Conclusion

Currently, most semiconductor foundries provide design kits with RF models for a range of active and passive devices such as MOSFET, inductors, capacitors, varactors, etc. though the models are typically accurate for frequencies below 20 GHz. At higher frequencies, the models are either extrapolated or need additional modeling effort from the designers. With the knowledge of device models, designers can optimize the circuit performance for their particular frequency of interest. This will help in achieving the specifications with a single tape-out and reduce the time to market for the RFICs.

From the device perspective, due to technology scaling, the f_T and f_{max} have consistently increased over the years enabling design of THz circuits in CMOS technology. In RFICs, the choice of the transistor is very important for the optimization of different circuits, and it is imperative to know the choice to optimize a particular transistor for a specific application. For example, for circuits like low noise amplifiers, the choice of transistor to optimize the f_T and NF_{min} is vital. For VCOs and power amplifiers, optimizing the active devices for f_{max} is crucial. For the passive circuits, understanding the material properties and geometrical parameters is very important for optimization of the quality factor to achieve a low loss and low power consumption.

References

1. K. J. Kuhn, Moore's Law past 32 nm: Future challenges in device scaling, *IEEE International Workshop on Computational Electronics*, pp. 1–6, 2009.
2. S. M. Sze and Kwok K. Ng, *Physics of Semiconductor Devices*, 3rd ed., John Wiley and Sons, Inc., 2007.
3. N. Paydavosi, S. Venugopalan, Y. S. Chauhan, J. P. Duarte, S. Jandhyala, A. M. Niknejad, and C. C. Hu, BSIM SPICE models enable FinFET and UTB IC designs, *IEEE Access*, vol. 1, pp. 201–215, 2013.
4. *Department of Electrical Engineering and Computer Science*, UC Berkeley, Berkeley, CA, USA. BSIM4v4.8.0 Technical Manual, 2013.
5. W. Liu, *MOSFET Models for SPICE Simulation: Including BSIM3v3 and BSIM4*, John Wiley and Sons, Inc., 2001.
6. W. Liu et al., BSIM3v3.3 MOSFET Model—Users' manual, *Department of Electrical Engineering and Computer Sciences*, UC Berkeley, 2005.
7. S. F. Tin, A. A. Osman, K. Mayaram, and C. Hu, A simple subcircuit extension of the BSIM3v3 model for CMOS RF design, *IEEE Journal of Solid-State Circuits*, vol. 35, no. 4, pp. 612–624, Apr. 2000.
8. T. E. Kolding, Calculation of MOSFET Gate Impedance, Technical Report R98–1009, ISSN 0908-1224.
9. K. S. Yeo, M. A. Do, and C. C. Boon, *Design of CMOS RF Integrated Circuits and Systems*, World Scientific Publishing Co. Pte. Ltd., 2010.
10. A. Scholten, G. D. J. Smit, B. De Vries, L. Tiemeijer, J. Croon, D. B. M. Klaassen, R. Van Langevelde, X. Li, W. Wu and G. Gildenblat, The new CMC standard compact MOS model PSP: Advantages for RF applications, *IEEE Journal of Solid-State Circuits*, vol. 44, no. 5, pp. 1415–1424, May 2009.
11. M. Miura-Mattausch, N. Sadachika, D. Navarro, G. Suzuki, Y. Takeda, M. Miyake, T. Warabino, Y. Mizukane, R. Inagaki, T. Ezaki, H. Mattausch, T. Ohguro, T. Iizuka, M. Taguchi, S. Kumashiro, and S. Miyamoto, HiSIM2: Advanced MOSFET model valid for RF circuit simulation, *IEEE Transactions on Electron Devices*, vol. 53, no. 9, pp. 1994–2007, Sep. 2006.
12. C. C. Enz, F. Krummenacher, and E. A. Vittoz, An analytical MOS transistor model valid in all regions of operation and dedicated to low-voltage and low-current applications, *Analog Integrated Circuits Signal Process*, vol. 8, no. 1, pp. 83–114, Jan. 1995.

13. Y. S. Chauhan, S. Venugopalan, M. A. Chalkiadaki, M. A. U. Karim, H. Agarwal, S. Khandelwal, N. Paydavosi, J. P. Duarte, C. C. Enz, A. M. Niknejad, and C. Hu, BSIM6: Analog and RF compact model for bulk MOSFET, *IEEE Transactions on Electron Devices*, vol. 61, no. 2, pp. 234–244, Feb. 2014.

14. Y. Gu and J. S. Yuan, Gate-oxide thickness effects on hot-carrier-induced degradation in n-MOSFETs, *Proceedings of SOUTHEASTCON*, pp. 665–669, Apr. 1996.

15. N. Wang, *Digital MOS Integrated Circuits*, Prentice-Hall, Englewood Cliffs, NJ, 1989.

16. M. Keshavarz Hedayati, A. Abdipour, R. Sarraf Shirazi, C. Cetintepe, and R. B. Staszewski, A 33 GHz LNA for 5G wireless systems in 28 nm bulk CMOS, *IEEE Transactions on Circuits and Systems II: Express Briefs*, vol. 65, no. 10, pp. 1460–1464, Oct. 2018.

17. A. Van der Ziel, *Noise in Solid-State Devices and Circuits*, John Wiley and Sons, Inc., 1986.

18. A. J. Scholten et al., Accurate thermal noise model for deep-submicron CMOS, *International Electron Devices Meeting, Technical Digest* (Cat. No. 99CH36318), pp. 155–158, Dec. 1999.

19. M. A. Chalkiadaki, A. Mangla, C. C. Enz, Y. S. Chauhan, M. A. U. Karim, S. Venugopalan, A. Niknejad, and C. Hu, Evaluation of the BSIM6 compact MOSFET model's scalability in 40 nm CMOS technology, *IEEE European Solid-State Circuits Conference*, pp. 34–37, Sep. 2012.

20. S. Venugopalan, K. Dandu, S. Martin, R. Taylor, C. Cirba, X. Zhang, A. Niknejad, and C. Hu, A non-iterative physical procedure for RF CMOS on using BSIM6, *IEEE Custom Integrated Circuits Conference*, pp. 1–4, Sep. 2012.

21. C. C. Enz and E. A. Vittoz, *Charge Based MOS Transistor Modeling: The EKV Model for Low-Power and RF IC Design*, John Wiley and Sons, Inc., 2006.

22. B. Dormieu, P. Scheer, C. Charbuillet, S. Jan, and F. Danneville, 4-port isolated MOS modeling and extraction for mmWave applications, *IEEE European Solid-State Device Research Conference*, pp. 54–57, Sep. 2012.

23. I. Kwon, M. Je, K. Lee, and H. Shin, A simple and analytical parameter extraction method of a microwave MOSFET, *IEEE Transactions on Microwave Theory and Techniques*, vol. 50, no. 6, pp. 1503–1509, Jun. 2002.

24. M. A. Chalkiadaki and C. C. Enz, Low-power RF modeling of a 40 nm CMOS technology using BSIM6, *IEEE International Conference on Mixed Design of Integrated Circuits and Systems*, pp. 57–62, Oct. 2013.

25. A. Mangla, M. A. Chalkiadaki, F. Fadhuile, T. Taris, Y. Deval, and C. Enz, Design methodology for ultra-low-power analog circuits using next generation BSIM6 MOSFET compact model, *Microelectronics Journal*, vol. 44, no. 7, pp. 570–575, Jul. 2013.

26. P. H. Woerlee et al., RF-CMOS performance trends, *IEEE Transactions on Electron Devices*, vol. 48, no. 8, pp. 1776–1782, Aug. 2001.

27. Y. Cheng, C. H. Chen, M. Matloubian, and M. J. Deen, High-frequency small signal AC and noise modeling of MOSFETs for RFIC design, *IEEE Transactions on Electron Devices*, vol. 49, no. 3, pp. 400–408, Mar. 2002.

28. C. H. Chen, M. J. Deen, Y. Cheng, and M. Matloubian, Extraction of the induced gate noise, channel noise, and their correlation in submicron MOSFETs from RF noise measurements, *IEEE Transactions on Electron Devices*, vol. 48, no. 12, pp. 2884–2892, Dec. 2001.

29. A. Scholten, L. F. Tiemeijer, R. van Langevelde, R. J. Havens, A. T. A. Zegers-van Duijnhoven, and V. C. Venezia, Noise modeling for RF CMOS circuit simulation, *IEEE Transactions on Electron Devices*, vol. 50, no. 3, pp. 618–632, Mar. 2003.

30. G. D. J. Smit, A. J. Scholten, R. M. T. Pijper, L. F. Tiemeijer, R. van der Toorn, and D. B. M. Klaassen, RF-noise modeling in advanced CMOS technologies, *IEEE Transactions on Electron Devices*, vol. 61, no. 2, pp. 245–254, Feb. 2014.

31. A. van der Ziel, Noise in solid-state devices and lasers, *Proceedings of the IEEE*, vol. 58, no. 8, pp. 1178–1206, Aug. 1970.

32. K. J. Kuhn, CMOS scaling for the 22 nm node and beyond: Device physics and technology, *IEEE International Symposium on VLSI Technology, Systems and Applications*, pp. 1–2, Apr. 2011.

33. K. Roy, S. Mukhopadhyay, and H. Mahmoodi-Meimand, Leakage current mechanisms and leakage reduction techniques in deep-submicrometer CMOS circuits, *Proceedings of the IEEE*, vol. 91, no. 2, pp. 305–327, 2003.

34. D. J. Frank, R. H. Dennard, E. Nowak, P. M. Solomon, Y. Taur, and H. S. P. Wong, Device scaling limits of Si MOSFETs and their application dependencies, *Proceedings of the IEEE*, vol. 89, no. 3, pp. 259–288, Mar. 2001.

35. C. Hu, Gate oxide scaling limits and projection, *IEEE International Electron Devices Meeting*, pp. 319–322, Dec. 1996.

36. Y. C. Yeo, T. J. King, and C. Hu, MOSFET Gate leakage modeling and selection guide for alternative gate dielectrics based on leakage considerations, *IEEE Transactions on Electron Devices*, vol. 50, no. 4, pp. 1027–1035, Apr. 2003.

37. J. Chen, T. Y. Chan, I. C. Chen, P. K. Ko, and C. Hu, Sub-breakdown Drain leakage current in MOSFET, *IEEE Electron Device Letters*, vol. 8, no. 11, pp. 515–517, Nov. 1987.

38. A. Cathelin, Fully depleted silicon on insulator devices CMOS: The 28 nm node is the perfect technology for analog, RF, mmWave, and mixed-signal system-on-chip integration, *IEEE Solid-State Circuits Magazine*, vol. 9, no. 4, pp. 18–26, Apr. 2017.

39. J. P. Noel et al., Multi-UTBB FDSOI device architectures for low-power CMOS circuit, *IEEE Transactions on Electron Devices*, vol. 58, no. 8, pp. 2473–2482, Aug. 2011.

40. O. Kononchuk and B. Y. Nguyen, *Silicon-On-Insulator (SOI) Technology: Manufacture and Applications*, Woodhead Publishing, pp. 286–287, 2014.

41. R. Guillaume, Millimeter-wave and terahertz frequency synthesis on advanced silicon technology, Master's Thesis, Université de Bordeaux, Dec. 2018. [Online]. Available: https://theses.hal.science/tel-02329132/.

42. X. Liu and S. Mourad, Performance of submicron CMOS devices and gates with substrate biasing, *IEEE International Symposium on Circuits and Systems*, pp. 9–12, May 2000.

43. P. Magarshack, P. Flatresse, and G. Cesana, UTBB FD-SOI: A process/design symbiosis for breakthrough energy-efficiency, *IEEE Design, Automation and Test in Europe Conference and Exhibition*, pp. 952–957, Mar. 2013.

44. L. Koshkinen, M. Hieenkari et al., Minimum-energy point design in FDSOI regular Vt, *IEEE SOI-3D-Subthreshold Microelectronics Technology Unified Conference*, pp. 1–2, Nov. 2015.

45. B. Razavi, *Design of Analog CMOS Integrated Circuits*, International Edition, McGraw Hill, 2001.

46. K. B. Ashby and I. A. Koullias, High Q inductors for wireless applications in a complementary silicon bipolar process, *IEEE Journal of Solid-State Circuits*, vol. 31, no. 1, pp. 4–8, Jan. 1996.

47. C. P. Yue and S. S. Wong, On-chip spiral inductors with patterned ground shields for Si-based RF IC's, *IEEE Journal of Solid-State Circuits*, vol. 33, no. 5, pp. 743–815, May 1998.

48. A. M. Niknejad and R. G. Meyer, Analysis, design, and optimization of spiral inductors and transformers for Si RF ICs, *IEEE Journal of Solid-State Circuits*, vol. 33, no. 10, pp. 1470–1481, Oct. 1998.

49. A. M. Niknejad and R. G. Meyer, Design, *Simulation and Applications of Inductors and Transformers for Si RF ICs*, Springer Link, 2000.

50. G. G. Rabjohn, Balanced Planar Transformers, US Patent 4,816,784, 1989.

51. J. R. Long, Monolithic transformers for silicon RF IC design, *IEEE Journal of Solid-State Circuits*, vol. 35, no. 9, pp. 1368–1382, Sep. 2000.

52. C. B. Sia, Device, design, characterization and modeling of inductors and interconnects for RFIC applications, Ph.D Thesis, Nanyang Technological University, Singapore, 2008.

53. K. Kamogawa, K. Nishikawa, I. Toyoda, T. Tokumitsu, and M. Tanaka, A novel high-Q and wide-frequency-range inductor using Si 3-D MMIC technology, *IEEE Microwave and Guided Wave Letters*, vol. 9, no. 1, pp. 16–18, Jan. 1999.

54. J. N. Burghartz, M. Soyuer, and K. A. Jenkins, Microwave inductors and capacitors in standard multilevel interconnect silicon technology, *IEEE Transactions on Microwave Theory and Techniques*, vol. 44, no. 1, pp. 100–104, Jan. 1996.

55. R. Yang, H. Qian, Ju. Li, Q. Xu, C. Hai, and Z. Han, SOI technology for radio frequency integrated circuit applications, *IEEE Transactions on Electron Devices*, vol. 53, no. 6, pp. 1310–1316, Jun. 2006.

56. A. Tekin, E. Zencir, D. Huang, W. Liu, and N. S. Dogan, A 700 MHz VCO using high-Q silicon on insulator (SOI) inductors, *IEEE Radio and Wireless Symposium*, pp. 427–429, Oct. 2006.

57. V. Malba, D. Young, J. J. Ou, A. F. Bernhardt, and B. E. Boser, High-performance RF coil inductors on silicon, *IEEE Electronic Components and Technology Conference* (Cat. No. 98CH36206), pp. 252–255, May 1998.

58. F. Mernyei, F. Darrer, M. Pardoen, and A. Sibrai, Reducing the substrate losses of RF integrated inductors, *IEEE Microwave and Guided Wave Letters*, vol. 8, no. 9, pp. 300–301, Sep. 1998.

59. C. P. Yue and S. S. Wong, On-chip spiral inductors with patterned ground shields for Si-based RF ICs, *IEEE Journal of Solid-State Circuits*, vol. 33, no. 5, pp. 743–752, May 1998.

60. I. J. Bahl, *Lumped Elements for RF and Microwave Circuits*, Norwood: Artech House, 2003.

61. S. R. Kythakyapuzha and W. B. Kuhn, Modeling of Inductors and transformers, *IEEE Radio Frequency Integrated Circuits Symposium* (IEEE Cat. No. 01CH37173), pp. 283–286, May 2001.

62. W. B. Kuhn and N. M. Ibrahim, Analysis of current crowding effects in multiturn spiral inductors, *IEEE Transactions on Microwave Theory and Techniques*, vol. 49, pp. 31–38, Jan. 2001.

63. J. D. Kraus, *Electromagnetics*, 4th ed., McGraw-Hill, 1991.

64. A. Weisshaar, H. Lan, and A. Luoh, Accurate closed-form expressions for the frequency-dependent line parameters of on-chip interconnects on lossy silicon substrate, *IEEE Transactions on Advanced Packaging*, vol. 25, no. 2, pp. 288–296, May 2002.

65. D. Melendy, Modelling of on-chip spiral inductors for silicon RFICs, Master's Thesis, Oregon State University, Nov. 2002.

66. A. C. Watson, Analysis and modeling of single-ended and differential spiral inductors in silicon-based RFICs, Master's Thesis, Oregon State University, Dec. 2003.

67. N. Mahalingam, K. Ma, K. S. Yeo, and W. M. Lim, K-band high-PAE wide tuning-range VCO using triple-coupled LC tanks, *IEEE Transactions on Circuits and Systems II, Express Briefs*, vol. 60, no. 11, pp. 736–740, Nov. 2013.

68. R. Long, Monolithic transformers for silicon RF IC design, *IEEE Journal of Solid-State Circuits*, vol. 35, no. 9, pp. 1368–1382, Sep. 2000.

69. H. C. Luong and J. Yin, Transformer design and characterization in CMOS process, in *Transformer-Based Design Techniques for Oscillators and Frequency Dividers*, Cham, Springer International Publishing, 2016.

70. S. G. Lee, Area efficient and symmetric design of monolithic transformers for silicon RF ICs, *IEEE TENCON*, vol. 2, pp. 880–882, Sep. 1999.

71. D. Kehrer, Design of monolithic integrated lumped transformers in silicon-based technologies up to 20 GHz, Ph.D Thesis, Technische Universität Wien—Fakultät für Elektrotechnik und Informationstechnik, Hacklweg, 2000.

72. K. T. Ng, B. Rejaei, and J. N. Burghartz, Substrate effects in monolithic RF transformers on silicon, *IEEE Transactions on Microwave Theory and Techniques*, vol. 50, no. 1, pp. 377–383, Jan. 2002.

73. T. Biondi, A. Scuderi, E. Ragonese, and G. Palmisano, Analysis and modeling of layout scaling in silicon integrated stacked transformers, *IEEE Transactions on Microwave Theory and Techniques*, vol. 54, no. 5, pp. 2203–2210, May 2006.

74. T. S. D. Cheung, and J. R. Long, Shielded passive devices for silicon-based monolithic microwave and millimeter-wave integrated circuits, *IEEE Journal of Solid-State Circuits*, vol. 41, no. 5, pp. 1183–1200, May 2006.

75. O. El-Gharniti, E. Kerherve, and J. B. Begueret, Characterization of Si-based monolithic transformers with patterned ground shield, *IEEE Radio Frequency Integrated Circuits Symposium*, pp. 261–264, Jun. 2006.

76. H. B. Liang, Y. S. Lin, C. C. Chen, and J. H. Lee, Optimization of PGS pattern of transformers/inductors in standard RF BiCMOS technology for RFIC applications, *IEEE Radio Frequency Integrated Circuits Symposium*, pp. 4–7, Jul. 2006.

77. S. M. Yim, T. Chen, and K. K. O, The Effects of a ground shield on the characteristics and performance of spiral inductors, *IEEE Journal of Solid-State Circuits*, vol. 37, no. 2, pp. 237–244, Feb. 2002.

78. B. Leite, E. Kerherve, J. B. Begueret, and D. Belot, Shielding structures for millimeter-wave integrated transformers, *IEEE International Conference on Electronics, Circuits and Systems*, pp. 239–242, Dec. 2009.

79. J. H. Ko, K. T. Kim, J. K. Byun, and H. S. Kim, Parameter extraction and optimal design of spiral inductor using evolution strategy and sensitivity, *IEEE Transactions on Magnetics*, vol. 46, no. 8, pp. 2831–2834, Aug. 2010.

80. S. Luo, C. C. Boon, L. Zhu, M. A. Do, and A. V. Do, A compact millimeter-wave CMOS bandpass filter using a dual-mode ring resonator, *International Symposium on Integrated Circuits*, pp. 270–272, Dec. 2011.

81. H. Ding, G. Wang, K. Lam, D. Jordan, A. Zeeb, A. Bavisi, E. Mina, and B. Gaucher, Modeling and implementation of on-chip millimeter-wave compact branch line couplers in a BiCMOS technology, *IEEE European Microwave Conference*, pp. 458–461, Oct. 2007.

82. H. Ding, K. Lam, E. Mina, J. Rascoe, D. Jordan, A. Zeeb, and B. Gaucher, On-chip millimeter-wave library device – scalable Wilkinson power divider/combiner, *IEEE Electronic Components and Technology Conference*, pp. 1246–1251, May 2007.

83. R. E. Collin, *Foundations for Microwave Engineering*, 2nd ed., NY: McGraw-Hill, 1992.

84. D. Pozar, *Microwave Engineering*, Wiley, 2004.

85. D. F. Williams and R. B. Marks, Accurate transmission line characterization, *IEEE Microwave and Guided Wave Letters*, vol. 3, no. 8, pp. 247–249, Aug. 1993.

86. T. Shibata and E. Sano, Characterization of MIS structure coplanar transmission lines for investigation of signal propagation in integrated circuits, *IEEE Transactions on Microwave Theory and Techniques*, vol. 38, no. 7, pp. 881–890, Jul. 1990.

87. F. Stellari and A. L. Lacaita, New formulas of interconnect capacitances based on results of conformal mapping method, *IEEE Transactions on Electron Devices*, vol. 47, no. 1, pp. 222–231, Jan. 2000.

88. I. Bahl, *Lumped Elements for RF and Microwave Circuits*, Norwood, USA: Artech IIouse Inc., 2003.

89. L. Lintao, W. Jinxiang, and F.-C. Lai, A new equivalent circuit model of MIM capacitor for RFIC, *International Conference on Microwave and Millimeter Wave Technology*, pp. 1–3, Apr. 2007.

90. T. Ytterdal et al., *Device Modeling for Analog and RF CMOS Circuit Design*, John Wiley and Sons, Inc., 2003.

91. M. N. Mariyatul, Design and modeling of on-chip planar capacitor for RF application, Master's Thesis, Universiti Teknologi Malaysia, Malaysia, May 2006.

92. R. Aparicio and A. Hajimiri, Capacity limits and matching properties of integrated capacitors, *IEEE Journal of Solid-State Circuits*, vol. 37, no. 3, pp. 384–393, Mar. 2002.

93. H. Le, J. Crossley, S. R. Sanders, and E. Alon, A sub-ns response fully integrated battery-connected switched-capacitor voltage regulator delivering 0.19 W/mm^2 at 73% efficiency, *IEEE International Solid-State Circuits Conference*, pp. 372–373, Feb. 2013.

94. J. Jiang, Y. Lu, C. Huang, W. Ki, and P. K. T. Mok, A 2-/3-phase fully integrated switched-capacitor DC–DC converter in bulk CMOS for energy-efficient digital circuits with 14% efficiency improvement, *IEEE International Solid-State Circuits Conference*, pp. 1–3, Feb. 2015.

95. N. Butzen and M. S. J. Steyaert, Design of soft-charging switched capacitor DC–DC converters using stage out-phasing and multi-phase soft-charging, *IEEE Journal of Solid-State Circuits*, vol. 52, no. 12, pp. 3132–3141, Dec. 2017.

96. J. Craninckx and M. Steyaert, A 1.8 GHz low-phase-noise CMOS VCO using optimized hollow spiral inductors, *IEEE Journal of Solid-State Circuits*, vol. 32, no. 5, pp. 736–744, May 1997.

97. L. S. Cutler and C. L. Searle, Some aspects of the theory and measurement of frequency fluctuations in frequency standards, *Proceedings of the IEEE*, vol. 54, no. 2, pp. 136–154, Feb. 1966.

98. S. M. Sze and K. K. Ng, *Physics of Semiconductor Devices*, 3rd ed., John Wiley and Sons, Inc., 2007.

99. N. Burghartz, M. Soyuer, and K. A. Jenkins, Integrated RF and microwave components in BiCMOS technology, *IEEE Transactions on Electron Devices*, vol. 43, no. 9, pp. 1559–1570, Sep. 1996.

100. P. Andreani and S. Mattisson, On the use of MOS varactors in RF VCOs, *IEEE Journal of Solid-State Circuits*, vol. 35, no. 6, pp. 905–910, 2000.

101. A. Porret, T. Melly, C. Enz, and E. Vittoz, Design of high-Q varactors for low-power wireless applications using a standard CMOS process, *IEEE Journal of Solid-State Circuits*, vol. 35, no. 3, pp. 337–345, 2000.

102. B. Razavi, *RF Microelectronics*, 2nd ed., Pearson Education, Inc., 2012.

103. D. Peterson, Varactor properties for wide-band linear-tuning microwave VCO's, *IEEE Transactions on Microwave Theory and Techniques*, vol. 28, no. 2, pp. 110–119, 1980.

104. R. Bunch and S. Raman, Large-signal analysis of MOS varactors in CMOS-Gm LC VCOs, *IEEE Journal of Solid-State Circuits*, vol. 38, no. 8, pp. 1325–1332, Aug. 2003.

105. E. H. A. Abidi, Varactor characteristics, oscillator tuning curves, and AM-FM conversion, *IEEE Journal of Solid-State Circuits*, vol. 38, no. 6, pp. 1033–1039, Jun. 2003.

106. C. Cao and K. O. Kenneth, Millimeter-wave voltage-controlled oscillators in 0.13 μm CMOS technology, *IEEE Journal of Solid-State Circuits*, vol. 41, no. 6, pp. 1297–1304, Jun. 2006.

107. C. M. Hung, Y. C. Ho, I. C. Wu, and K. O., High-Q capacitors implemented in a CMOS process for low-power wireless applications, *IEEE Transactions on Microwave Theory and Techniques*, vol. 46, no. 5, pp. 505–511, May 1998.

108. H. Wang, K. Y. Lin, Z. M. Tsai, L. H. Lu, H. C. Lu, C. H. Wang, J. H. Tsai, T. W. Huang, and Y. C. Lin, MMICs in the millimeter-wave regime, *IEEE Microwave Magazine*, vol. 10, no. 2, pp. 99–117, Feb. 2009.

Competition divides us.
Collaboration connects us.
Sharing benefits all of us.

—Kiat Seng Yeo

Collaboration and innovation go hand in hand.
Collaboration creates joy. Joy brings hope.
Innovation leads to excellence.
Excellence brings prosperity.

—Kiat Seng Yeo

Chapter 2

Transceiver Building Blocks

Recently, radio frequency (RF) energy has been prevalent around us so much so that present-day research is targeting to utilize this energy to charge autonomous Internet-of-Things (IoT) nodes seamlessly using RF energy harvesters. This is the extent of the availability of RF sources around us which is possibly from most of the modern-day appliances at home and offices, RF base stations, wireless routers, automobiles, etc. Most interestingly, the IoT nodes themselves also incorporate RF transceivers. Now, we are trying to find out what the RF transceiver is and how it is designed to enable such a remarkable trend in the modern world [1–5].

An RF transceiver is a transducer device that is used in wireless communication to transmit and receive signals as RF electromagnetic (EM) energy which can be between two mobile devices and between the baseband station and the mobile devices. The growth of the RF as integrated circuit (IC) design research is aided by the evolution of the international standards including the industrial scientific and medical (ISM) band. They are built using semiconductor technology and the scope of this book is to design RF ICs in CMOS and BiCMOS technology.

As the CMOS or BiCMOS technology nodes advance and shrink, two main advantages become evident—one is the operating

CMOS RF and mm-Wave Transceivers and Synthesizers
Bharatha Kumar Thangarasu, Nagarajan Mahalingam, Kaixue Ma, and Kiat Seng Yeo

ISBN 978-981-5129-80-9 (Hardcover), 978-1-003-67356-9 (eBook)
www.jennystanford.com

frequency is increased with an equivalent RF transceiver system bandwidth that increases based on the corresponding fractional bandwidth and the reduced wavelength is moving towards complete integration as a system-on-chip (SoC) solution.

Most wireless communication needs to support two-way communication. It also needs the system to be integrated with a receiver (RX), a transmitter (TX) and a frequency synthesizer. By integrating these three sub-systems as a single chip, the RF transceiver is complete and enables an efficient system with minimum uncertainties of the interface connections.

We try to understand each of the sub-systems in detail in the subsequent chapters. This chapter mainly focuses on the various building blocks that contribute to the development of the sub-systems of the transceiver.

A generalized RF transceiver architecture that is typically used in several wireless applications today ranging from communications to radar applications is shown in Fig. 2.1. The external interface between the transmitter and the receiver can be wireless using antennas and in wired applications, the antenna is replaced with the impedance-controlled transmission lines—instead of a dedicated antenna for transmitter and receiver front ends, by incorporating either a transmit/receive (T/R) switch and the duplexer to direct the signals from the transmitter output to the antenna while isolating that signal from the receiver input which is operating in the time-domain duplexing (TDD) and frequency-domain duplexing (FDD) modes, respectively [6–9].

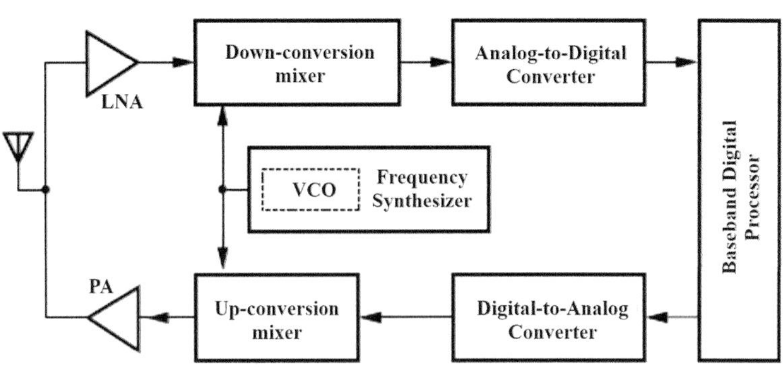

Figure 2.1 Generic RF transceiver for RF communication.

The impedance matching network (MN) elements and other passive frequency-selective circuit elements such as transformer baluns, directional couplers and filters are used in the RF transceiver system as shown in Fig. 2.1.

The received signal from the antenna is directed via the T/R switch or the duplexer to the receiver chain path. The first stage of the receiver is usually called low-noise amplifier (LNA) based on its gain, amplifies the received signal before the down-conversion mixer, and suppresses the noise contribution of the subsequent stages in the receiver chain. The receiver chain mixer frequency down-converts the received signal by mixing it with a stable internally generated signal called local oscillator (LO) signal to obtain the low-frequency baseband (BB) or intermediate frequency (IF) signal. In principle, the receiver suppresses all the unwanted noise and undesired signals such as interferences. In the transmitter chain, a baseband low-frequency signal is up-converted and amplified by a power amplifier (PA) before transmitting signals to the antenna. The transmitter hence drives the power and translates the frequency of a baseband signal to that as required by the RF transmitter to establish communication with the dedicated receiver [10–15].

To accomplish the required transmitter and receiver functions, RF and microwave components must be developed either individually or as part of an integrated circuit. The remainder of this section will examine issues related to individual components that comprise the radio. The details about the frequency synthesizer will be covered in Chapters 6 and 7.

2.1 Amplifiers

Amplifiers are active blocks needed in the RF transceivers to enhance the strength of the signal for better further processing of the signals. The amplifiers are circuits which require supply voltage to bias the active devices which enables them to amplify the output signals in accordance with the input signals.

The main performance parameter of an amplifier is the gain which is the amplification factor that depends on the source and load impedances. Hence, to have a generalized model of the amplifier becomes difficult as the gain is a function of two different variables namely the source and load impedances. Generalized amplifiers can be better realized as a black box with a two-port network with four terminals as shown in Fig. 2.2—two terminals at the input with voltage across the terminals and the current through them. Likewise, there are two terminals at the output. Hence, we can have two simultaneous equations for each output variable which are individually dependent on the two input independent variables.

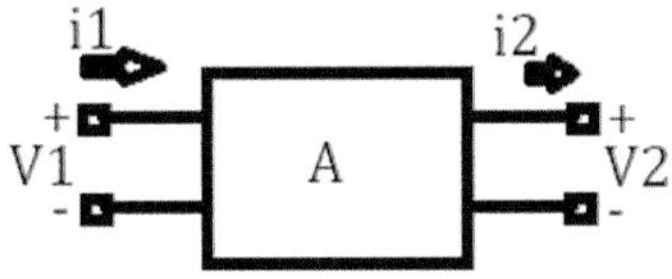

Figure 2.2 Generalized representation of an amplifier.

$$v_2 = a \cdot v_1 + b \cdot i_1 \tag{2.1}$$

$$i_2 = c \cdot v_1 + d \cdot i_1 \tag{2.2}$$

We find an interesting insight into the amplifier characteristics based on Eqs. (2.1) and (2.2). By suitable substitution, the amplifier transfer function can be determined as follows:

$$a = \left(\frac{v_2}{v_1}\right)\Big|_{i_1=0} \text{ is the voltage gain} \tag{2.3}$$

$$b = \left(\frac{v_2}{i_1}\right)\Big|_{v_1=0} \text{ is the transimpedance} \tag{2.4}$$

$$c = \left(\frac{i_2}{v_1}\right)\Big|_{i_1=0} \text{ is the transconductance} \tag{2.5}$$

$$d = \left(\frac{i_2}{i_1}\right)\Big|_{v_1=0} \text{ is the current gain} \tag{2.6}$$

Based on these four *a*, *b*, *c*, and *d* transfer function parameters as shown in Eqs. (2.3) to (2.6), the amplifiers can be classified as voltage amplifiers, transimpedance amplifiers (TIA), transconductance amplifiers (TCA), and current amplifiers, respectively. One exception when we consider RF amplifiers is the PA in which both voltage amplification and current amplification are possible simultaneously, based on an optimum load impedance condition that improves either the efficiency or the linearity performance and a detailed study. Alternatively, an LNA is an RF amplifier based on an optimum source impedance that improves the noise figure and gain performance. These two RF amplifiers, namely the PA and LNA are predominantly used in the RF front end mainly in the transmitter and the receiver chain, respectively. The third type of amplifiers seen in the RF transceivers are the variable gain amplifiers and by having the possibility to vary their gain based on a control signal enables them to be used as reconfigurable amplifiers that can vary the overall gain of the transmitter and receiver systems. These three RF amplifiers are discussed in detail in the next sub-sections.

2.1.1 Low-Noise Amplifier

A receiver with a wide input power range also called its dynamic range is desirable. Hence, the lower limit on this dynamic range is the sensitivity which is also called the minimum detectable received signal. To enable the receivers to operate at the small input powers around its sensitivity, the input is to be amplified to suppress the noise of the subsequent stages in the receiver chain along with its reduced noise contributions. Hence, the two main requirements of the LNA are the high gain and the low noise figure (NF). The LNA is the most crucial block in determining the overall receiver chain performance in a wireless RF transceiver and is the receiver front end amplifier. The LNA's noise figure has the predominant contribution on the overall receiver's NF and eventually on the receiver chain's sensitivity. The LNA, in addition to minimizing the overall receiver sub-system NF, provides a significant gain that will minimize the receiver chain non-linearities.

2.1.1.1 General considerations of the LNA design

LNA being the receiver front end amplifier has to be designed with the following considerations to be effective in the receiver chain.

(a) NF

The NF of the LNA design without any scaling will be directly added to the receiver's overall NF and becomes the most dominant contributor in addition to the input devices such as the antenna feed network, T/R switch or duplexer, filter such as the image reject filters, hybrid couplers, etc., and finally the scaled noise contributed by the subsequent blocks after LNA in the receiver chain. The NF of LNA has two main contributions, one is the active device's noise which is usually represented as the input-referred noise and the second contribution is from the source resistance as provided from the amplifier's input path. Hence from Fig. 2.3 below we can observe these two main noise contributors as two voltage sources as $\overline{V_{n,\text{in}}^2}$ and $4kTR_s$, respectively.

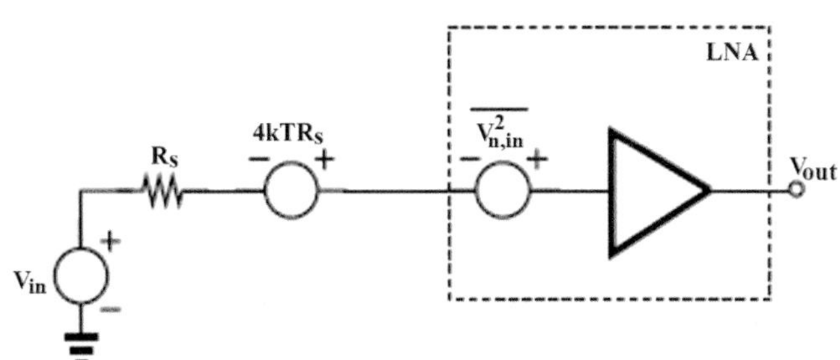

Figure 2.3 Noise contributors in LNA.

$$\text{NF} = \frac{\overline{V_{n,\text{out}}^2}}{A_v^2} \cdot \frac{1}{4kTR_s} = 1 + \frac{\overline{V_{n,\text{in}}^2}}{4kTR_s} \tag{2.7}$$

From Eq. (2.7), we can observe that for a low NF of the LNA, the input should be low loss and the transistor is to be selected with a low noise figure.

(b) Gain

The LNA is desired to have a large gain to suppress the noise contributed by the subsequent receiver chain blocks such as mixer, baseband amplifiers and filters as shown in Fig. 2.4. The LNA's gain is determined based on suitable tradeoff between the noise figure and the linearity performance of the receiver chain. Larger gain will affect the linearity but will improve the NF performance and in contrary, lower gain will result in improved linearity, but degraded NF. In low-power receiver design topology, the LNA is directly interfaced to the down-conversion mixer without any intermediate impedance matching networks. In this case, it is more suitable to analyze the receiver performance based on the LNA's voltage gain instead of power gain.

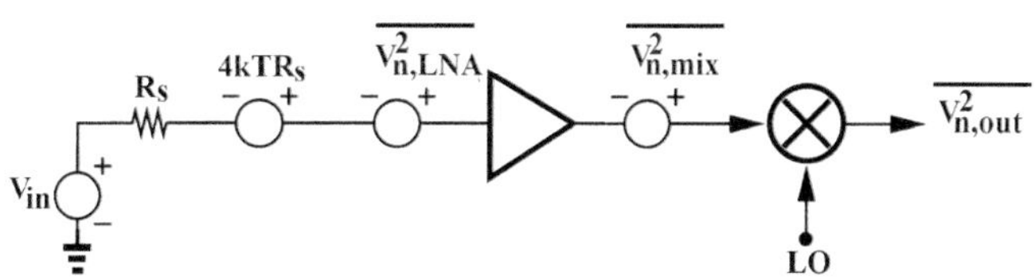

Figure 2.4 NF of the RX chain.

Hence, the noise figure of the LNA together with the mixer is considered in the NF calculations such that, $NF_{total} = NF_{LNA} + NF_{mixer}/(\text{Voltage gain}_{LNA})^2$ given as

$$\mathrm{NF_{tot}} = \frac{A_{v1}^2(\overline{V_{n,\mathrm{LNA}}^2} + 4kTR_s) + \overline{V_{n,\mathrm{mix}}^2}}{A_{v1}^2}\frac{1}{4kTR_s}$$

$$= \mathrm{NF_{LNA}} + \frac{\overline{V_{n,\mathrm{mix}}^2}}{A_{v1}^2}\cdot\frac{1}{4kTR_s}. \tag{2.8}$$

Similarly, for the linearity performance is shown as

$$\frac{1}{\mathrm{IP}_{3,\mathrm{tot}}^2} = \frac{1}{\mathrm{IP}_{3,\mathrm{LNA}}^2} + \frac{\alpha_1^2}{\mathrm{IP}_{3,\mathrm{mixer}}^2}, \tag{2.9}$$

where α_1 is the voltage gain of the LNA.

at the intercept point of the fundamental output component and the third-order non-linearity output component.

(f) Bandwidth

The LNA being the front end amplifier must provide a wide –3 dB bandwidth to ensure the gain variation in the operating frequency range is below 1 dB and provide a flat response will minimal passband ripple, The LNA design is to be implemented as a band-pass filter response designed for a fractional bandwidth over the center frequency which also determines the quality factor (Q) of the design defined as the ratio of the center frequency to the bandwidth.

2.1.1.2 Example design of a low-noise amplifier

We will study one example to better understand the design concepts involved in the LNA design which is a cascode CS stage with inductive degeneration shown in Fig. 2.5. The input cascode stage with inductive load provides the flexibility of lower VDD as compared to the resistive loaded CS stage. Additionally, the degeneration inductor and the cascode topology together ensure the design becomes unconditionally stable over any variation of the source impedance. The degeneration inductor also ensures an improved noise figure performance.

The design of the low-noise amplifier with the operation frequency in the 2.4 GHz ISM band starts with the determination of the transconductance g_m.

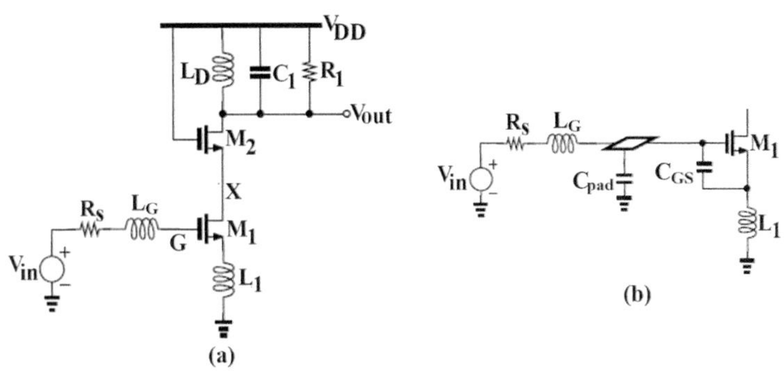

Figure 2.5 (a) Cascode CS LNA with inductive degeneration and (b) Inclusion of parasitics.

1. For low-power design using a 28 nm CMOS process, the drain current I_d is used to bias as 1 mA. This is based on the process selected will provide a g_m = 20 mS, for an input device selected as RF NMOS transistor, the aspect ratio (w/l) is determined as the over-drive voltage = V_{dsat} = 100 mV, and K_P = 250 μA/V^2.

$$V_{ds,sat} = \sqrt{\frac{(2 \cdot I_d)}{\left(K_n \cdot \frac{w}{l}\right)}} \tag{2.13}$$

From this we can get the transistor size as

$$\frac{w}{l} = \frac{(2 \cdot I_d)}{(K_n \cdot V_{ds,sat}^2)} = \frac{(2 \times 1\ \text{mA})}{(250\ \mu\text{A}/\text{V}^2 \times (0.1\ \text{V})^2)} \approx 800.$$

For a conservative design, we can set w/l = 1000.

For l_{min} = 0.03 μm, w = 30 μm

$$\frac{w}{l} = \frac{30\ \mu\text{m}}{0.03\ \mu\text{m}}$$

2. Next step, we check the input parasitic capacitance

$$C_{gs} = 9\frac{\text{fF}}{\mu\text{m}^2} \cdot w \cdot l = 9\text{fF} \times 60\mu\text{m} \times 0.06\mu\text{m} = 32.4\,\text{fF a}$$

$$C_{ds} \approx 0.3\ \text{fF} \times 60\ \mu\text{m} = 18\ \text{fF}$$

Total capacitance with layout parasitics = C_{gs} + C_{ds} ≈ 50 fF × 2 = 100 fF

3. Next, we determine the degeneration inductance, L_S, using

$$g_m \cdot \frac{L_s}{C} = 50\ \Omega$$

$L_s = 50\ \Omega \times \frac{C}{g_m} = 50\ \Omega \times \frac{100\ \text{fF}}{20\ \text{mS}} = 0.25\ \text{nH}$, which is possible to include as an on-chip inductor.

4. Next step is to determine the series input inductance

$$f_o = 2.45\ \text{GHz} = \frac{1}{\left[2\Pi \cdot \sqrt{(L_{\text{ext}} + L_S) \cdot C}\right]}$$

$$= \frac{1}{\left[2\Pi \cdot \sqrt{(L_{\text{ext}} + 0.25\ \text{nH}) \cdot 100\ \text{fF}}\right]}$$

We find that L_{ext} = 42 nH which is quite large for an on-chip inductor.

As a practical LNA design, we consider including the ESD bond pad capacitance of 100 fF, the bond-wire inductance of about 1 nH and package capacitance of 100 fF along with the input transistor's parasitic capacitance of 100 fF. By neglecting the bond-wire inductance, the three parasitic capacitances are effectively in parallel resulting in an overall capacitance of 300 fF.

Now, this practical consideration will result in a L_{ext} of 15 nH which can be practically implemented as on-chip inductor. This will provide a good first-order estimation of the LNA design which will further need to be optimized to cater for the higher-order uncertainties in the design such the device models and the design parasitics.

2.1.2 Power Amplifier

Transmitters are required to send out signals either through a wireless or a wired channel with larger power levels to the transmitting antenna or cables, respectively, and hence extend the communication distance. In this process, the PA design is to be such that there are no additional spurs in the transmitter output and the signal should also be with reduced distortions, causing the signals to be entirely delivered without any disruptions. In addition, this transmitter's front end amplifier is the most DC power consuming building block in the entire RF transceiver. Altogether, the PA must include high power added efficiency (PAE) and high linearity which are the complementary requirements. Hence, the PA design is needed to have a better tradeoff between efficiency and linearity performance.

Typically, a PA is an amplifier in the transmitter front end with several constraints as follows,

1. As an example, if a transmitter is required to deliver a 0.5 W (+27 dBm) of RF power to a 50 Ω load as the transmit antenna. The output peak voltage amplitude denoted as V_p and the peak-to-peak voltage swing as V_{pp}, at the antenna by calculation reaches a maximum of 14 V and the peak current through the load will be 280 mA.

   For a CMOS-based CS amplifier stage as shown in Fig. 2.6(a) when used to drive such a resistive load, it requires a supply voltage to be greater than V_{pp}. However, if the load at the drain terminal is implemented using an inductor connected to the supply voltage V_{DD} as shown in Fig. 2.6(c), the drain voltage swing's peak exceeds the V_{DD} voltage and can reach twice the V_{DD}. The voltage is dropped across the V_{DS} of the device which may exceed the device's breakdown voltage. One immediate solution is to reduce the V_{DD} supply voltage at the expense of reduced transition frequency (f_T) and reduced output power, but there is still the exceeded voltage from the 50 Ω load which has to be addressed.

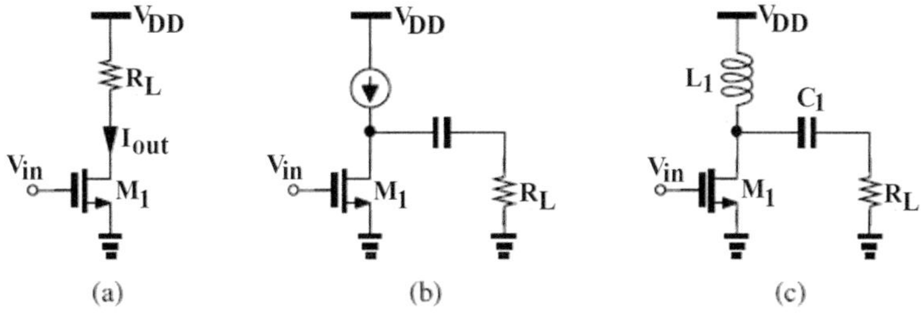

Figure 2.6 (a) Basic CS amplifier (b) Current sourced CS amplifier (c) Inductive loaded CS amplifier.

2. The efficiency of a PA is critical in the overall RF transceiver design as the PA is the most power consuming block. There are two metrics to define the efficiency of PA namely the drain efficiency and PAE. The main difference between them is that in the latter metric, the gain of the PA is also considered which provides a better insight of the overall

performance. Achieving ideal efficiencies close to 100% is practically difficult and for a linear PA it is usually about 50%, hence for an output power of 0.5 W the DC power consumption will be about 1 W. Hence, the PA is to be designed to minimize the losses to have better efficiency performance.

3. Linearity of PA is the second critical complementary consideration in which intermodulation distortion and gain compression are the predominant quantities. The inter-modulation distortion is mainly determined using the adjacent channel power rejection ratio (ACPR) using two-tone test and for the gain compression the test is based on single tone input signal by gradually increasing the signal amplitude and observing the output signal spectrum.
4. As we can see, by reducing supply voltage to say about 1 V, the maximum voltage swing is limited to a maximum of about twice the V_{DD} which is 2 V. Therefore, a voltage swing of 14 V with a load current of 280 mA is required to drive the output power of 0.5 W across a 50 Ω load. This difference in the voltage can be mitigated by an output matching network with a voltage transformation ratio of 1:7. In order to simultaneously achieve optimal power transfer, conserve energy, and disregard the matching network's insertion loss, the reciprocal of the same current ratio from the secondary to the transformer's primary coil will yield a drain current of around 2 A. This large current may need to be conducted by the transistor channel that may result in device reliability issues.

As a PA designer, these above design constraints are to be carefully considered and a suitable design topology along with bias condition that determines the operating class of PA is to be selected depending on the requirement specifications.

The PA designs have been mainly classified based on improved linearity such as classes A, B, AB and C and high efficiency namely the classes D, E, and F as shown in Table 2.1. The class F is an extension of the linear PAs with harmonic suppression. The former classes of PA are grouped as linear PA and the latter are considered as switching PA.

Figure 2.7 shows the bias conditions of comparison of linear PA classes against conduction angle and power conversion efficiency.

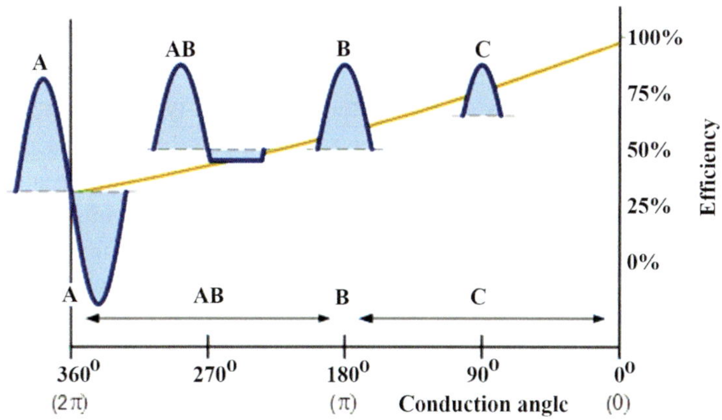

Figure 2.7 Classes of linear PA.

Table 2.1 Comparison of PA classes including both linear and switching PAs

	Linear mode PAs	Switching mode PAs
PA classes	A, B, AB, C, J	D, D^{-1}, E, F^{*}, F^{-1*}
Transistor operation	RF current source	RF switch - ON/OFF
Linearity	Good	Limited
Drain efficiency	Medium-low	High (~100%)
Modulations	Varying amplitude (ASK, QPSK, QAM, etc.)	Constant amplitude (BPSK, GFSK, etc.)
Harmonics	Suppressed	Waveform shaping
Gain	High	Low
Operating frequency	High	Low-medium

*Extended to linear mode PAs

A sample Class AB PA design procedure steps is as follows:

1. The PA design starts with the determination of the class of PA based on the specifications, mainly the linearity and the efficiency. If the linearity is more critical than the

efficiency, then we can select the linear PA starting from class A and subsequently depending on the exact efficiency requirement we can select the PA class towards class AB, class B and class C. In the example design, the P_{out} is +10 dBm and efficiency > 50%. Hence, we select a linear PA as class AB.

2. From this we determine the transistor type, aspect ratio and the bias condition based on the output power level requirement such as transition frequency (f_T) to be larger than the operating center frequency, threshold voltage (V_{TH}) based on the operating voltage signal swing to be less than the device breakdown voltage, and large transconductance (g_m).
3. From the DC operation point, we can determine the g_m as 100 mS, and the optimum load R_{Lopt} = 60 Ω based on a supply voltage of 1.1 V.
4. To determine the bias condition there are three notable methods namely the class of PA, for third-order term g_{m3} = 0, and constant first-order g_m region along the load line. Hence the bias voltage is in the range from 0.45 V to 0.7 V depending on the above-mentioned criteria.
5. Then we can check the stability factors and ensure the main PA device is unconditionally stable.
6. Then design the output matching network based on load pull analysis to get the most suitable network with minimum matching network components.
7. To design the input-matching network with minimum components and is matched to 50 Ω source impedance.
8. Subsequently more practical design constraints such the parasitics of the layouts, ESD bond pads and the bond-wires are to be included in the simulation test bench.

2.1.3 Variable Gain Amplifier

A variable gain amplifier (VGA) is one of the key RF transceiver front end building blocks that enable wireless mobile communication. The VGA gain control range is determined by the receiver dynamic range that can provide regulated output power to the baseband chipset. The research towards the recent VGA design is to target for high data rate communication in giga-bit per second (Gbps) range. Hence the VGA must support a wide bandwidth. As there is constant improvement of the state-of-the-art performance, the applications as well as the design capacity for overall RF transceiver system integration progresses gradually towards an SoC solution.

For the VGA to be integrated in a single-chip RF transceiver SoC, it is desirable to have certain improved features such as lower DC power consumption, smaller die area, better input and output impedance matching, insensitive to process-voltage-temperature (PVT) variations, linear gain control input, broad bandwidth, and seamless digital baseband interface independent of the gain control setting.

Based on the gain control, the VGA are classified as the discrete or digital gain control VGA (DVGA) or programmable gain amplifier (PGA) and analog gain control (VGA). In analog VGA the gain control is to be provided as an analog signal and the digital baseband has to integrate a digital-to-analog converter (DAC) to achieve gain control. Unlike the DVGA in which the digital baseband can be directly interfaced without any DAC.

By using binary weighted digital control, a DVGA can be implemented by switching between several fixed gain and attenuation stages or by using an array of passive or active circuit components selected discretely by using switches. The switching gain stage DVGA consumes a large die area due to unit circuit multiplication, while the DVGA with switching circuit components is alternatively compact and also consumes less current consumption with smaller gain control linearity errors.

Usually, the application demands the desired accuracy resolution of the reconfigurable gain, and the type of gain variation as either discrete (digital) or continuous (analog) characteristics. It can be achieved by DVGA or VGA, respectively, in which the parameter control is used as a switch and operated in either the saturation or cutoff regions while for the analog control the parameter control is in the linear active region as shown in Fig. 2.8.

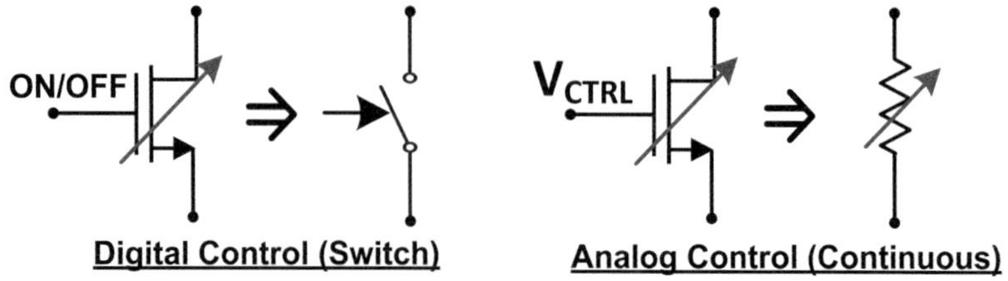

Figure 2.8 Types of gain control modes.

A well-known gain control option is the VGA and was designed initially before being implemented as DVGA. The VGA gain control can be obtained by varying the device parameters that affect the amplifier gain as shown in Fig. 2.9.

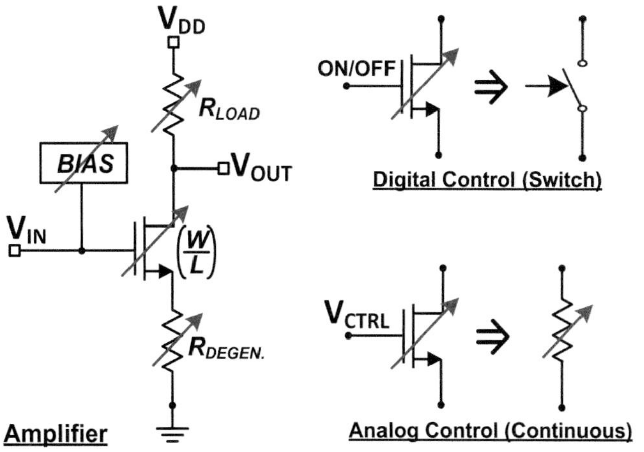

Figure 2.9 Amplifier gain variation options to achieve the variable gain.

The gain of such a typical amplifier is defined as

$$A_V = \frac{g_m \cdot R_{\text{LOAD}}}{(1 + g_m \cdot R_{\text{DEGEN}})}, \tag{2.14}$$

where g_m is the transconductance, R_{LOAD} the load resistance and R_{DEGEN} the degeneration resistance. The transconductance, g_m, is related to the amplifying transistor size and the bias operating condition given as

$$g_m = f\left(\text{BIAS}, \left(W/L\right)\right). \quad (2.15)$$

The gain control can be achieved by DVGA or PGA or as continuously varying VGA one or more of the amplifier's device parameters using continuous controllable or discrete switches that can tune or select or either the biasing current or voltage, load resistance, degeneration resistance or the transistor aspect ratio, that affect the amplifier gain performance.

Practically, the gain is stated in decibel (dB) scale which has a logarithmic function defined as

$$A_V\big|_{\text{dB}} = 20 \cdot \log_{10}(A_V) = 20 \cdot \log_{10}\left[\frac{g_m \cdot R_{\text{LOAD}}}{\left(1 + g_m \cdot R_{\text{DEGEN}}\right)}\right]. \quad (2.16)$$

By proper design selection, we can achieve $g_m \cdot R_{\text{DEGEN}} << 1$ such that with R_{LOAD} as constant then g_m is an exponential function of a parameter such as "a", then the gain in dB ($A_V|_{\text{dB}}$) will become linearly proportional to the exponential parameter, "*a*." This property of gain control is called linear-in-decibel or dB-linearity which is most preferred in VGA or DVGA design.

2.2 Mixers

Since passive mixers are a common feature in transceivers today. This chapter will examine and comprehend the fundamentals of their operation. We examine transceivers with passive mixers that are powered by clocks with 50% and 25% duty cycles. It is demonstrated that the passive mixer possesses an impedance transformation attribute as a result of the absence of reverse isolation between the RF and IF ports.

This feature of the passive mixers is useful to develop blocker-resilient receivers because it can be used to frequency-shift low-*Q* baseband impedances and create on-chip high-*Q* filters with center frequencies that are accurately controlled by the LO clock. It is also disclosed how issues like *I*/*Q* cross talk and disparate high- and low-side conversion gains can result from this absence of reverse isolation. The optimal gain and linearity

These two tones with comparable magnitudes are also present at point V_B, which is the RF side of the Q channel's passive mixer. The issue arises from the components at sites V_A and V_B being 180° out of phase at the picture frequency, $f_{LO} - f_m$. This increases the difference in high- and low-side up-conversion gains by causing an image current to flow between the two channels via the two series capacitors C. The size of the series capacitors, C, determines how much current is seen in the image. As a result, a smaller C would result in a lower image current but a smaller conversion gain penalty. The circulating picture issue can be avoided by designing the transmitter in a different approach that uses a 50% passive mixer. In Fig. 2.13, the architecture is displayed. First, two independent transconductors (G_m) units with their outputs shorted together are used to buffer the two RF outputs in the I and Q channels. The PA driver units in this architecture would be more complex than those in Fig. 2.12, having four RF inputs instead of two.

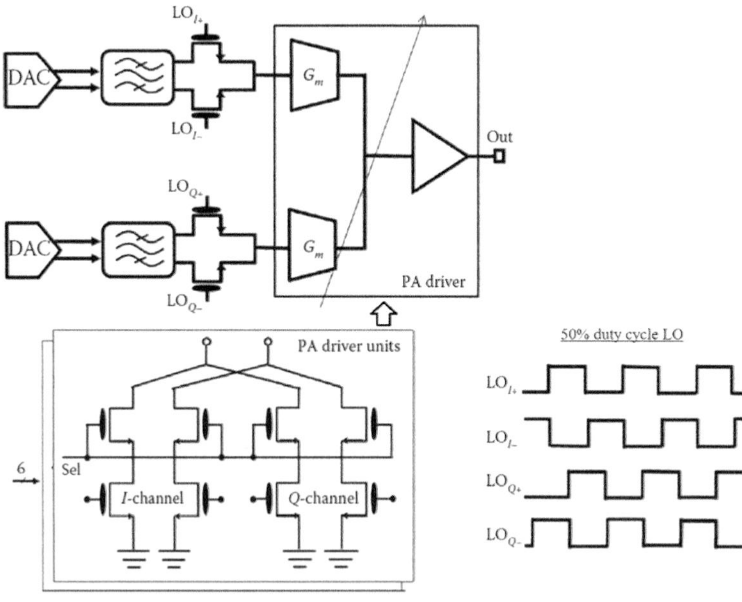

Figure 2.13 Transmitter design with 50% passive mixer.

Given the possibility of a random mismatch between the two transconductors in the PA driver units' *I* and *Q* channels, the IRR would now fluctuate across a range of gain levels. Because of this, every gain setting needs a different calibration process. Likewise, it might be necessary to calibrate the LO feedthrough across a range of RF gain values. The transmitter architecture with passive mixers maintains low noise and low supply voltage applicability despite the aforementioned disadvantages.

2.2.1.2 Design of transmitters with 25% passive mixer

As seen in Fig. 2.14, designing the transmitters with a 25% passive mixer is another method of removing the picture current. The transmitter with the 25% duty-cycle passive mixer is actually the most ideal design overall, with the exception that it requires 25% LO clocks instead of the more common 50% duty-cycle clocks. Its architecture results in the fewest possible routing parasitics and is straightforward, particularly when it comes to floor planning layout.

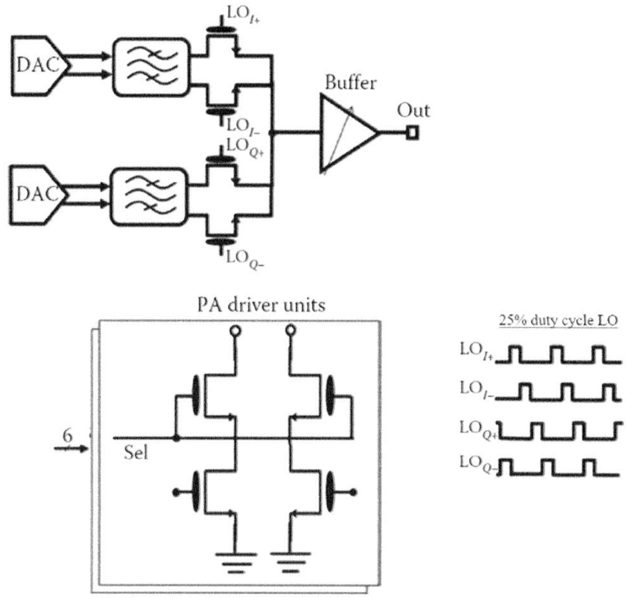

Figure 2.14 TX design with 25%-based passive mixer.

Additionally, the passive mixer is better than the active mixer in terms of noise, particularly flicker noise, and power consumption because its switches don't carry any DC currents. Furthermore, because there is no device stacking, the passive mixer-based architecture is ideal for low-voltage applications. Moreover, LO feedthrough and IRR only need to be calibrated for a single gain level if gain control is done on the RF side because the passive mixer remains unchanged throughout a range of gain settings. LO feedthrough and IRR performance can be achieved with acceptable gain settings by applying the calibration results. The calibration time is reduced with a single-point calibration like this. Additionally, with only two RF inputs and two RF outputs (assuming differential implementation), the PA driver's units are extremely basic. The sole disadvantage, as previously indicated, is the requirement for low-noise, 25% duty-cycle nonoverlapped clocks, which are easily produced from 50% LO clocks by utilizing four AND gates.

We may simplify the transmitter as shown in Fig. 2.15 to better understand how the transmitter functions with the 25% passive mixer. The I and Q low-pass filters can be swapped out for their Thevenin equivalents on the baseband side using the Thevenin theorem. The single-ended output impedance visible at the low-pass filters' outputs is known as Z_{out}. When the low-pass filter output is disconnected from the mixer switches, the voltage at that point is known as the open-circuit voltage of the I channel, or $\pm v_{\text{BB},I}$.

This voltage is obviously the DAC voltage in the I channel after the transfer function of the low-pass filter was applied. Comparably, the open-circuit voltage at the Q channel's output is represented by $\pm v_{\text{BB},Q}$. It is a reasonable assumption based on practical experience that the input impedance observed from the PA driver is capacitive on the RF side. This capacitor is the same size as C_L. We now assume the switches can be treated as linear time-variant (LTV) system when they are in the OFF state, which is an open circuit, and when they are in the ON mode, which is a resistor of size R_{SW}.

For LTV systems, the Thevenin theorem is still true. Because there are no longer any series baseband voltage sources in the switching system, which consists of four switches and four identical baseband impedances Z_{out}, the four baseband voltages can be replaced with a single RF voltage in series (the circuit on the right-hand side of Fig. 2.15). It is easy to determine the Thevenin voltage on the RF side using the following equation:

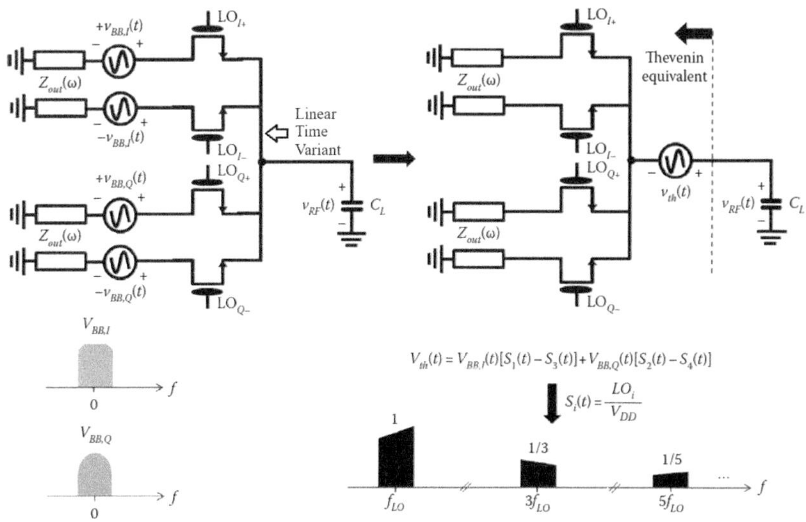

Figure 2.15 Simplified representation of the TX with 25% passive mixer.

$$v_{th}(t) = v_{BB,I}(t)[S_{I+}(t) - S_{I-}(t)] + v_{BB,Q}(t)[S_{Q+}(t) - S_{Q-}(t)] \quad (2.17)$$

where $S_{I+}(t)$ is defined as '1' if the associated switch is ON and '0' in the other case. $S_{I+}(t)$ is therefore a periodic signal with a 25% duty cycle. The definitions of $S_{I-}(t)$, $S_{Q+}(t)$, and $S_{Q-}(t)$ are similar. A frequency-domain graphic explanation of the Thevenin voltage spectrum, $v_{th}(t)$, is provided as shown in Fig. 2.16(a). It is assumed that the baseband signals in the Q and I channels have distinct frequency spectrums.

Naturally, we are aware that the I and Q channels' baseband signals are usually statistically separate signals. Additionally shown is the spectrum for the equivalent RF Thevenin voltage. The intended up-converted signal is in the region of the spectrum surrounding the LO.

Nevertheless, there are certain unpleasant up-converted components surrounding higher-order odd harmonics. The frequency components that are up-converted about $3f_{\text{LO}}$ are the same as those around f_{LO}, but they are mirrored around the center and scaled by a third. The scaling factor is 1/5 and there is no mirroring at $5f_{\text{LO}}$. The remaining higher-order odd harmonics of the LO follow a similar pattern. We shall return to this later. The transmitter's linearity performance is negatively affected by the mirrored spectrum surrounding $3f_{\text{LO}}$.

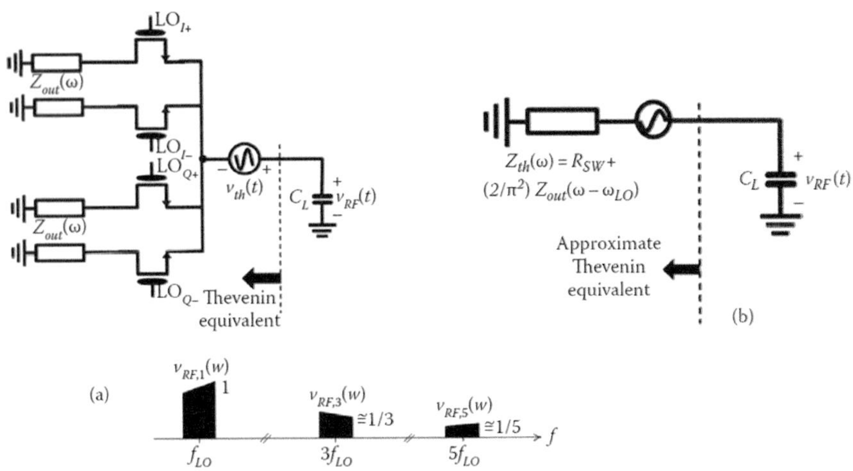

Figure 2.16 System transfer function from BB to RF (a) Thevenin's voltage equivalent circuit (b) basic Thevenin's voltage equivalent circuit by ignoring higher harmonics.

The nature of the LTV system now allows all the frequency components of the Thevenin voltage around f_{LO}, $3f_{\text{LO}}$, $5f_{\text{LO}}$, ... to contribute to the final RF signal across Z_L about f_{LO}, as illustrated in Fig. 2.16(a). It can be demonstrated that the following expression represents the ultimate RF voltage across the RF impedance near f_{LO}:

$$V_{RF,I}(\omega)=\frac{\sqrt{2}}{\pi}\frac{Z_L(\omega)}{Z_L(\omega)+R_{\text{SW}}} \times\frac{e^{j\pi/4}V_{\text{BB},I}(\omega-\omega_{\text{LO}})+e^{-j\pi/4}V_{\text{BB},Q}(\omega-\omega_{\text{LO}})}{1+\frac{2}{\pi^2}Z_{\text{out}}(\omega-\omega_{\text{LO}})\sum_{p=-\infty}^{+\infty}\frac{1}{(4p+1)^2\left[Z_L(4p\omega_{\text{LO}}+\omega)+R_{\text{SW}}\right]}},\tag{2.18}$$

where R_{SW} is the switch resistance and Z_L is the impedance of the RF load, represented by the capacitor C_L in Fig. 2.16(a). Furthermore, the RF voltage at $3f_{LO}$ is provided by

$$V_{RF,3}(\omega)=\frac{\sqrt{2}}{3\pi}\frac{Z_L(\omega)}{Z_L(\omega)+R_{SW}}$$
$$\times\frac{e^{-j\pi/4}V_{BB,I}(\omega-3\omega_{LO})-e^{-j\pi/4}V_{BB,Q}(\omega-3\omega_{LO})}{1+\frac{2}{\pi^2}Z^*_{out}(\omega-3\omega_{LO})\sum_{p=-\infty}^{+\infty}\frac{1}{(4p+1)^2\left[Z_L(4p\omega_{LO}-\omega)+R_{SW}\right]}}.$$

(2.19)

This is the frequency components surrounding f_{LO} mirrored. For the highest conversion gain for the transfer function, the switch resistance R_{SW} and $(2/\pi^2)$ Z^*_{out} must be significantly smaller than the impedance of Z_L at the LO frequency because Z_L is essentially a capacitor and there is no resonance (inductor in shunt with C_L). This indicates that the baseband filter's output impedance has to be low.

R_{SW} and Z^*_{out} could have values of 10 Ω or less as typical values. It should be noted that the baseband low-pass filter's Z_{out} was supposed to be low pass with a cutoff frequency significantly lower than the LO frequency in the calculations of Eqs. (2.18) and (2.19). This estimate is accurate for many real-world designs.

In addition to being difficult to work with, Eqs. (2.18) and (2.19) lack intuition. Figure 2.16(b) provides a good approximation, where a linear time-invariant (LTI) impedance is used to simulate the switching section of the Thevenin equivalent in Fig. 2.16(a). The following equation approximates this new Thevenin impedance:

$$Z_{TH}(\omega)=R_{SW}+\frac{2}{\pi^2}Z_{out}(\omega-\omega_{LO}) \tag{2.20}$$

Therefore, the corresponding $Z_{out}(\omega-\omega_{LO})$ is a frequency-shifted baseband output impedance that is frequency-shifted to the LO with a scaling factor of $(2/\pi^2)$. The venin impedance is equal to R_{SW} in series with Z_{out}. The computations are made simpler because this approximation is based on LTI and the model no longer includes a switching system. This LTI technique predicts the transfer function from the baseband to the RF voltage across C_L, which is given by

$$V_{\mathrm{RF},I}(\omega) \cong \frac{\sqrt{2}}{\pi}\left\{e^{j\pi/4}V_{\mathrm{BB},I}(\omega-\omega_{\mathrm{LO}})+e^{-j\pi/4}V_{\mathrm{BB},Q}(\omega-\omega_{\mathrm{LO}})\right\}$$
$$\times\frac{Z_L(\omega)}{Z_L(\omega)+R_{\mathrm{SW}}+\frac{2}{\pi^2}Z_{\mathrm{out}}(\omega-\omega_{\mathrm{LO}})}. \quad (2.21)$$

When the input frequency is swept from 1.5 GHz to 2.5 GHz (the LO frequency is 2 GHz), and Fig. 2.16(a) depicts the transfer function derived from the simulation results and compares it against the LTV forecast of Eq. (2.18) and the LTI-based prediction of Eq. (2.21). The LTI-based approximation has an inaccuracy of less than 1 dB, whereas the LTV-based approach predicts the transfer function with the maximum precision.

Considering how easy it is to use, it is worthwhile. A zoom into the frequency band of ±20 MHz over the LO is shown in Fig. 2.16(b). It appears that the conversion gain from baseband to RF can vary by up to 1 dB between −20 and +20 MHz. The low-pass filter's ripple would be increased by this ripple. Certain restrictions are placed on the allowed ripple by the constraints on the error-vector magnitude (EVM) of each transmission standard.

As a result, in contrast to the active mixer design, the high- and low-side conversion gains may differ in the passive mixer design. We now understand that the absence of reverse isolation between the RF and IF ports is the cause of the discrepancy between the high- and low-side conversion gains. The inquiry pertains to how the transmitter's high-side and low-side conversion gains might be reduced.

There are two answers to the conversion gain formula in Eq. (2.18) if we look at it: (1) The baseband filter's output impedance can be made extremely low over the appropriate baseband signal bandwidth, such as, in the case of WCDMA, from DC to 2 MHz. Baseband frequencies make this quite feasible because we can always use feedback. (2) The output impedance, Z_{out}, can be designed to be predominantly real over the bandwidth of the baseband signal.

To guarantee that the low-pass filter's output impedance, Z_{out}, is as low as possible (less than 5 Ω) throughout the baseband signal's bandwidth—for instance, from DC to 2 MHz in the case of WCDMA—a buffer can be placed after it, as illustrated in Fig. 2.17. To avoid negatively affecting the transmitter's linearity performance, this buffer should also be very linear. In order to maintain the transmitter's out-of-band noise requirements—particularly in the receive band, which typically imposes demanding requirements—the buffer must have extremely low noise.

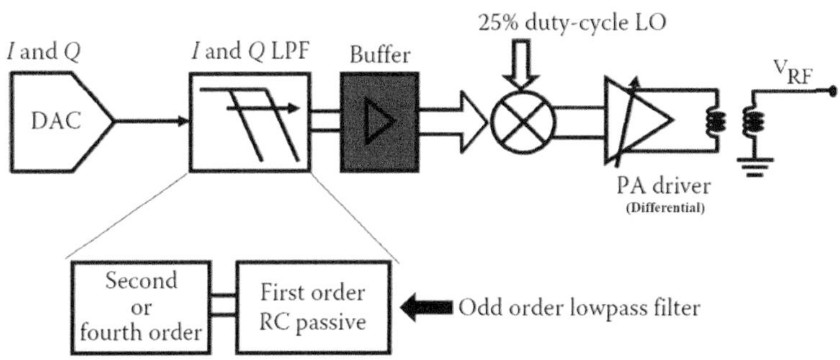

Figure 2.17 TX chain with LPF, buffer and the 25% passive mixer.

First-order RC passive filters are typically used as the final level of baseband filtering to reduce noise from the DAC and earlier low-pass filter stages (complex poles). Thus, when the baseband portion is taken into account, the buffer stage is the only one that significantly contributes to the receive-band noise. The buffer needs to be built with little noise from the start in order to achieve this. Creating a buffer with a very low output impedance over the intended baseband signal bandwidth and a high output impedance at higher frequencies would be a very good design strategy. In this manner, the up-conversion gain for the desired baseband signal is maintained high, but the up-conversion gain for far-out noise components would be lowering Z_{out} in Eq. (2.21) is significant for these components.

It was clarified that in addition to the intended signal at f_{LO}, there are up-converted signals at $3f_{LO}$, $5f_{LO}$, and so forth on the RF side of the passive mixer, which is the PA driver input. The up-converted signal is one third of the desired signal in magnitude and is flipped over its center with respect to the desired signal at f_{LO} around $3f_{LO}$. The signal is not inverted and has a one fifth magnitude around $5f_{LO}$.

If the mixer outputs contained solely the intended signal, spectrum regrowth, as measured by the adjacent channel leakage ratio (ACLR), would result from the PA driver's third-order non-linearity. The ACLR is defined as the power of the leaked signals to the adjacent channels divided by the power of the targeted signal. The expanded signal produces frequency components in the adjacent channels.

Thus, we can have upper and lower ACLR numbers based on this criterion. What is the effect of the existence of up-converted frequency components around $3f_{LO}$, $5f_{LO}$, etc. on ACLR? Here we simply consider the frequency components in the vicinity of $3f_{LO}$. The desirable signal and the undesired components at $3f_{LO}$ would combine and produce more undesired components around f_{LO} because of the third-order non-linearity of the PA driver. Some of these elements would overlap the intended channel, and the remainder would produce frequency components in the channels that are next to it, which would reduce the effectiveness of the ACLR. Mathematically, it can be demonstrated that the generated signal owing to the PA driver non-linearity around f_{LO} is completely different from the desired signal with different statistics since the up-converted signal around $3f_{LO}$ is mirrored in the frequency domain. Only EVM would be impacted by components that fell over the acceptable signal spectrum, but even so, the influence on EVM is usually minimal. However, ACLR can be severely deteriorated by frequency components falling on neighboring channels. This degradation, for instance, is roughly 6 dB in a WCDMA transmitter. This means that, for a particular PA driver non-linearity, when the input consists just of the intended signal and all three LO components are involved, the ACLR that can be preserved will be deteriorated by 6 dB (Fig. 2.18). The result increases the difficulty of meeting the PA driver's already strict linearity criteria. Before the final signal is applied to the PA driver in an active mixer configuration, a tuned LC load can be used to filter out the first three LO components and beyond; this is not as simple to accomplish in a passive mixer situation.

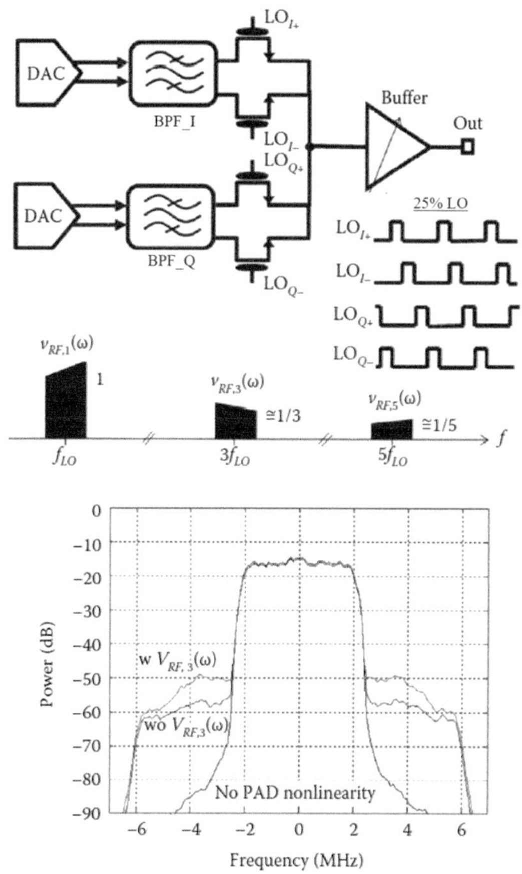

Figure 2.18 TX chain PA driver stage linearity requirements.

2.2.2 Down-Conversion Mixer

A zero- or low-IF receiver's typical construction is depicted in Fig. 2.19. The targeted signal and a few undesirable in-band or out-of-band interferers make up the received signal in the antenna. There is a possibility that the out-of-band blockers will be millions of times stronger than the desired signal. The weak desired signal needs the gain of the LNA to be set as high as possible to overcome the noise of the subsequent stages.

Strong out-of-band blockers are traditionally attenuated and kept from compressing the LNA by placing an external SAW filter in front of the on-chip LNA. Apart from a passband loss known as insertion loss, the SAW filter passes the signal along with the intended in-band blockers without causing any appreciable attenuation.

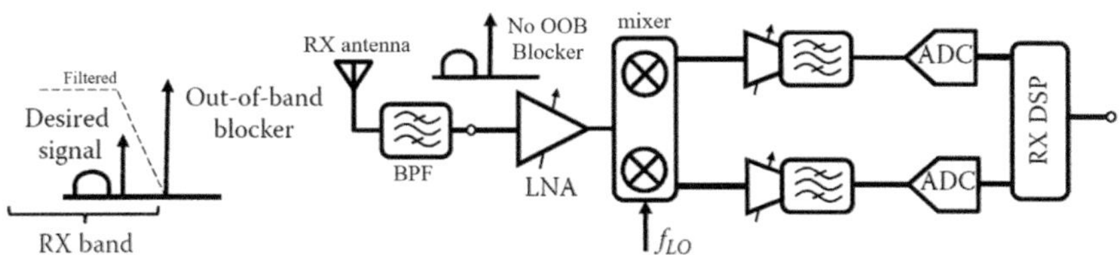

Figure 2.19 Down-conversion mixer in RX chain.

After that, an I/Q down-conversion mixer processes the desired signal into zero or low-IF. Using low-Q baseband low-pass filters, the in-band blockers that were nearly hard to attenuate in the RF stage can now be removed in the IF stage. After passing through this low-pass filter, the composite down-converted signal and blockers are attenuated to the point where two I/Q analog-to-digital converters (ADCs) with appropriate resolution and power consumption can digitize the baseband signals. This includes both in-band blockers and residues of out-of-band blockers.

The realm of digital signal processing (DSP) handles the remaining receiver functionality. In order to ease the filtering processes, RF filters that are unduly sharp are not necessary thanks to the down-conversion operation.

As a result, this receiver's core invariably has a mixer that translates frequencies, moving the chosen channel from the RF to the baseband. Any mixing system must be either non-linear or linear; if linear, it must be a time-variant system since no additional frequency components are formed at the output in an LTI system. Present-day receivers employ LTV mixers, which come in active or passive varieties, as shown in Fig. 2.20.

The well-known active mixer is Fig. 2.20(a). The RF voltage originating from the LNA output serves as the input. An RF current is created by converting the RF voltage and superimposing it over the bias current I_{BB}. The RF signal is down-converted to baseband as a result of the mixer's switches commutating both the bias current and the RF.

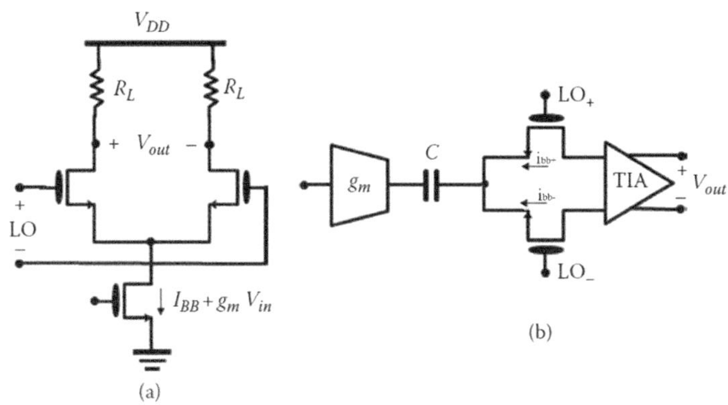

Figure 2.20 Down-conversion mixers in RX chain (a) active and (b) current-based passive mixers.

There are two known problems with this active mixer. First, it has been demonstrated in the literature that the bias current commutated by the switching pair is directly proportional to the output noise of an active mixer caused by the flicker noise of the MOS switches. As a result, the bias current amplifies the flicker noise that comes from the switches, which is usually significant. Typically, flicker noise causes issues for zero-IF receivers, particularly when receiving signals with limited bandwidth like GSM.

This active mixer's high LO feedthrough (LO feedthrough), which is again proportionate to the bias current, is another problem. Although the LO feedthrough is lower with a double-balanced mixer architecture, it is still usually higher than with a passive mixer. The requirement for additional voltage-to-current conversion in addition to the LNA's is another disadvantage of this active mixer. Particularly in SAW-less receivers where the linearity requirements are quite strict, this additional *V*-to-*I* conversion reduces the receiver linearity. The last and last disadvantage is the requirement for a high supply voltage because of the devices stacked between the ground and the supply voltage.

The near-infinite isolation of the active mixer between the RF input and baseband output, which facilitates simpler design and comprehension, is one of its unique characteristics. We can reach mixer topology in Fig. 2.20(b), the so-called current-driven passive mixer, if we are ready to give up this isolation. We get a mixer that only commutates the AC signal current when we set the commutated DC current to zero. The triode operation of the mixer switches and the absence of DC currents are ensured by the capacitive coupling of the input transconductance stage to the switching pair.

The switches must turn on in the deep triode region and be strongly driven by powerful LO voltages to have a minimal ON resistance. The triode MOS switch uses its ON resistance to directly connect the input transconductor to the output load when the LO is high. Transimpedance amplifiers, often known as current buffers, or TIAs, buffer the switches on the baseband side. All internal nodes in the receiver chain become low impedance due to this current buffer, which ideally has zero input impedance.

Consequently, all internal nodes experience lower voltage fluctuations, which improves linearity. Furthermore, flicker noise from the switches does not show up at the baseband output because there is no bias current that is commuted by the switches. The LO feedthrough is also significantly smaller for the same reason. This architecture is also highly conducive to low-voltage applications because it does not involve a stack of devices.

The IF and RF ports of the passive mixers are not reverse isolated, in contrast to the active mixers. This chapter will demonstrate how, because of the absence of reverse isolation, the mixer uses simple frequency shifting to reflect the baseband impedance to the RF and vice versa.

2.2.2.1 Design of receivers with 50% passive mixer

A receiver front end with current-driven passive mixers powered by 50% duty-cycle clocks is shown in Fig. 2.21. Typically, the LNA design features large transconductance to reduce the noise contribution of the subsequent stages and is inductively degraded for greater IIP_3. Furthermore, as can be seen, there are no intermediate transconductance stages between the LNA and the I/Q mixers; instead, the mixers are directly connected to the LNA.

If not, the current-to-voltage and voltage-to-current conversions caused by these transconductance stages would negatively impact the receiver's non-linearity. Z_L, the LNA load, is an LC load that is parallel. Due to the two series capacitors C connecting the mixers for the I and Q channels to the LNA, the mixer switches do not carry any DC currents. The two quadrature channels each get a portion of the LNA's RF current.

Large shunt capacitors are connected between the differential inputs of the op-amps in the RC feedback-based current buffers to filter out the components of the received signal that are at high-frequency offsets from the intended signal. Strong blockers are filtered to reduce voltage swings at all internal nodes and enhance linearity. The low-frequency IM_2 components produced inside the LNA are likewise blocked by the series capacitors C. Otherwise, these parts can leak to the baseband and degrade the receiver IIP_2 because of mismatches in the mixer switches.

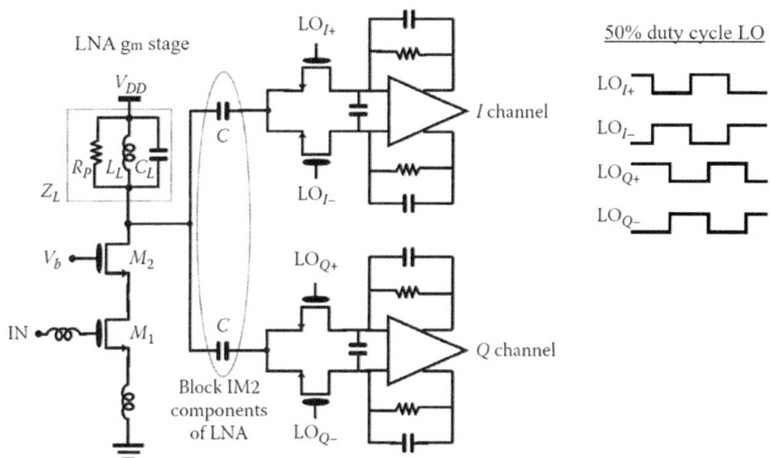

Figure 2.21 RX front end with 50% LO-based current-driven passive mixer and g_m-based LNA.

Assume that the switches of a mixer timed at a rate f_{LO} with rail-to-rail square-wave differential LOs commutate an RF current IRF in order to comprehend the operation of the mixing system, as seen in Fig. 2.22. After flowing into low-pass filtered baseband impedances Z_{BB}, the differential down-converted currents are transformed into baseband voltages, V_{BB}, across the baseband impedances.

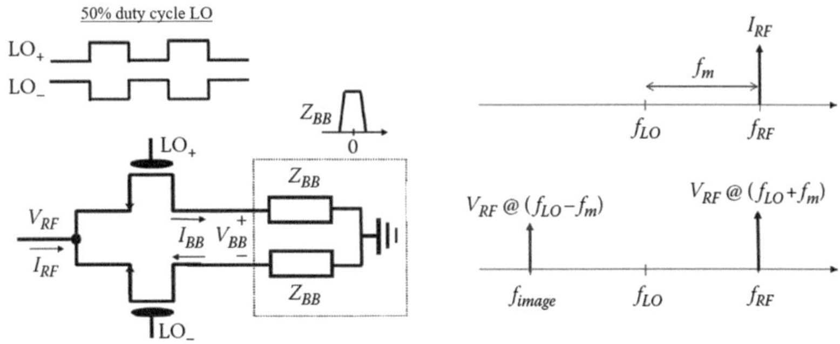

Figure 2.22 Mathematical analysis of 50% LO-based passive mixer.

These baseband voltages are up-converted back to the RF due to the passive mixer's lack of reverse isolation, creating an RF voltage around f_{LO} on the RF side of the switches. Assuming that there is a single tone in the RF current at $f_{\text{LO}} + f_m$, where f_m is a little frequency offset, $f_{\text{LO}} + f_m$, sometimes referred to as the main frequency at the incident RF frequency, and $f_{\text{LO}} - f_m$, often referred to as the image frequency, are the two primary frequency components of the resultant RF voltage around f_{LO}, as can be demonstrated. Various formulas provide the phasors of various RF voltage components:

$$V_{\text{RF}}@(f_{\text{LO}}+f_m)=\left[R_{\text{SW}}+\frac{2}{\pi^2}Z_{\text{BB}}(f_m)\right]I_{\text{RF}} \tag{2.22}$$

$$V_{\text{RF}}@(f_{\text{LO}}-f_m)=\frac{2}{\pi^2}Z_{\text{BB}}(-f_m)I^*_{\text{RF}}, \tag{2.23}$$

where the sign of the conjugate term is *. The image component of the resulting RF voltage can have a magnitude equal to or greater than the main component with a tiny switch resistance.

Furthermore, it is easily demonstrable that the down-converted baseband currents vs. the stimulus RF current are

$$I_{\text{BB}}@f_m=\begin{cases}\frac{1}{\pi}I_{\text{RF}} & f_m>0\\ \frac{1}{\pi}I^*_{\text{RF}} & f_m<0\end{cases}. \tag{2.24}$$

Given an ideal RF current stimulus (Fig. 2.23) at the incident (main) frequency, the input impedance as observed from the RF side can be expressed as follows:

$$Z_{\text{in}} \cong R_{\text{SW}} + \frac{2}{\pi^2}\{Z_{\text{BB}}(s - j\omega_{\text{LO}}) + Z_{\text{BB}}(s + j\omega_{\text{LO}})\} \quad (2.25)$$

It is equivalent to a band-pass filter (BPF) connected in series with the switch resistance. This BPF is a high-Q BPF that is created by shifting the frequency of the low-pass filter to the RF. More significantly, this high-Q BPF is clock adjustable, meaning that its center may be precisely adjusted based on the clock frequency.

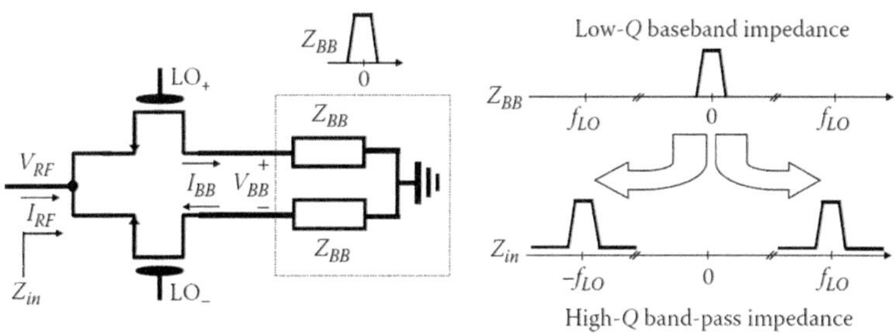

Figure 2.23 50% LO-based passive mixer impedance transformation.

The resulting components of the RF voltage at both the main and image frequencies would be low by the same amount determined by this filtering, as seen in Fig. 2.24, if the input RF current is present at a frequency offset from f_{LO} greater than the bandwidth of the baseband filter. However, both the main and image voltages that arise would be high if the RF current from the f_{LO} is located in the passband of this filter. In other words, this integrated high-Q BPF affects both the main and image voltage components. Unwanted blocks can be lessened by using this feature. By lowering the voltage swing across the switches, this high-Q filtering would improve linearity.

Let's now consider a second passive mixer circuit, Fig. 2.25, whose LOs are phase-shifted 90° from the first one. The image components will be 180° out of phase but have the same magnitudes if the same RF current is injected into the two mixing systems, but the RF voltages at the main frequency will be the same.

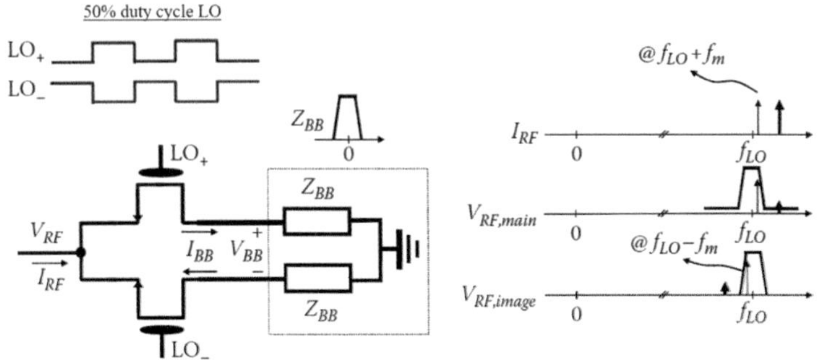

Figure 2.24 High-Q BPF of 50% passive mixer and the image component.

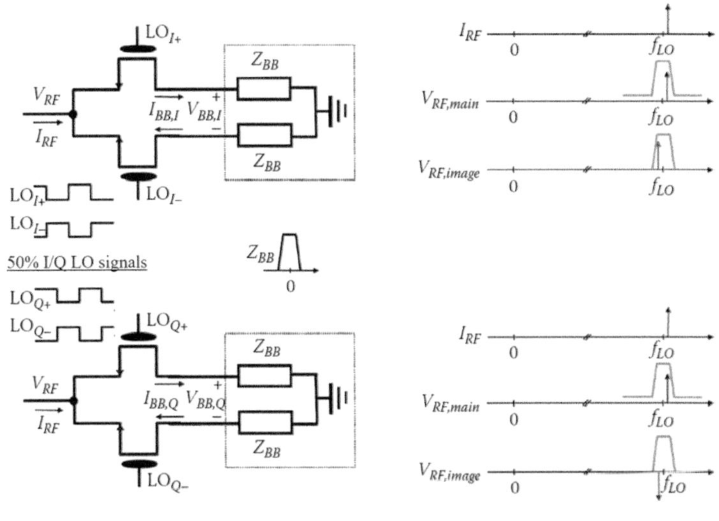

Figure 2.25 50% I/Q LO-based current-driven passive mixer.

This technique causes a current to flow between the two channels (between nodes A and B) at the picture frequency in a quadrature down-conversion, as seen in Fig. 2.26. Z_L receives no picture current because, at the image frequency, V_A + V_B equals zero. The primary baseband currents are superposed with the undesirable picture current after it has been down-converted. The image current's magnitude is inversely proportional to the series capacitors C's impedance at f_{LO} and proportionate to the baseband impedance Z_{BB}.

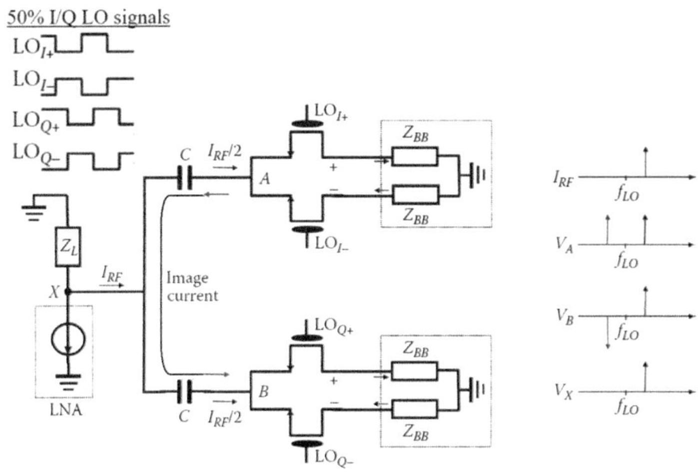

Figure 2.26 Image current flow in the I/Q LO-based current-driven passive mixer.

The performance of the I/Q receiver utilizing this 50% current-driven passive mixer, which will be discussed next, will be negatively impacted by the huge image current if the input impedance of the TIA is not low enough.

Let's ignore the higher-order harmonics for the time being. In theory, the RF current I_{RF} should have been split inversely proportional to the impedances seen from the three routes between Z_L and the two I and Q channels if there was no image current. The following expression could represent a relationship between the corresponding baseband currents:

$$I_{\mathrm{BB},I} = -jI_{\mathrm{BB},Q} = \frac{\frac{1}{\pi} Z_L(\omega_{\mathrm{LO}} + \omega_m) I_{\mathrm{RF}}}{2Z_L(\omega_{\mathrm{LO}} + \omega_m) + Z_C(\omega_{\mathrm{LO}} + \omega_m) + R_{\mathrm{SW}} + \frac{2}{\pi^2} Z_{\mathrm{BB}}(\omega_m)} \tag{2.26}$$

Nevertheless, the transfer function for the baseband currents is changed to the following because of the circulating image current:

$$I_{\mathrm{BB},I} = -jI_{\mathrm{BB},Q}$$

$$= \frac{\frac{1}{\pi} Z_L(\omega_{\mathrm{LO}} + \omega_m)\left[Z_C^*(\omega_{\mathrm{LO}} - \omega_m\right] I_{\mathrm{RF}}}{\left[2Z_L(\omega_{\mathrm{LO}} + \omega_m) + Z_C(\omega_{\mathrm{LO}} + \omega_m) + R_{\mathrm{SW}} + \frac{2}{\pi^2} Z_{\mathrm{BB}}(\omega_m)\right]\left[Z_C^*(\omega_{\mathrm{LO}} + \omega_m) + R_{\mathrm{SW}} + \frac{2}{\pi^2} Z_{\mathrm{BB}}(\omega_m)\right] - \left[\frac{2}{\pi^2} Z_{\mathrm{BB}}(\omega_m)\right]^2} \quad (2.27)$$

The transfer function Eq. (2.27) has these two additional components in comparison to Eq. (2.26): $[Z_C^*(\omega_{\mathrm{LO}} - \omega_m) + R_{\mathrm{SW}}]$ both in the numerator and $[Z_C^*(\omega_{\mathrm{LO}} - \omega_m) + R_{\mathrm{SW}} + (2/\pi^2) Z_{\mathrm{BB}}(\omega_m)]$ within the denominator. These additional components have a significant impact on a number of interesting and undesirable outcomes, including: (i) different high- and low-side conversion gains; (ii) different high- and low-side IIP_2 and IIP_3 values; (iii) non-linearity-generated component leakage from one channel's baseband side to the other; and (iv) decreased receiver conversion gain. Each of the previously stated consequences is explained in brief below:

(a) Variable high-side and low-side mixer conversion gains

One of the image current's earliest effects is its differential high-side and low-side mixer conversion gains. Alternatively, as illustrated in Fig. 2.26—A path, the RF current's frequency component at $f_{\mathrm{LO}} + f_m$ has a distinct conversion gain than the one at $f_{\mathrm{LO}} - f_m$. For a typical design, Fig. 2.26—B path shows the simulated high- and low-side conversion improvements against the well-matched expectations of Eq. (2.27).

The high- and low-side conversion gains in Eq. (2.27) might differ greatly because, as you can see, Z_{BB} is a complex conjugate of its values at positive frequencies. Furthermore, it can be demonstrated from Eq. (2.27) that if the baseband impedance—that is, a pure resistive input impedance—was symmetric about DC, the high- and low-side conversion gains would not differ. Even without the picture current, these various high- and low-side conversion increases would have also occurred.

The effect is amplified by the picture current, resulting in even larger disparities between the high- and low-side conversion gains.

(b) Variations in IIP_2 and IIP_3 levels on the high-side and low-side mixer conversion gains

Additionally, it stands to reason that the high- and low-side IIP_3 or IIP_2 figures would differ as well, given the differences in the high- and low-side conversion gains.

(c) Baseband components produced by non-linearity leaking from one channel to another

The leakage of low-frequency components produced by non-linearity on the baseband side of the mixer switches from one quadrature channel to the other is another effect of I/Q cross talk. Assume, as seen in Fig. 2.27(a), that all baseband loads in the I and Q pathways are the same, with the exception of the I channel's (Fig. 2.27(a)'s upper channel) baseband load, which contains second- or third-order non-linear components.

Let's now assume that the two tones at f_1 and f_2 make up the RF input. Following a down-conversion, they split into two baseband tones at $|f_{LO} - f_1|$ and $|f_{LO} - f_2|$ levels. On the matching baseband side, low-frequency intermodulation components are produced as a result of the non-linearity of the current buffer in the I channel. The same intermodulation components exist and, if the receiver is not constructed correctly, may be as strong when the Q-side voltage is examined even when its load is entirely linear.

IIP_2 and IIP_3 readings are typically unexpected because of the effect. Mathematically, it can be demonstrated that in the case of IM_2 (receiver IIP_2), the IM_2 component that leaks from the I channel to the Q channel and vice versa adds up constructively in one channel to the IM_2 component that is already generated in that channel and destructively in the other in Fig. 2.27(b). Consequently, the IIP_2 of the receiver measured for one channel is improved and that of the other channel is degraded due to the leaking effect.

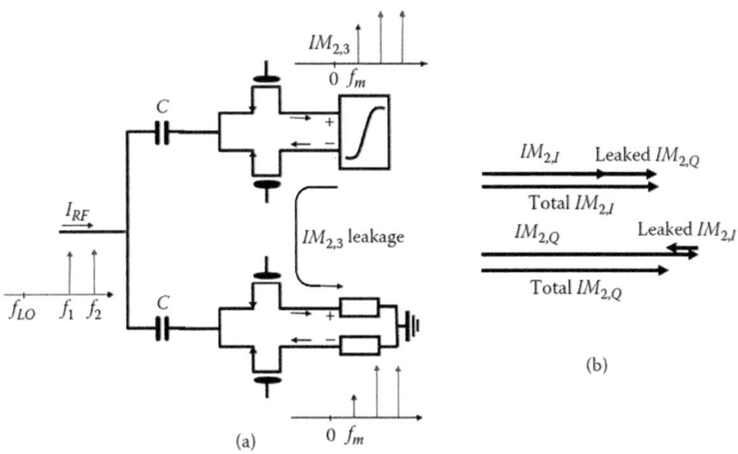

Figure 2.27 Issues with the I/Q passive mixers (a) I/Q cross talk due to leakage of non-linear BB components from one channel to another (b) I/Q cross talk effects on IIP_2.

(d) Decreased receiver conversion gain

The picture current also has the effect of lowering the receiver conversion gain. Stated differently, the conversion gain is lower when the image current is present. It can be demonstrated if we refer to the conversion gain with the image current included Eq. (2.27) as I_{BB} (actual) and the conversion gain without it Eq. (2.26) as I_{BB} (ideal).

$$\left|\frac{I_{BB}(\text{actual})}{I_{BB}(\text{ideal})}\right| = \begin{cases} 1 & \frac{2}{\pi^2} Z_{BB} \to 0 \\ \left|\frac{Z_C^*(\omega_{LO}) + R_{SW}}{2Z_L(\omega_{LO}) + 2R_{SW}}\right| \leq 1 & \frac{2}{\pi^2} Z_{BB} \to \infty \end{cases} \tag{2.28}$$

$$\left|\frac{I_{\mathrm{BB}}(\text{actual})}{I_{\mathrm{BB}}(\text{ideal})}\right|=\begin{cases}1 & C\rightarrow 0\\ \left|\dfrac{R_{\mathrm{SW}}}{R_{\mathrm{SW}}+\dfrac{2}{\pi^2}Z_{\mathrm{BB}}(\omega_m)}\right|<1 & C\rightarrow\infty\end{cases}\tag{2.29}$$

Therefore, the conversion gain is generally smaller due to the picture current, and the receiver's noise figure degradation can be significant.

2.2.2.2 Design of receivers with 25% passive mixer

Research has shown that receivers with 25% duty-cycle passive mixers operate better than those with 50% passive mixers. This section explains the rationale behind this. A receiver front end with a current-driven passive mixer powered by 25% duty-cycle clocks is shown in Fig. 2.28. To reduce the noise contribution of the stages that follow, the LNA is inductively degraded and built with a high transconductance in the signal band.

There are no intermediate transconductance stages separating the LNA and the complicated I/Q mixer; instead, the mixer is directly linked to the LNA using a single capacitor of size C^*. The complicated I/Q mixer is coupled to the LNA via the series capacitor C. The LNA load, Z_L, is a parallel LC load. The baseband buffers are based on the op-amp with feedback RC. A large shunt capacitor is placed between the differential inputs of the op-amp in each quadrature channel to filter out components that are located at high-frequency offsets, which are places where out-of-band blockers can be particularly strong. This enhances the receiver linearity by reducing the voltage swings at all internal nodes.

Once more, because of the nature of the passive mixer, the LNA from the complex mixer sees the impedance as a high-Q BPF, assuming the baseband impedances are low-pass in Fig. 2.28. This integrated high-Q on-chip filter can be used to attenuate undesirable blockers by functioning as an on-chip SAW filter.

If device mismatches in the mixer switches result in a degradation of the receiver IIP_2, the series capacitor C prevents any potential IM_2 components formed inside the LNA. Additionally, common-mode feedback within the LNA may be used in place of the series capacitor *C*, assuming, for the purpose of illustration, that the LNA is just an inverter.

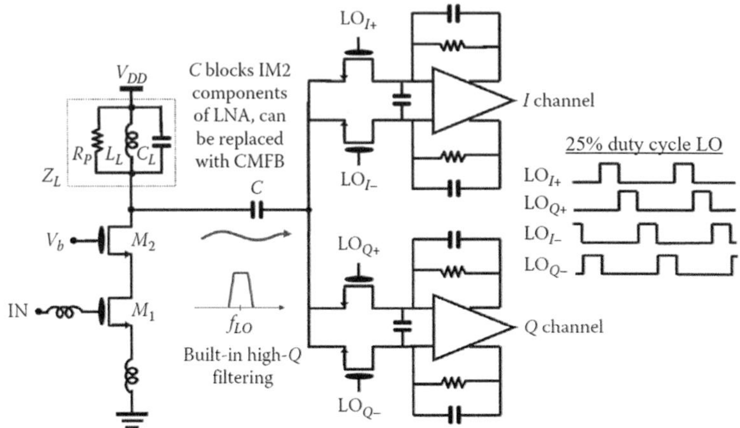

Figure 2.28 25% LO current-driven passive mixer in the RX front end.

The switches of the mixer, clocked at a rate f_{LO} with rail-to-rail 25% duty-cycle quadrature LO clocks, commutate an RF current I_{RF}, as seen in Fig. 2.29. After flowing into baseband impedances Z_{BB} and being low-pass filtered, the differential down-converted currents become baseband voltages across the baseband impedances. These baseband voltages are up-converted back to the RF around the LO on the RF side of the switches because the passive mixers lack reverse isolation.

It can be demonstrated that the input impedance as seen from the RF side is provided by the following expression, assuming that the input is an ideal RF current source:

$$Z_{\mathrm{in}} \cong R_{\mathrm{SW}} + \frac{2}{\pi^2}\{Z_{\mathrm{BB}}(s - j\omega_{\mathrm{LO}}) + Z_{\mathrm{BB}}(s + j\omega_{\mathrm{LO}})\}, \tag{2.30}$$

which indicates that when a BPF is connected in series with it, the baseband impedance is equal to the switch resistance. This BPF is really the same low-pass filter that has had its frequency moved to the RF to create an integrated high-*Q* BPF. This technique has the primary advantage over the 50% duty-cycle mixer in that no picture component is produced. As a result, the recipient does not experience severe *I*/*Q* cross talk.

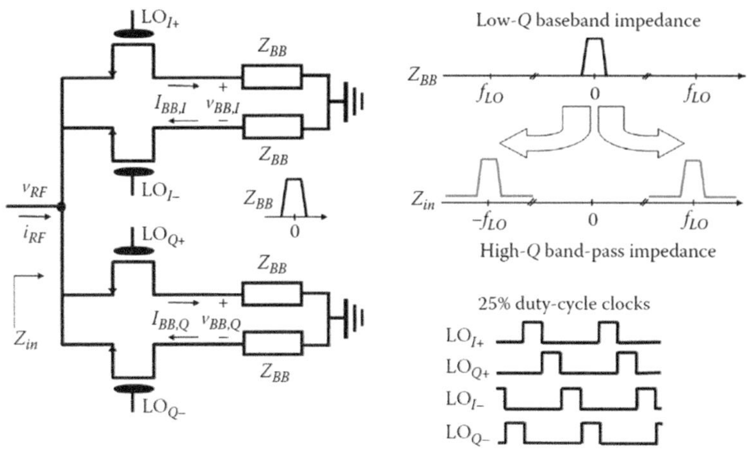

Figure 2.29 25% LO passive mixer's impedance transformation.

A receiver with a 25% passive mixer can be modeled as shown in Fig. 2.30. The output impedance of the LNA is determined by Z_L, and it is represented with a current source I_{RF}. Z_{BB} represents the single-ended input impedance observed from the current buffer. Figure 2.30 also shows the quadrature 25% rail-to-rail clocks that power the mixer switches. There is no noticeable I/Q cross talk since there is only one switch that is ON at any given time, either from the I or the Q channel, and the LNA current flows to the associated channel. Stated differently, no picture current can flow from the I channel to the Q channel.

The impedance observed from the LNA load and the impedance observed from the switching system in series with the capacitor C split the LNA current. Therefore, the following expression represents the RF current that the switches commutate:

$$I_C \cong \frac{Z_L(\omega)}{Z_L(\omega)+Z_C(\omega)+R_{\mathrm{SW}}+\frac{2}{\pi^2}Z_{\mathrm{BB}}(\omega-\omega_{\mathrm{LO}})} I_{\mathrm{RF}} \tag{2.31}$$

This expression is merely an estimate; the precise equation, which considers the effects of higher-order harmonics, will shortly be determined. As was already observed, there is little I/Q cross talk when using the 25% mixer. As a result, high- and low-side conversion gains are better matched because there is less picture current. Additionally, the IIP_2 and IIP_3 values are better, the conversion gain is larger, and the receiver noise figure is lower because there is no picture current.

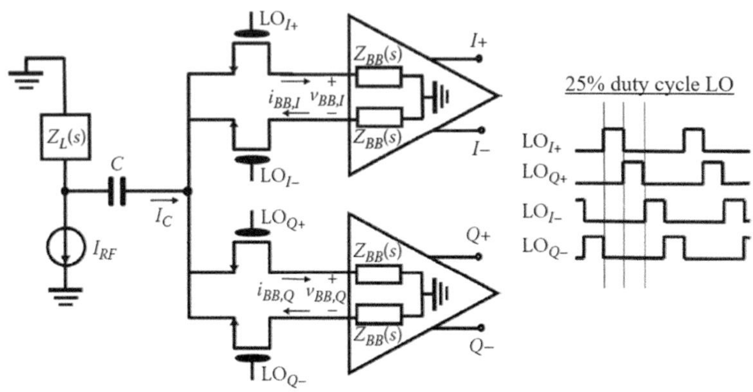

Figure 2.30 25% LO-based passive mixer RX model.

The LNA load in the standard design is an LC tank tuned at f_{LO}, and the series capacitor C is chosen just large enough to exhibit a low impedance at f_{LO} in Fig. 2.31(a). In this manner, depending on the amount of R_p, the RF current (LNA output current) is simply sent to the mixer with no gain and, in fact, with considerable attenuation. (or the inductor Q).

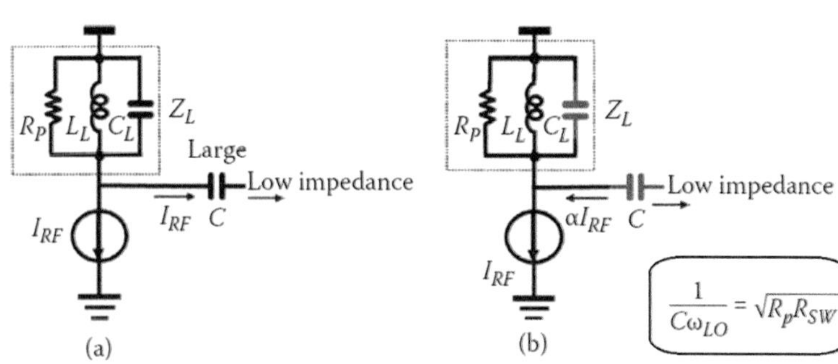

Figure 2.31 Components sizing in a narrowband RX with 25% passive mixer (a) conventional (b) optimized designs.

The fact that the current of an inductor or a capacitor in any parallel RLC tank is Q times greater than the overall tank current at the resonance, allows us to optimize the design in Fig. 2.31(b). As a result, there is an ideal series capacitor C size that depends on the switch resistance R_{SW} and the inductor loss R_p to achieve maximum conversion gain. Clearly, to achieve the highest conversion gain, the impedance of the series capacitor, C, needs to be equal to $R_p \cdot R_{SW}$.

The magnitude of C_L is also determined by the resonance at f_{LO} of the parallel between the LC load, Z_L, and the series capacitor, C. With this selection of parts, $R_p \cdot R_{SW}$, a linear, low-noise gain of considerable magnitude magnifies the current of the LNA signal.

Higher-order harmonics have not yet been incorporated into the transfer function derivation process. The transfer functions from the RF current to the baseband voltages at the input of TIAs are provided by the following equation if we do not add harmonic up- and down-conversions:

$$V_{BB,I} = -jV_{BB,Q} = \frac{Z_L(\omega_{LO}+\omega_m)Z_{BB}(\omega_m)}{Z_L(\omega_{LO}+\omega_m)+Z_C(\omega_{LO}+\omega_m)+R_{SW}+\frac{2}{\pi^2}Z_{BB}(\omega_m)}\frac{2\sqrt{2}}{\pi}e^{j\frac{\pi}{4}}I_{RF} \tag{2.32}$$

When it is assumed that the input RF current at $\omega_{LO} + \omega_m$ is a tone. Nevertheless, the transfer function becomes more complex if we incorporate higher-order harmonics, as shown by

$$V_{BB,I} = -jV_{BB,Q} = \frac{Z_L(\omega_{LO}+\omega_m)Z_{BB}(\omega_m)}{Z_L(\omega_{LO}+\omega_m)+Z_C(\omega_{LO}+\omega_m)+R_{SW}} \times \frac{\frac{2\sqrt{2}}{\pi}e^{j\frac{\pi}{4}}I_{RF}}{1+\frac{2}{\pi^2}Z_{BB}(\omega_m)\sum_{k=-\infty}^{+\infty}\frac{1}{(4k+1)^2\left[Z_L((4k+1)\omega_{LO}+\omega_m)+Z_C((4k+1)\omega_{LO}+\omega_m)+R_{SW}\right]}} \tag{2.33}$$

Equation (2.33) lets us know that we are now working with an infinite denominator sum. Overall, the conversion gain is anticipated to be somewhat smaller because of the harmonic up- and down-conversions.

As can be seen, the simplified transfer function provides a reliable and practical approximation that eliminates the need to deal with harmonics and the complex equations that go along with them, except for a negligible error around the LO.

2.3 Filters

The RF filter is a frequency-selective device that can be constructed to allow or block frequencies based on requirements in the complex domain of signal processing-based communication. The RF filters are made to satisfy certain amplitude criteria for the signals, such as roll-off rate, stop-band attenuation, and passband ripple, in addition to the frequency characteristics. These inactive but essential parts have the capability to allow or block signals according to their frequency properties, and they are essential in many kinds of applications in many different sectors of the economy.

Gaining an understanding of the subtleties of RF filters—their various forms, functions, and essential features—opens a world of opportunities for the most accurate signal management. Each form of filter—from band-pass and notch filters to high-pass and low-pass filters—offers a special capacity to customize signal transmission, guaranteeing accuracy, effectiveness, and integrity in electronic systems. Examining these aspects reveals the complexity and adaptability of RF filters, highlighting their essential role in the development of contemporary technology.

What are RF filters and why do we need RF filters?

When it comes to controlling the flow of signals according to their frequencies, RF filters are essential. These passive devices act as gatekeepers, permitting or prohibiting communications based on the properties of their frequencies. Essentially, filters are frequency-dependent elements that determine which signals to pass based on the needs of a certain application. Their importance is best illustrated by the fact that they keep undesired signals from getting to the receiver, which helps to avoid problems like jamming.

Purpose of RF filters

Effectively blocking unwanted transmissions is the main goal of RF filters. Passive devices function smoothly by either allowing or preventing the transmission of signals according to their frequency. When undesired signals potentially disrupt the regular operation of electronic devices or communication systems, the crucial nature of this function becomes evident. Essentially, RF filters serve as vital protectors, keeping signals that are outside of the intended frequency range from interfering with the integrity and effectiveness of signal transmission.

How are the filters implemented?

(a) Low loss and frequency-dependent passive LC components;

(b) Resonant cavity-based filters;

(c) Planar filters on printed circuit boards (PCBs): Selecting the right housing complies with the needs of applications. Some PCBs filters may require surface mount technologies (SMT), while others may perform best with SMA connectors or specialized enclosures;

(d) Ceramic filters;

(e) Waveguide-based filters; and

(f) Distinct filter methods have different properties, such the BAW (bulk acoustic wave) and SAW. For example, SAW filters might be appropriate in situations where a wider frequency rejection is enough, whereas BAW filters provide tighter frequency rejection, which is crucial for applications like cellular and wireless systems.

The basic idea behind how RF filters work is that they are frequency-dependent passive parts. When these filters are exposed to incoming radio frequency signals, they detect the frequencies and take appropriate action, either permitting some frequencies to flow through or blocking them. The workings of RF filters are supported by a variety of technologies, each exhibiting unique design subtleties and performance characteristics. Among these technologies, waveguide filters, planar filters built into PCBs, cavity filters, ceramic filters, and LC filters—which combine inductors and capacitors—are worth mentioning. Though they both aim to manipulate signals, each technology has unique design features of its own.

LC filters are fundamental elements in the field of electronic engineering that help to explain the fundamental ideas behind filtering processes. An inductor in a straightforward series arrangement, for example, can serve as a low-pass filter, and a capacitor connected in series can create a high-pass filter. These simple filter configurations are examples of standard solutions used in daily life. The frequency dependency of RF filters is fundamental to their operation. The frequency of the input signal they come across determines how well they work.

To visualize this, think of a power vs. frequency graph that shows several signals at different power levels. Differential output responses are produced when a particular filter is positioned in the signal path, demonstrating the filter's special capacity to modify the signal according to its frequency properties as shown in Fig. 2.32.

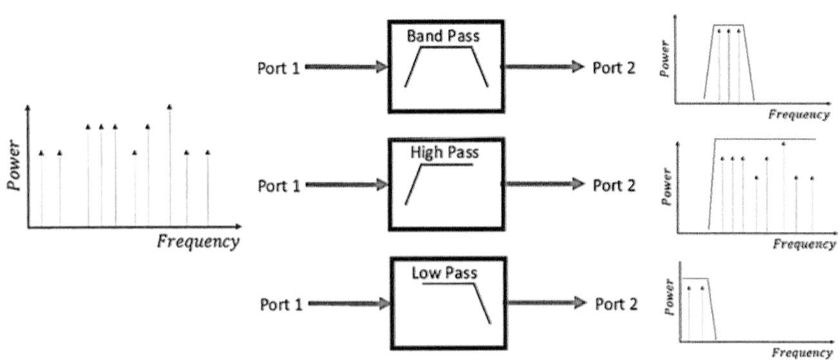

Figure 2.32 Generic filter characteristics.

2.3.1 Low-Pass Filters

On the other hand, low-pass filters allow low-frequency signals to flow through more effectively while blocking the advancement of higher-frequency components. Because of this, the output shows that lower frequencies are transmitted, whereas higher frequencies encounter resistance and are prevented from passing through the filter. By adjusting signal properties for maximum performance, these various filter types play a crucial role in many electronic applications. They are all excellent instruments for customizing signal passage based on frequency needs.

2.3.2 Band-Pass Filter

Band-pass filters allow just a specific range of frequencies to pass, which fine-tunes signal transmission. A certain range of frequencies is isolated and allowed while others are attenuated by this selective filtration. When one looks at the output at port 2 of this kind of filter, they can see which frequencies pass through while a variety of others are blocked. It's important to remember that in practical applications, the blocked frequencies experience attenuation beyond the cutoff point even though they don't completely disappear. Because of this attenuation, which drastically lowers their power levels, interference is reduced and a clearer signal visualization free of undesirable frequency disturbances is made possible.

2.3.3 High-Pass Filter

High-pass filters (HPFs) work by allowing high frequencies to get through while preventing lower frequencies from being transmitted. The intended high-frequency range is successfully carved out while lower-frequency components are suppressed thanks to this selective action. This filter's port 2 best illustrates this behavior by allowing higher frequencies to flow through beginning at the designated cutoff point. On the other hand, it limits the passage of lower frequencies such that they don't go past this set point.

Furthermore, the notch filters might be an extension of the BPF in which the notch filters allow and block a particular frequency. In contrast to band-pass filters, notch filters block a particular small frequency range. These filters are very good at attenuating a single frequency or a small range, which helps them suppress undesirable sounds in that frequency range.

In addition to these basic filter types, other parts like diplexers and duplexers are included. Duplexers are mainly used in FDD systems. They filter signals by allowing different frequencies to either pass through or block to meet different frequency needs. Like RF filters in theory, diplexers operate on several frequencies and display frequency-dependent features. While they are not technically speaking RF filters, their frequency dependence enables them to behave as such, which makes them useful in a variety of applications where accurate frequency manipulation is essential.

These many types of filters and associated parts are essential instruments for controlling signal frequencies in a variety of applications, improving signal quality and performance in many electronic systems.

2.3.4 Characteristics of RF Filters

The following essential qualities, each of which is vital to the operation of RF filters, determine their applicability and effectiveness in a variety of applications:

(a) Frequency range and cutoff frequency

Crucial factors that affect the filter's capacity to accept or reject frequencies on a selective basis are its frequency range and cutoff frequency. Sharp cutoff filters are excellent in isolating desired frequencies and attenuating those that are close to one another. Broader cutoff filters are more economical, but they may also provide less precise results.

(b) Insertion loss ensuring signal integrity

The attenuation of signal power as it passes through the filter is shown as insertion loss within the passband. It is essential to keep insertion loss minimal during transmission to maintain signal integrity and strength.

(c) Attenuation outside the passband frequencies for mitigating interference

To reduce interference, high attenuation or rejection outside of the assigned passing frequencies is necessary. Optimal rejection levels guarantee undesired frequencies are greatly reduced, enhancing the signal's purity. They are frequently described in data sheets as S1 or comparable values.

(d) Power handling and matching power requirements

It's crucial to make sure the filter can manage the incoming electricity. Evaluating the filter's power-handling capacity ensures that it will withstand changes in power input without breaking down.

The field of radio frequency filters is broad and includes many different kinds, including cavities, PCB, integrated, and LC filters, all of which are designed for particular uses. Comprehensive data sheets offer a thorough understanding of their features, facilitating well-informed choices that are in line with the various requirements of various applications. A search for these filters reveals a plethora of choices, each specifically engineered to satisfy a range of requirements, highlighting the components' versatility in a variety of complex technical environments.

2.4 Couplers

The RF signals have to be carefully routed to the multiple paths. In the process of routing or guiding the signals, the power division ratio can be equal or unequal depending on the requirement and the phase characteristics can be converted such as 0° with no phase difference (in-phase) at the outputs called power dividers, and 180° and 90° phase difference as hybrid couplers. In addition, impedance matching, isolation, directivity and coupling have to be considered in the coupler's design.

An ideal transmission line has two parameters namely the characteristic impedance, Z_0 and the electrical length for a given single frequency. When transformed to the practical transmission line, the parameters for the design have physical dimensions such as length, width, thickness, diameter, etc., and electrical parameters—resistivity, dielectric constant, magnetic coupling, etc. The transmission lines can be implemented as a coplanar wave-guide (CPW), microstrip and stripline.

This can be characterized by the designers using electro-magnetic (EM) simulators. The EM simulators will provide a result as the *S*-parameters depending on the number of ports of the passive network. Most of the passive designs can then be analyzed using the *S*-parameter matrix which will consider the design as a black box with the parameters associated with the ports. The network properties such as insertion loss, return loss, reciprocity, symmetricity, isolation, etc. can be determined from the *S*-parameter matrix.

We can study the various types of RF passive structures based on the number of ports.

2.4.1 One-Port Network

The simplest is the one-port which are called stubs and are usually associated with electrical length in units of quarter wavelength with the other end as either open circuit or short circuit to ground as illustrated in Fig. 2.33.

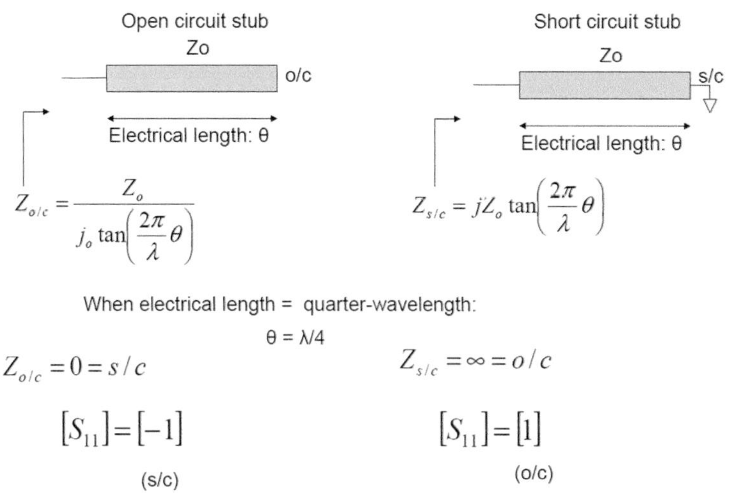

Figure 2.33 One-port network characteristics: open-circuit and short-circuit stubs.

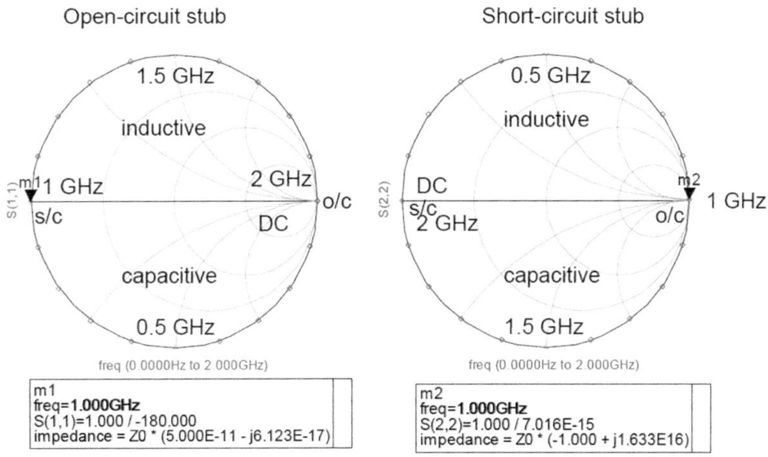

Figure 2.34 Quarter wavelength stub characteristics: open-circuit and short-circuit stubs.

The quarter-wavelength open-circuit stub presents a short circuit and the quarter-wavelength short-circuit stub presents an open circuit. The characteristics of a one-port stub will become complementary at double the operating frequency as depicted in Fig. 2.34.

2.4.2 Two-Port Network

The quarter wavelength impedance transformer using a two-port network is shown in Fig. 2.35. In order to achieve impedance matching, the impedance transformer's characteristic impedance Z_0 is to be designed based on the input and load impedances as indicated in Fig. 2.35.

2.4.3 Three-Port Network

(a) Power dividers and directional couplers

The design can be used as power dividers and power combiners. The design can be ideally lossless. The three-port networks are T-junction or power dividers. The power dividers provide in-phase output signals with an equal power division or uneven power division is also possible. We can use the scattering matrix to derive the basic parameters of the n-port networks. By using the S-matrix, we can define several critical parameters isolation, coupling, and directivity, which are important quantities for the characterization of couplers and hybrids.

For example, a three-port coupler has two inputs and one output, forming a power combiner. This structure can be considered as a T-junction. The scattering matrix of such a three-port network will have in total nine independent elements as defined in the 3×3 S-matrix as

$$[S]=\begin{bmatrix} S_{11} & S_{12} & S_{13} \\ S_{21} & S_{22} & S_{23} \\ S_{31} & S_{32} & S_{33} \end{bmatrix}. \tag{2.34}$$

If the three-port network is implemented as a passive structure, then it can be reciprocal and its S-matrix is suggested to be symmetrical with ($S_{ij} = S_{ji}$) with $i \neq j$. Usually, it is preferred to have a network which is lossless or with low loss and has to be impedance matched at all its ports. It is impossible to build such a three-port network which is simultaneously lossless, symmetrical, and impedance matched at all its ports. If all the network ports are matched, then the parameter S_{ii} = 0, and additionally if the network is symmetrical, but cannot achieve lossless network, then the S-matrix reduces to

$$[S]=\begin{bmatrix} 0 & S_{12} & S_{13} \\ S_{12} & 0 & S_{23} \\ S_{13} & S_{23} & 0 \end{bmatrix}. \tag{2.35}$$

By considering the conservation of energy in the network and setting the condition for a lossless network, then $\sum|S_{ij}|^2 = 1$ with $i \neq j$, suggests the network with all the ports matched, cannot achieve reciprocity. Alternatively, considering the three-port network, which will be lossless and reciprocal with a physical constraint that only two ports are impedance matched resulting in the S-matrix as follows:

$$[S]=\begin{bmatrix} 0 & S_{12} & S_{13} \\ S_{12} & 0 & S_{23} \\ S_{13} & S_{23} & S_{33} \end{bmatrix} \tag{2.36}$$

Finally, considering the three-port circulator, the network can achieve lossless and matched at all the three ports but cannot have reciprocity as shown in Fig. 2.36 and Eq. (2.37).

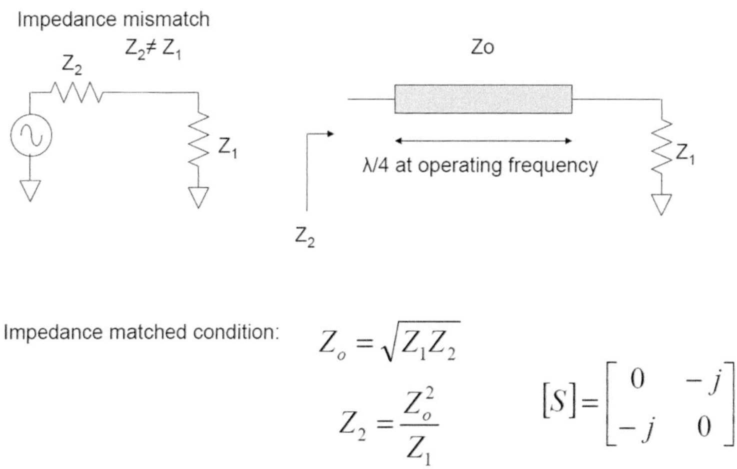

Figure 2.35 Quarter wavelength impedance transformer characteristics.

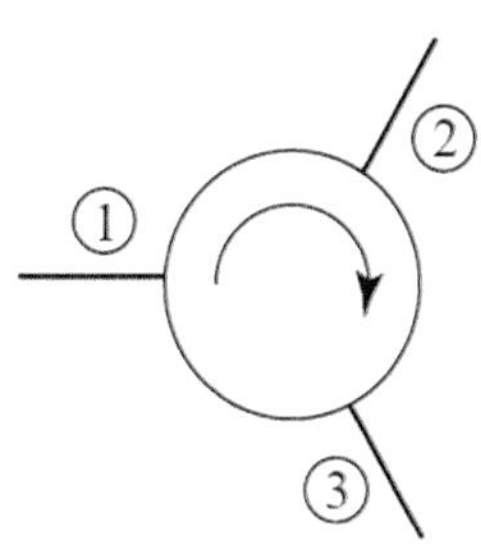

Figure 2.36 Three-port circulator.

$$[S]=\begin{bmatrix} 0 & 0 & 1 \\ 1 & 0 & 0 \\ 0 & 1 & 0 \end{bmatrix} \tag{2.37}$$

(b) T-junction

It is the simplest form of power divider or combiner which has one input with two outputs or two inputs with one output, respectively. A typical T-junction is shown below in Fig. 2.37 with port 1 as input and port 2 and port 3 as outputs. From the derived S-matrix Eq. (2.38), it is evident that all the ports are not matched with $S_{ii} \neq 0$. This network is lossless with $\sum|S_{ij}|^2 = 1$ and is also reciprocal with $S_{ij} = S_{ji}$.

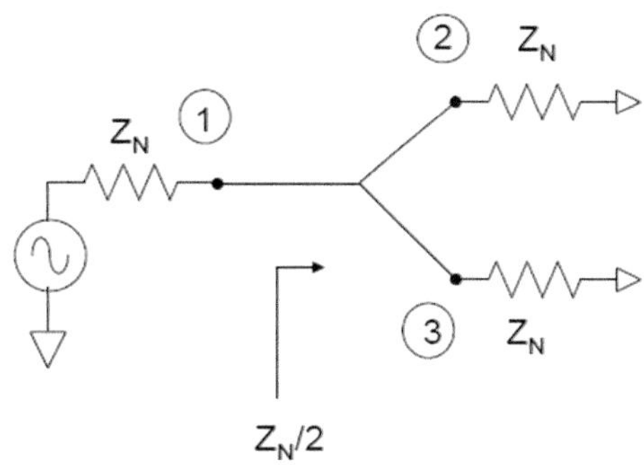

Figure 2.37 Three-port power combiner/splitter.

$$[S] = \begin{bmatrix} -\frac{1}{3} & \frac{2}{3} & \frac{2}{3} \\ \frac{2}{3} & -\frac{1}{3} & \frac{2}{3} \\ \frac{2}{3} & \frac{2}{3} & -\frac{1}{3} \end{bmatrix} \tag{2.38}$$

Example: To design a lossless T-junction with 50 Ω source impedance power divider with characteristic impedances such that the two output powers are in a 3:1 ratio. We compute the S-matrix for this T-junction network.

Solution: The input power to the T-junction with the voltage V_0 at the junction is,

$$P_{\text{in}} = \frac{1}{2} \times \frac{V_o^2}{Z_0} \tag{2.39}$$

The two output powers are,

$$P_1 = \frac{1}{2} \times \frac{V_o^2}{Z_1} = \frac{1}{4} \times P_{\text{in}} = \frac{1}{4} \times \frac{1}{2} \times \frac{V_o^2}{Z_0}$$

$$P_2 = \frac{1}{2} \times \frac{V_o^2}{Z_2} = \frac{3}{4} \times P_{\text{in}} = \frac{3}{4} \times \frac{1}{2} \times \frac{V_o^2}{Z_0}$$

Based on this, the characteristic impedances of the output ports are,

$$Z_1 = 4 \times Z_0 = 200\ \Omega$$

$$Z_2 = \frac{4}{3} \times Z_0 = 66.6\ \Omega$$

The effective impedance as seen at the junctions is $Z_1 \,||\, Z_2$ = 50 Ω which is the same as the source impedance.

The impedances looking into the two outputs are

1. Z_{out1} = 50 Ω || 66.6 Ω = 28.56 Ω
2. Z_{out2} = 50 Ω || 200 Ω = 40 Ω

The reflection coefficients at the output ports are

$$\Gamma_{\text{out1}} = \frac{(28.56 - 50)}{(28.56 + 50)} = -0.272$$

$$\Gamma_{\text{out2}} = \frac{(40 - 50)}{(40 + 50)} = -0.111$$

The S-matrix = $[S]$ = $\begin{bmatrix} 0 & 0.25 & 0.75 \\ 0.25 & -0.272 & 0.863 \\ 0.75 & 0.425 & -0.111 \end{bmatrix}$

Hence, this network is lossless as there are no resistive components. However, it is evident that it is not symmetric and not matched at the output ports. It also does not have isolation at the output ports, i.e., $S_{23} \neq 0$ and $S_{32} \neq 0$.

(c) Quarter wavelength power divider

For achieving matching at the input port, the output ports can be included with quarter wavelength transmission line components as described in Fig. 2.38. From the S-matrix Eq. (2.40), we can observe that this network is lossless, but the output isolation is not achieved. It is evident that output ports are not matched. Although this network is lossless with $\sum|S_{ij}|^2 = 1$, it is reciprocal with $S_{ij} = S_{ji}$ when $i \neq j$.

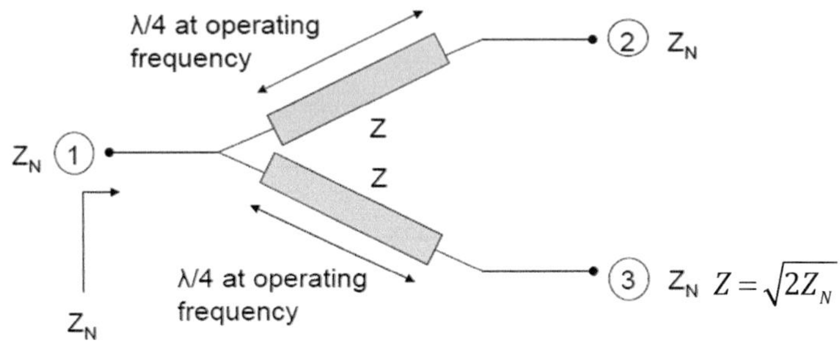

Figure 2.38 Analysis of three-port power combiner/splitter.

$$[S] = \begin{bmatrix} 0 & \frac{-j}{\sqrt{2}} & \frac{-j}{\sqrt{2}} \\ \frac{-j}{\sqrt{2}} & \frac{1}{2} & \frac{1}{2} \\ \frac{-j}{\sqrt{2}} & \frac{1}{2} & \frac{1}{2} \end{bmatrix} \tag{2.40}$$

(d) Wilkinson power divider

Just by adding a resistor between the output ports, the three-port junction achieves matching at all the ports, along with isolation of output ports. This network shown in Fig. 2.39 is called Wilkinson power divider. However, the network becomes lossy. This power divider can be made with equal or unequal output power division.

$$[S] = \begin{bmatrix} 0 & \frac{-j}{\sqrt{2}} & \frac{-j}{\sqrt{2}} \\ \frac{-j}{\sqrt{2}} & 0 & 0 \\ \frac{-j}{\sqrt{2}} & 0 & 0 \end{bmatrix} \tag{2.41}$$

By using the following even and odd-mode analysis, such complex networks can be better examined.

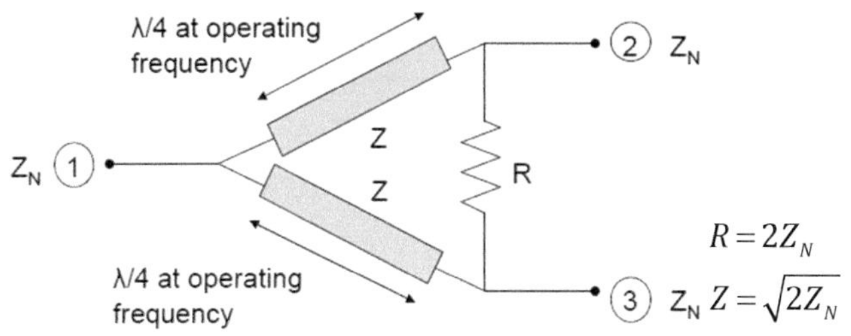

Figure 2.39 Three-port Wilkinson's power divider.

We normalize all the impedances to the characteristic impedance Z_o and include voltage sources at both the output ports as shown in Fig 2.40. By drawing a line splitting this network symmetrically across the midplane, the two output resistors with halved value will provide a resistance of normalized impedance value of 1, indicating output impedance matched. Similarly, by doubling the source impedance at the input port using shunt connection as shown in Fig. 2.40, we can individually analyze one half of the network under even and odd excitation.

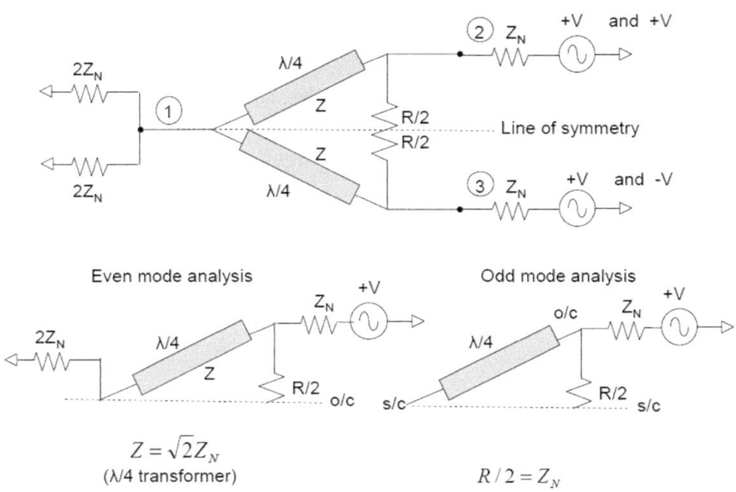

Figure 2.40 Even- and odd-mode analysis of three-port Wilkinson's power divider.

From the analysis, we can summarize that the S-matrix for the network as shown in Eq. (2.41) suggests that the input port is matched, output ports are matched for both even and odd modes, there is symmetry achieved and most importantly there is output port isolation.

2.4.4 Four-Port Networks Take the Form of Directional Couplers and Hybrids

Directional couplers can be designed for even as well as uneven power division while hybrid junctions usually have equal power division either with 90° or 180° phase shifts between the output ports. A general scattering matrix with all the four ports matched and with reciprocity is shown,

$$[S]=\begin{bmatrix} 0 & S_{12} & S_{13} & S_{14} \\ S_{12} & 0 & S_{23} & S_{24} \\ S_{13} & S_{23} & 0 & S_{34} \\ S_{14} & S_{24} & S_{34} & 0 \end{bmatrix} \tag{2.42}$$

By suitable substitutions and equalizing the parameters in the above S-matrix, for a lossless network based on conservation of energy suggests that $S_{14} = S_{23} = 0$, $|S_{13}| = |S_{24}|$ and $|S_{12}| = |S_{34}|$. By further substitution of the phase and power distributions of the output ports with reference to the input ports we can obtain two types of couplers namely the 180° hybrid coupler and the 90° quadrature hybrid coupler.

2.4.4.1 180° hybrid

The magic-T hybrid also called the 180° hybrid coupler with 180° phase separation between the output ports is a special case of the generic directional coupler and the corresponding S-matrix is,

$$[S]=\frac{1}{\sqrt{2}}\begin{bmatrix} 0 & 1 & 1 & 0 \\ 1 & 0 & 0 & -1 \\ 1 & 0 & 0 & 1 \\ 0 & -1 & 1 & 0 \end{bmatrix} \tag{2.43}$$

The 180° hybrid coupler as the name suggests is a Four-port network with a 180° phase difference between the output ports.

Based on the port input either port 4 or port 1, the output port 2 and port 3 will have a phase difference of 180° or in-phase, respectively, as shown in Fig. 2.41. Similarly, when a coupler is operated as a power combiner, with input signals provided at ports 2 and 3, the difference signal is seen at port 4 while the sum signal of the inputs will be obtained at port 1.

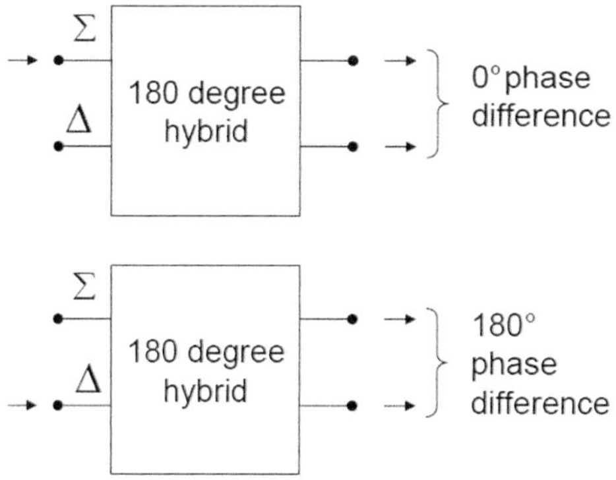

Figure 2.41 Rat-race hybrid.

2.4.4.2 90° hybrid

Another type of hybrid coupler is known as the quadrature coupler with a 90° phase difference at the output port. The quadrature coupler hybrids are usually built using microstrip lines or stripline structures which are also known as branch-line hybrid. Another quadrature coupler type is called the coupled line coupler or the Lange coupler.

With reference to Fig. 2.42, the basic operation of the branch-line coupler is that, with all the four ports matched, the input power provided at port 1 is equally power divided between port 2 and port 3, with a phase shift of 90° between them. Port 4 is the isolated port. The scattering matrix of this quadrature hybrid is as follows,

$$[S]=\frac{1}{\sqrt{2}}\begin{bmatrix} 0 & 1 & j & 0 \\ 1 & 0 & 0 & j \\ j & 0 & 0 & 1 \\ 0 & j & 1 & 0 \end{bmatrix} \tag{2.44}$$

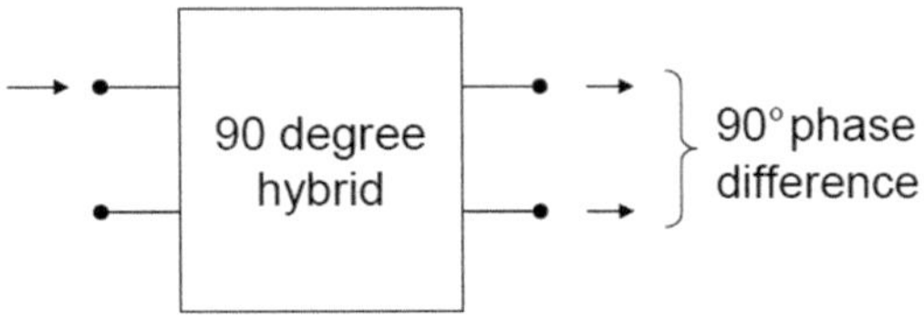

Figure 2.42 Branch-line hybrid directional coupler.

2.5 Matching Networks

For analog circuits, the impedances are mainly the resistive components which is the real part. The signal transfer occurs mainly as either voltage signal if it is a high-impedance node and a current signal for the low-impedance path. In RF circuits, the impedance consists of real and imaginary parts, and the signal is mostly considered to be transferred as power signals.

Why are impedance matching networks needed?

If the source and load impedances have the mismatched real part, it will result in power loss, and if there is a mismatch with the imaginary part, it will result in signal phase variations. Hence, to achieve maximum power transfer, the real part of the source and load impedance must be equal and have zero phase difference. There should also be conjugate imaginary parts which will cancel each other. Therefore, by achieving conjugate impedance matching, the RF circuits can achieve maximum power transfer. This is achieved by additional circuits in the RF signal paths called impedance matching networks. By achieving proper impedance matching, the reflection coefficients can be reduced, and the signal transmission will be lossless resulting in more power delivered to the load from the source. The MN ensures the impedances at both ends are unaffected and enables maximum power transfer.

As shown in Fig. 2.43, by using impedance transformation method, a series RC or RL network can the transformed into its equivalent parallel network. Clearly, R_P is larger than R_S and X_P is larger than X_S, i.e., $L_P > L_S$ and $C_P < C_S$.

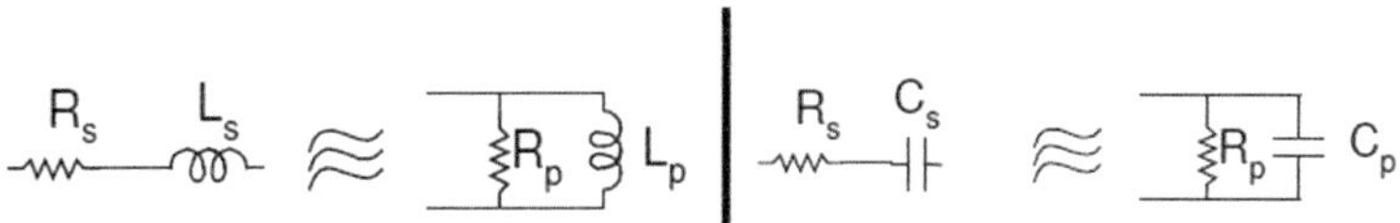

$$R_p = (1+Q^2)R_s \qquad L_p = \frac{(1+Q^2)}{Q^2}L_s \qquad Q = \frac{\omega L_s}{R_s} = \frac{R_p}{\omega L_p}$$

$$R_p = (1+Q^2)R_s \qquad C_p = \frac{Q^2}{(1+Q^2)}C_s \qquad Q = \frac{1}{\omega R_s C_s} \qquad Q = \omega R_p C_p$$

Figure 2.43 Series to shunt impedance transformations.

Figures 2.44 and 2.45 show the transformer characteristics with unload and load conditions, respectively. Broadband design is achieved using transformers. However, by including a shunt capacitor across the primary and the secondary coils, the frequency response can be made as a narrowband design.

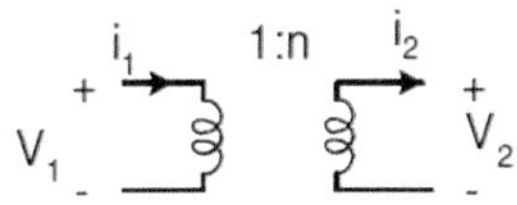

$$\left.\begin{array}{l} v_2 = nv_1 \\ i_2 = \frac{1}{n}i_1 \end{array}\right\} \Rightarrow p_2 = v_2 i_2 = nv_1 \frac{1}{n} i_1 = v_1 i_1 = p_1$$

Figure 2.44 Transformer characteristics with unloaded condition.

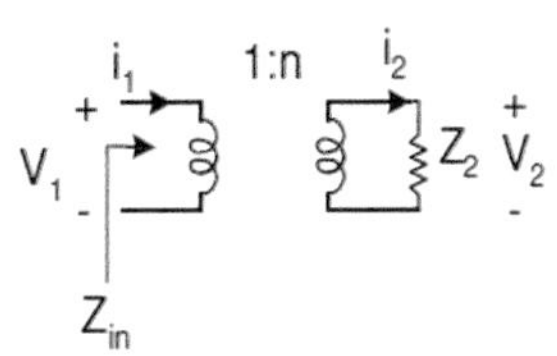

$$i_2 = \frac{v_2}{Z_2}$$

$$i_1 = ni_2 = n\frac{v_2}{Z_2} = n^2\frac{v_1}{Z_2}$$

$$Z_{in} = \frac{v_1}{i_1} = \frac{Z_2}{n^2}$$

Figure 2.45 Transformer characteristics with loaded condition.

2.5.1 L-Type Matching Network

The LC components can be arranged in four structures as indicated in Fig. 2.46, in which first two are low pass and next two are high pass depending on the source and load resistance values.

From these L-matching networks, it is seen that the resistance can be transformed from a large to small value and vice versa using LC components. This can be analyzed as follows:

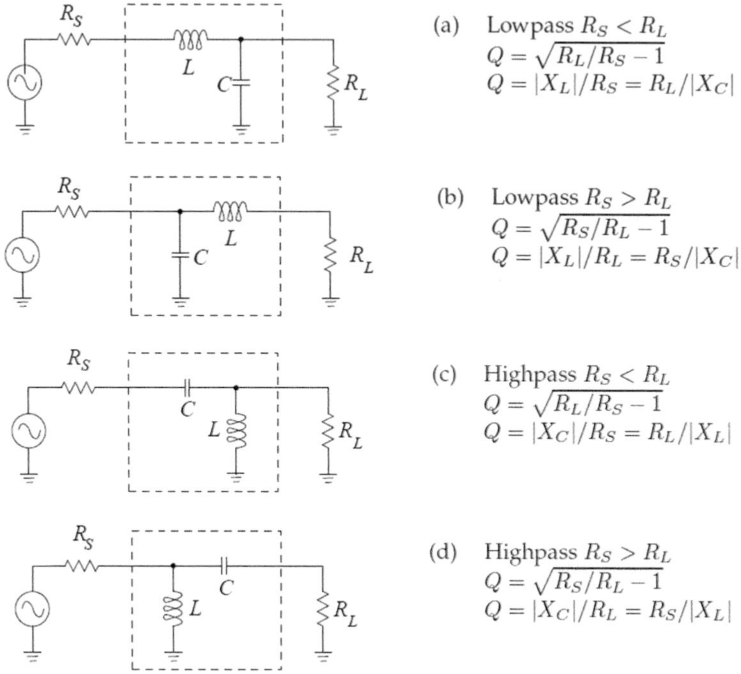

Figure 2.46 L-matching network characteristics with one shunt reactive and one series reactive element based on source and load condition.

An example of the L-matching network is shown below:

- $R_S = 200\ \Omega$
- $R_L = 1\ \text{k}\Omega$
- Frequency = 2.4 GHz

We need to provide DC voltage isolation from the source to load.

Solution: We can deduce the following as $R_S < R_L$ and we need to have high-pass characteristics. Based on this, the L-matching structure for a generic reactance as shown in Fig. 2.47 looks as follows:

1. The structure requires X_S in series to R_S and X_L as shunt to the load.
2. Since we need to block the DC, X_S should be a capacitor and X_L will be an inductor.

The values of the LC components can be determined as follows:

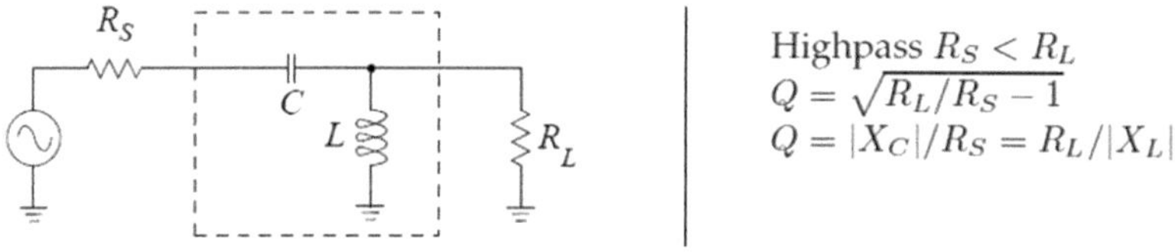

Figure 2.47 L-matching network characteristics for $R_S < R_L$ condition.

$$Q_S = Q_P = \sqrt{\frac{R_L}{R_S} - 1} = \sqrt{\frac{1000}{200} - 1} = \sqrt{4} = 2$$

$$\frac{X_C}{R_S} = 2,$$

$$X_C = 2 \times R_S = 400\ \Omega$$

$$\frac{1}{(2\Pi \times f \times C)} = 400\Omega$$

$$C = \frac{1}{(2\Pi \times f \times 400)} = 165\ \text{fF}$$

$$\frac{R_L}{X_L} = 2$$

$$X_L = \frac{R_L}{2}$$

$$L = \frac{R_L 1}{(2\Pi \times f \times 2)} = 33.1\ \text{nH}$$

L-matching networks are a two-element MN and the Q is fixed based on the R_S and R_L values. This limits the choice of Q of the MN and provides a limitation on the MN design. To enable an additional degree of freedom in the MN design, for choosing Q based on the application determining the bandwidth the 3-element matching network is needed. There are two types of

3-element networks, namely the T-network and the Pi-network as shown in Fig. 2.48.

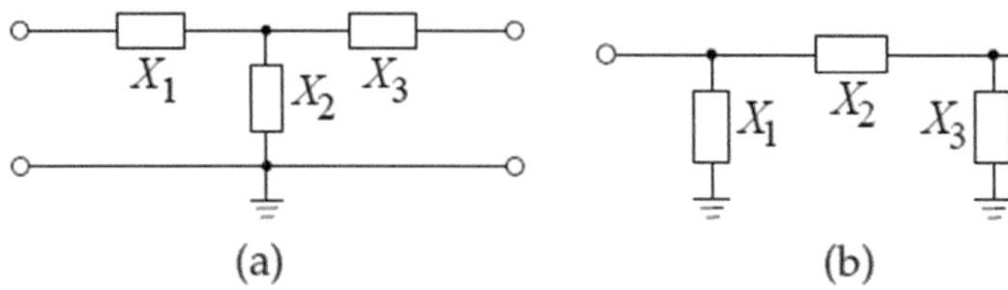

Figure 2.48 3-element matching network (a) T-network and (b) Pi-network.

2.5.2 T-Type Matching Network

The T-network can be considered as two back-to-back connected L-networks that can be independently designed to match the load and the source impedance conditions based on an imaginary virtual resistance, R_V, placed at the center in between the two L-networks. Since the R_V resistance as shown in Fig. 2.49 is in shunt to both the L-network to be selected as a larger resistance than both R_S and R_L. R_V is chosen according to the equation,

$$Q=\sqrt{\frac{R_V}{\min(R_S,R_L)}-1}, \tag{2.45}$$

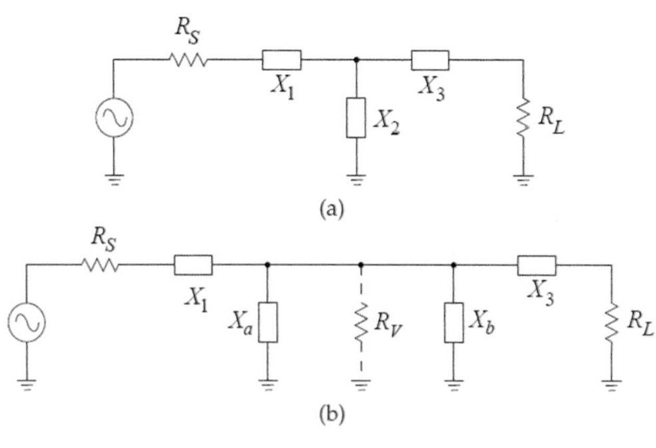

Figure 2.49 Design analysis (a) T-network and (b) virtual intermediate resistor-based two L-networks.

where Q is the chosen loaded Q of the overall T-network. Each L-network is calculated separately based on R_S/R_V and R_V/R_L, respectively. This will result in four possible designs for the T-network, based on the given R_S, R_L, and Q.

2.5.3 Pi-Type Matching Network

The Pi-network can also be considered as two juxtaposed back-to-back connected L-network as shown in Fig. 2.50. The virtual resistor R_V is in series with both the L-network and hence its value can be chosen as a smaller value than the minimum of R_S and R_L. The value of R_V is determined for a given R_S, R_L and the desired Q of the network using the equation as:

$$Q = \sqrt{\frac{\max(R_S, R_L)}{R_V} - 1} \tag{2.46}$$

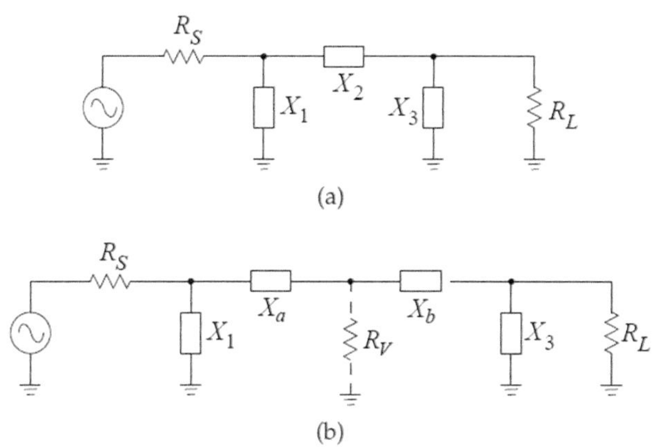

Figure 2.50 Design analysis (a) Pi-network and (b) virtual intermediate resistor-based two L-networks.

An example of the design of Pi-network based on R_S = 500 Ω, R_L = 1 kΩ, Q = 12 is illustrated in Fig. 2.51 and Fig. 2.52. We can determine the value of R_V using Eq. (2.46) above,

$$R_V = \frac{\max(R_S, R_L)}{Q_2 + 1} = \frac{1\ \text{k}\Omega}{145} = 6.85\ \Omega$$

$$Q_a = \sqrt{\frac{R_S}{R_V} - 1} = \sqrt{\frac{500}{6.85} - 1} = 8.48 = \frac{X_a}{R_V} = \frac{R_S}{X_1}$$

We solve for X_a and X_1.

$$Q_b = 12 = \frac{X_b}{R_V} = \frac{R_L}{X_3}$$

We then find X_b and X_3.

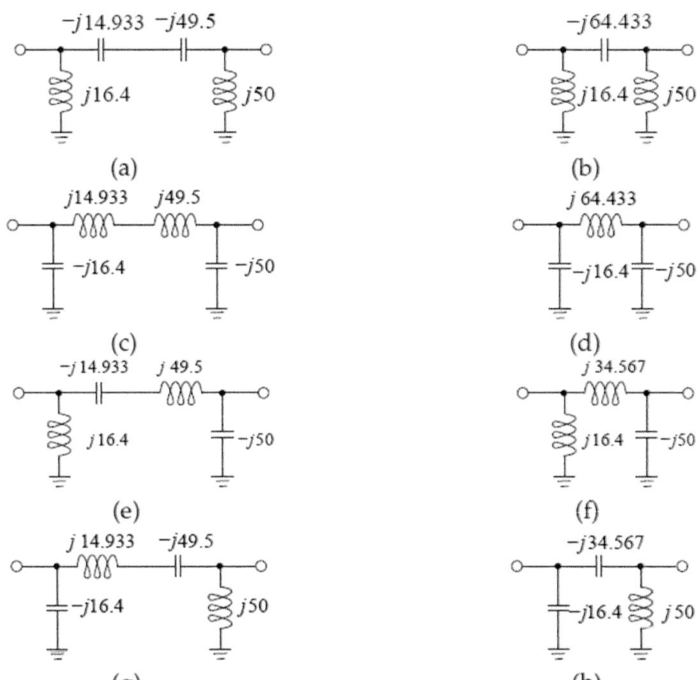

Figure 2.51 Pi-network's four example arrangements (a, (c), (e), (g) networks and (b), (d), (f), (h) final equivalent reduced Pi-networks.

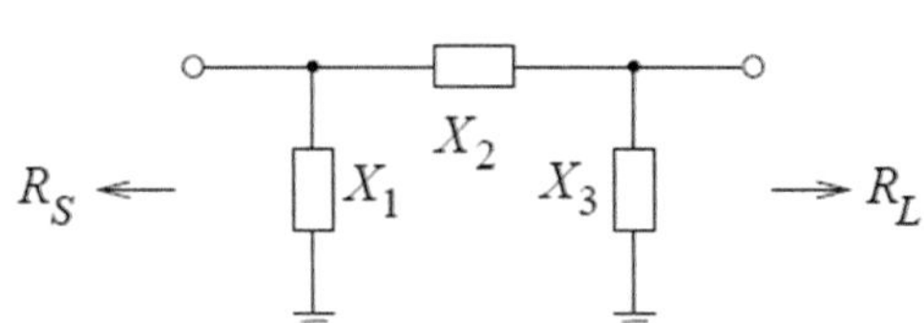

Figure 2.52 3-element matching Pi-network.

As an extension to the three-element MN is the multistage MN which can have broadband low-Q MN

The L-network connected in cascade can be arranged based on R_S and R_L values as follows:

- If $R_S > R_L$, then $R_S > R_V > R_L$
- If $R_S < R_L$, then $R_S < R_V < R_L$

The minimum Q to achieve a maximum matching bandwidth is

$$Q_1 = Q_2 = \sqrt{\frac{R_V}{R_{\min}} - 1} = \sqrt{\frac{R_{\max}}{R_V} - 1}\sqrt{\frac{R}{6.85} - 1} = 8.48$$

$$R_V = \sqrt{R_S \times R_L}.$$

For n stages,

$$R_{\min} < R_{V1} < R_{V2} < \ldots < R_{Vn-1} < R_{\max}$$

$$\frac{R_{V1}}{R_{\min}} = \frac{R_{V2}}{R_{V1}} = \ldots = \frac{R_{\max}}{R_{Vn-1}}.$$

2.5.4 Broadband (Low *Q*) Matching

It is not possible to choose the circuit Q and thus the bandwidth when using the L-network matching. But if the chosen Q is more than what can be produced with a L-network, Pi-network and T-network matching allow the circuit Q to be chosen regardless of the source and load impedances. Hence, compared to an L-network, the Pi and T-networks produce designs with narrower bandwidths.

Using two (or more) series-connected L sections instead of the Pi or T-networks' back-to-back layout is one design method for broadband matching (see Fig. 2.54). The virtual resistor, or R_V, that is supposed to be positioned at the intersection of the two L-networks is still the foundation of the design (see Fig. 2.53), but this time, R_V is selected to be between R_S and R_L, or mathematically:

$$R_{\min} \le R_V \le R_{\max},\tag{2.47}$$

where $R_{\min} = \min(R_L, R_S)$ with the network will have

$$Q_1 = \sqrt{\frac{R_V}{R_{\min}} - 1}.\tag{2.48}$$

And for $R_{\max} = \max(R_L, R_S)$, the other network will have

$$Q_2 = \sqrt{\frac{R_{\max}}{R_V} - 1}.\tag{2.49}$$

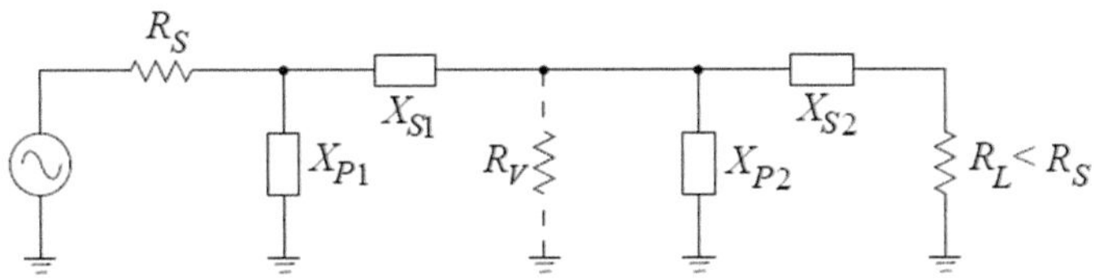

Figure 2.53 Wide bandwidth matching network with cascaded L-networks.

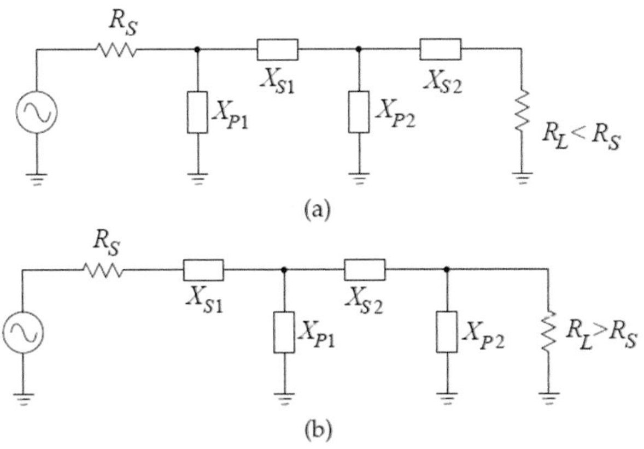

Figure 2.54 Wide bandwidth matching networks.

The maximum bandwidth obtained for a minimum Q as available from this network is given as

$$Q_1 = Q_2 = \sqrt{\frac{R_V}{R_{\min}} - 1} = \sqrt{\frac{R_{\max}}{R_V} - 1}.\tag{2.50}$$

Hence, the maximum matching network bandwidth attained when R_V is designed as the geometric mean of R_S and R_L as shown:

$$R_V = \sqrt{R_L R_S} \tag{2.51}$$

Broader bandwidths can be achieved by cascading multiple L-matching networks as

$$R_S < R_{V1} < R_{V2\cdots} < R_{Vn-1} < R_L. \tag{2.52}$$

For an optimum bandwidth, the impedance ratios should be equal to

$$\frac{R_V}{R_S} = \frac{R_{V2}}{R_{V1}} = \frac{R_{V3}}{R_{V2}} = \cdots = \frac{R_L}{R_{Vn-1}}, \tag{2.53}$$

and the Q value is given by

$$Q = \sqrt{\frac{R_{V1}}{R_S} - 1} = \sqrt{\frac{R_{V2}}{R_1} - 1} = \cdots = \sqrt{\frac{R_L}{R_{Vn-1}} - 1}. \tag{2.54}$$

2.6 Conclusion

This chapter provides a foundation for the various building blocks utilized in the RF transceiver. The building blocks include the various amplifiers, mixers, filters, couplers and the matching networks. The various types of each of the building blocks based on their application with performance parameters are defined to get a better understanding of the design. In addition, this chapter also emphasizes the design procedures and the design considerations for the building blocks to enable them to be subsequently integrated as sub-systems and eventually as the complete RF transceiver SoC. The building blocks discussed in this chapter are the main ingredients in the RF transceiver sub-systems design namely the receiver, the transmitter, and the frequency synthesizer. The subsequent chapters progress towards the design and implementation of the RF transceiver sub-systems.

References

1. B. Razavi et al., Multiband UWB transceivers, *IEEE Custom Integrated Circuits Conference*, Sep. 2005, pp. 141–148.
2. B. Razavi, Design considerations for direct-conversion receivers, *IEEE Transactions on Circuits and Systems I: Regular Papers*, vol. 44, no. 6, pp. 428–435, Jun. 1997.
3. A. A. Abidi, Direct-conversion radio transceivers for digital communications, *IEEE Journal of Solid-State Circuits*, vol. 30, no. 12, pp. 1399–1410, Dec. 1995.
4. R. Hartley, Modulation system, US Patent 1,666,206, Apr. 1928.
5. D. K. Weaver, A third method of generation and detection of single-sideband signals, *Proceedings of the IRE*, vol. 44, no. 12, pp. 1703–1705, Dec. 1956.
6. J. Rudell et al., A 1.9 GHz wideband IF double conversion CMOS receiver for cordless telephone applications, *IEEE Journal of Solid-State Circuits*, vol. 32, no. 12, pp. 2071–2088, Dec. 1997.
7. L. Der and B. Razavi, A2 GHz CMOS image-reject receiver with LMS calibration, *IEEE Journal of Solid-State Circuits*, vol. 38, no. 2, pp. 167–175, Feb. 2003.
8. M. Gingell, Single-sideband modulation using sequence asymmetric polyphase networks, *Electrical Communication*, vol. 48, pp. 21–25, 1973.
9. S. Lerstaveesin and B. S. Song, A complex image rejection circuit with sign detection only, *IEEE Journal of Solid-State Circuits*, vol. 41, no. 12, pp. 2693–2702, Dec. 2006.
10. J. Crols and M. S. J. Steyaert, A single-chip 900 MHz CMOS receiver front end with a high-performance low-IF topology, *IEEE Journal of Solid-State Circuits*, vol. 30, no. 12, pp. 1483–1492, Dec. 1995.
11. F. Behbahani et al., CMOS mixers and polyphase filters for large image rejection, *IEEE Journal of Solid-State Circuits*, vol. 36, pp. 873–887, Jun. 2001.
12. J. Crols and M. S. J. Steyaert, Low-IF topologies for high-performance analog front ends of fully integrated receivers, *IEEE Transactions on Circuits and Systems II: Express Briefs*, vol. 45, no. 3, pp. 269–282, Mar. 1998.
13. K. Feher, *Wireless Digital Communications*, New Jersey: Prentice-Hall, 1995.

14. R. Steele, *Mobile Radio Communications*, New Jersey: IEEE Press, 1992.
15. B. Razavi, A 900 MHz/1.8 GHz CMOS transmitter for dual-band applications, *IEEE Journal of Solid-State Circuits*, vol. 34, no. 5, pp. 573–579, May 1999.

STEM (Science, Technology, Engineering and Mathematics)
is everywhere, but nowhere the same because it requires
different Skill sets, Trainings, Experiences and Mindsets (STEM).

—Kiat Seng Yeo

From STEM to STEAM

The conventional STEAM: Science, Technology, Engineering,
Arts and Mathematics
The modern STEAM: Science, Technology, Engineering,
Artificial Intelligence and Mathematics

—Kiat Seng Yeo

Chapter 3

Receiver Sub-System

3.1 Receiver Architectures

Based on the study of the quality factor and the fractional bandwidth of the channel-selection filtering, we can see that it is difficult to achieve good filtering at high carrier frequencies. Hence, we must find alternative ways for shifting the desired signal channel to a lower frequency band. By this, the channel selection is possible by a using filter with a realistic quality (Q) factor as shown in Fig. 3.1(a), The frequency shifting is achieved by the down-conversion mixer which was studied in the previous chapter. The mixer frequency down-conversion operation is illustrated as a signal multiplication with a stable local oscillation as a sinusoidal signal. Transforming a sinusoid to the frequency domain is equivalent to an impulse at $\pm\omega_{\mathrm{LO}}$ and the multiplication operation of a sinusoid in the time domain is the convolution with an impulse signal in the frequency domain. From this we can observe that the frequency band is modulated as two components with frequency shifting at the sum and difference frequencies as $\pm(\omega_{\mathrm{in}} \pm \omega_{\mathrm{LO}})$. The desired down-converted components is at $\pm(\omega_{\mathrm{in}} - \omega_{\mathrm{LO}})$ and can be easily retrieved by using a low-pass filter (LPF) as shown in Fig. 3.1(a). From this LPF, all the undesired

---

CMOS RF and mm-Wave Transceivers and Synthesizers
Bharatha Kumar Thangarasu, Nagarajan Mahalingam, Kaixue Ma, and Kiat Seng Yeo

ISBN 978-981-5129-80-9 (Hardcover), 978-1-003-67356-9 (eBook)
www.jennystanford.com

higher frequency components will be attenuated and only the down-converted frequency component will be allowed to pass through. This entire operation is termed as down-conversion. A low-noise amplifier (LNA), which was also studied in the previous chapter, mainly is used to suppress the noise in the receiver chain as shown in Fig. 3.1(c). Thus, a complete receiver requires these four blocks namely LNA, down-conversion mixer, LPF and the LO generator. The down-converted center frequency depends on the difference between the RF input frequency and the LO frequency. If the difference is almost close to 0 Hz, then the receiver is called direct-conversion receiver or homodyne receiver and if the difference is non-zero then the down-converted signal is called the intermediate frequency (IF) component. In the latter case, the receiver requires multiple frequency down-conversions to achieve the desired receiver operation and such receivers are called heterodyne receivers. Heterodyne receivers require multiple LO frequency signals thus increasing the complexity of the receiver design [1].

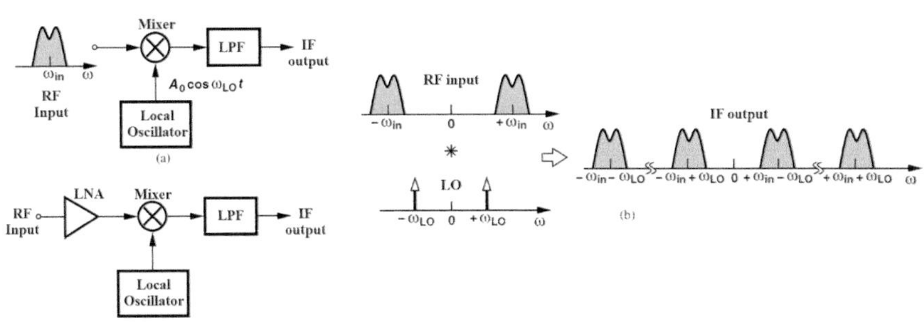

Figure 3.1 (a) Frequency down-conversion by mixer, (b) frequency spectra, and (c) LNA first to suppress noise.

To achieve a seamless down-conversion of all the N channels in a band to one single IF band, the LO signal can be made as a variable such that its frequency is varied in accordance with the channel frequency to result in same IF frequency as shown in Fig. 3.2. This variable LO frequency is achieved by the sub-system called the frequency synthesizer.

Now we will study the various issues related to the receiver design and the possible solutions to the issues.

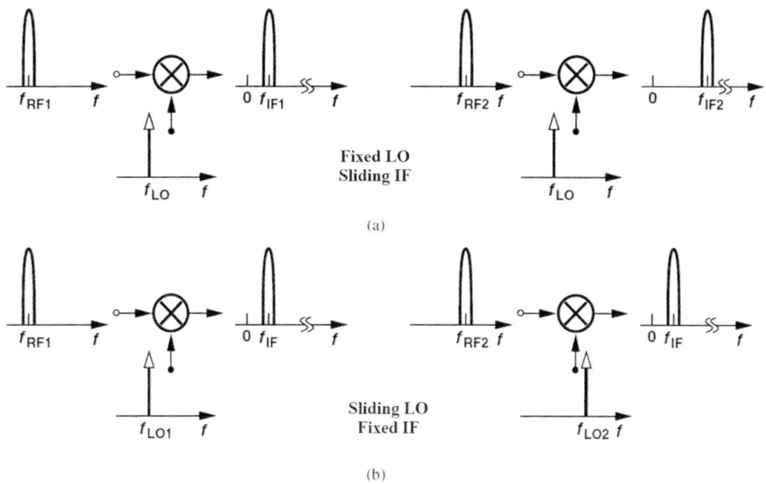

Figure 3.2 (a) Fixed LO frequency; (b) fixed IF down-conversion frequency.

3.1.1 Problem of Image Component

This is a problem related to the heterodyne receivers in which a RF frequency band which is on the other side of the LO with the same spacing as the IF frequency, may also be down-converted to the IF frequency and this RF frequency band is called the image of the desired signal. From this, we can observe that when the input RF band is on the higher or lower side of the LO, it will be down-converted to the IF component as shown in Fig. 3.3.

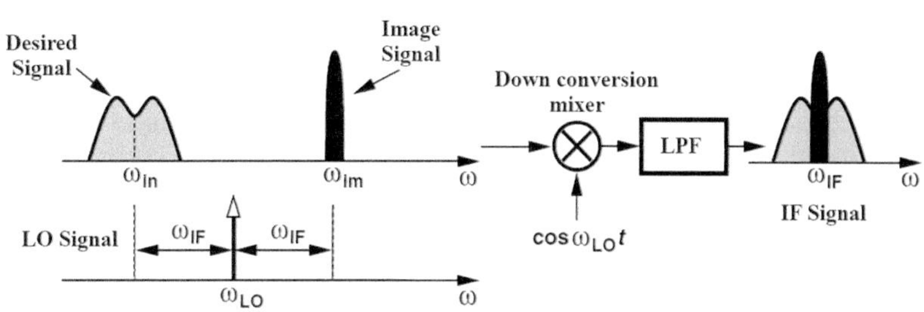

Figure 3.3 Heterodyne down-conversion mixing issue with the image.

Since there are many such RF sources that can be positioned at the image frequency band with varying power levels, these image components may result in the down-converted IF signal corruption and are considered as interference signals [2]. These interferers can be valid signals for other receivers but will interfere with this receiver's desired IF signal. Hence, to mitigate this issue, the receiver should implement a suitable image rejection technique.

3.1.2 High-Side and Low-Side Injection

From the discussion of the image signal, we can see that with the same LO frequency, there is a possibility of two RF input frequency bands which can be down-converted to the same IF frequency. If the LO frequency is larger than the RF input band center frequency, it is called the high-side injection, and the low-side injection is when the LO signal frequency is less than the RF input frequency. The selection of the desired injection method is mainly determined by the complexity of designing the frequency synthesizer with a higher LO signal and also on the image signal strength and the required image rejection capability.

3.1.3 Image Rejection

As we have seen, the image component will become an interferer, and the receiver must handle this by selectively suppressing the image component. In general, the predominant method followed in many receiver designs is to incorporate an image rejection filter at the receiver input. Since such filters will have passive insertion loss and to minimize the effect on the overall noise figure of the receiver, the image-reject filter is preferably positioned after the LNA as shown in Fig. 3.4. By this filtering method, the desired signal alone is passed to the mixer and the image component will be sufficiently attenuated by the stopband of the filter. The attenuation level also depends on how far the image component is from the desired component which in turn is based on the IF frequency which is the larger.

In addition to the noise consideration, the receiver linearity and frequency selectivity of the image-reject filter suggest it to be a passive filter with 50 Ω impedance at its input and output ports. This will push the LNA design to drive a smaller load which will result in high power consumption. Hence, alternatively to mitigate most of these concerns, the image-reject filter may need to be implemented as an external off-chip component with high Q-factor, low insertion loss and sharp frequency roll-off.

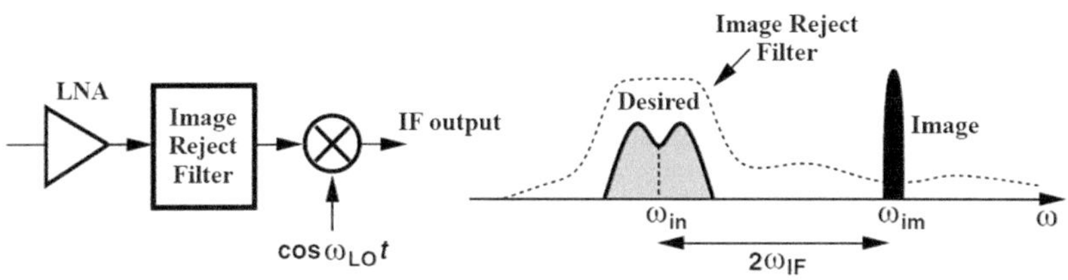

Figure 3.4 Image suppression by band-pass filtering.

3.1.4 Image Rejection versus Channel Selection

There is a system design trade-off when considering the image-reject and the channel selection. A larger ω_{IF} is preferred for better image rejection and also relaxes the requirement of the image reject filter. However, it will result in a greater number of down-conversion stages in the heterodyne architecture and impose a larger Q-factor requirement of the image-reject filter which together will increase the overall design complexity.

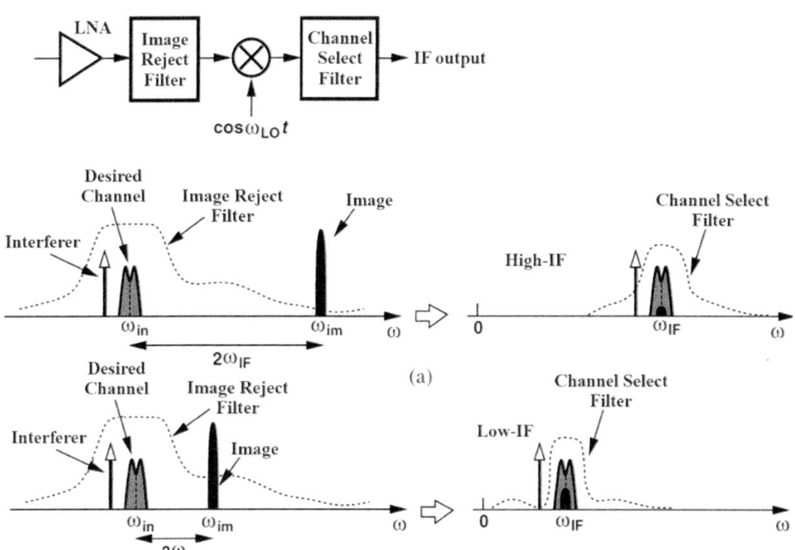

Figure 3.5 Compromise of the band-pass filtering between the image suppression and the channel selection for (a) high-IF and (b) low-IF conditions.

Figure 3.5 shows the two contrasting scenarios of having a larger ω_{IF} against a smaller ω_{IF}. We can summarize that a high ω_{IF}, results in a better image rejection and a low ω_{IF} will attenuate the in-band interferences but with a degraded image rejection. This trade-off is particularly critical for the heterodyne receivers and the designers are to pay attention to it.

3.1.5 Dual Down-Conversion

To provide a better compromise between the two scenarios of the IF values, and to provide a better circuit implementation, the alternative option is to implement the design with multiple down-conversions by multiple LO signals. A special case is to use dual down-conversion as shown in Fig. 3.6. Here, the IF is also large, and it also mitigates each stage image rejection requirement and also the Q-factor requirement of the image rejection filters.

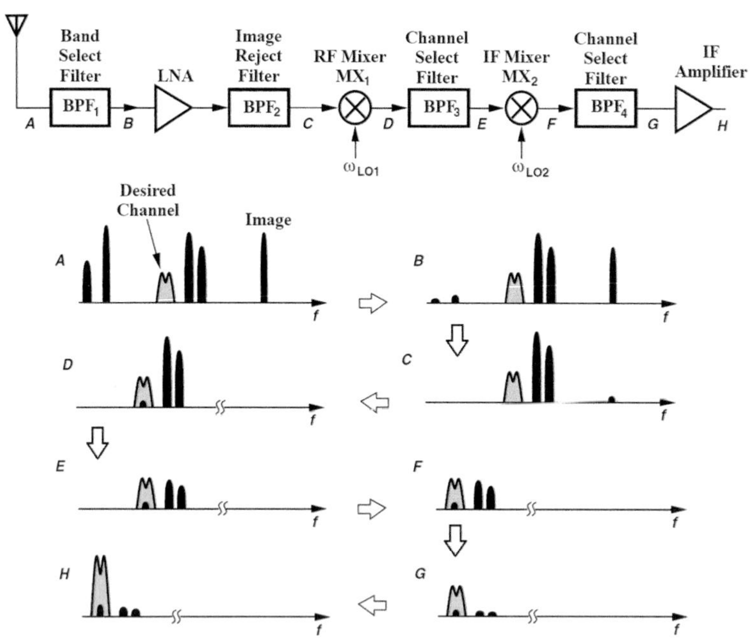

Figure 3.6 Two-step down-conversion (dual)-IF receiver.

From Fig. 3.6 we can see the spectrum of the signals as it passes through the heterodyne receiver chain from point A to point H. The final spectrum at H suggests that with enough gain of the desired signal, a better rejection of the image and other interferer signals as well as improved noise performance is achieved in the dual down-conversion receiver design [1].

In practice, as the number of receiver's down-conversion stages increase, the overall performance will be improved. But as the additional stages increase, this will increase the overall receiver complexity and hence creating a limitation on the number of mixers on the receiver chain. Hence, as a safe estimation, the heterodyne receivers are usually limited to only two or dual down-conversion receiver design.

3.1.6 Modern Heterodyne Receivers

As we studied in the previous sub-section, the dual down-conversion receiver as shown in Fig. 3.6 has incorporated many passive bulky filters and two LO signals. Hence, this system design method due to its complexity becomes outdated. Recently, the receiver system and the circuit design do not include any such bulk filters except for the external front-end band-select image-reject filter. By eliminating all the other filters in the receiver chain, the channel selection is not possible, and additionally there will be a second image component which cannot be eliminated. In this case, a zero second IF receiver design is an option as shown in Fig. 3.7. In this case, the second LO signal LO2 is planned such that it is at the center of the first IF signal and hence the second IF becomes zero thus one half of the first IF will become the image for the second stage which will not affect the receiver functionality, One of the biggest advantages here is that there will be no other interferers at any other frequencies resulting due to zero second IF down-conversion.

We will further try to understand the effects of an image signal which is the desired signal itself. For this, we have to know two terms namely the symmetrically modulated and asymmetrically modulated signals.

In case of amplitude modulation (AM) signal as shown in Fig. 3.8(a), the transmitted baseband signal will have a symmetrical spectrum which will be up-converted by a transmitting carrier signal, and the resulting output transmitter spectrum will also retain the symmetric spectrum with respect to the carrier frequency which is the transmitter's f_{LO}. From this, we can confirm that the AM signals after modulation remain symmetric in the spectrum since the AM spectra has the exact same copy of the information on both the side bands. Hence AM AM-based receivers will not have any image-related concerns.

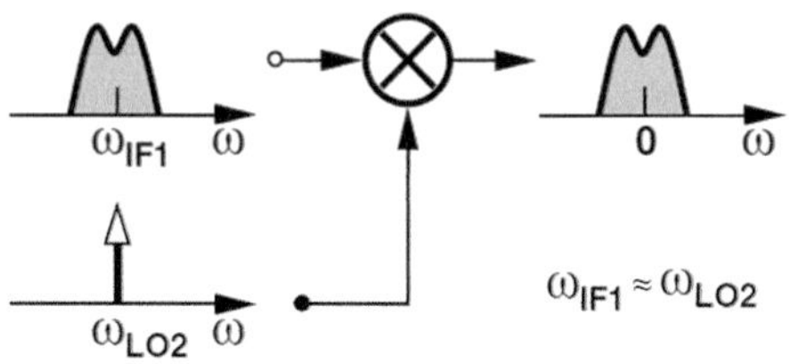

Figure 3.7 Selection of the LO2 frequency to mitigate the formation of the second image.

Alternatively, when we consider the frequency modulation (FM)-based receivers, in which the information is transmitted based on different frequency information. Hence, we cannot achieve a symmetric spectrum on both sides of the carrier as shown in Fig. 3.8(b).

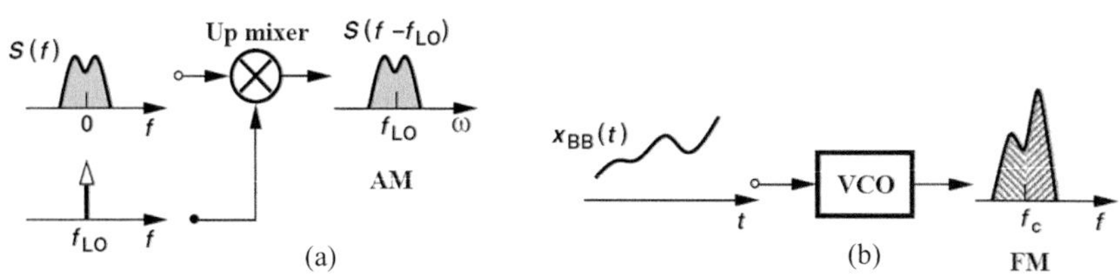

Figure 3.8 (a) Varying amplitude modulation signal; (b) fixed amplitude frequency modulation signals.

Increasing the data rate requirement of the recent day communication standard is mainly possible by having more complex modulation schemes such as FSK, GFSK, QPSK, and QAM. All these complex modulation schemes present an asymmetric spectrum with reference to their carrier frequencies. The asymmetry will not only be in the amplitude response but will also be with the phase response in the frequency domain. We can further check the effect of the asymmetric spectrum on the zero-IF down-converted spectrum which will result in the down-converted signal's corruption as shown in Fig. 3.9. This is also called self-corruption.

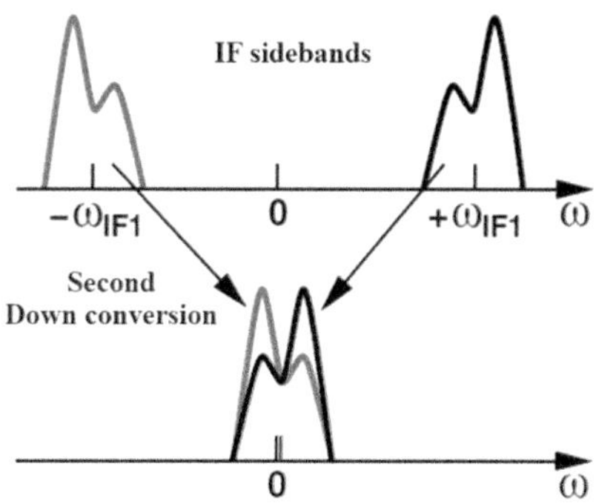

Figure 3.9 After dual down-conversion, overlay of the signal sidebands in the baseband.

We can try to see how to resolve this signal corruption. For this, we will create or split the input modulated signal by mixing them separately using two LO2 signals which have a phase difference of 90° between them. This is shown in Fig. 3.10, and this stage is called quadrature down-conversion. Here, the mixing is performed by the same LO2 frequency signals, but they are differentiated in phase by 90° resulting in two baseband signals which themselves have a phase difference of 90°. The resulting baseband outputs, $x_{BB,I}(t)$ and $x_{BB,Q}(t)$, are known as the quadrature baseband signals. One interesting aspect of this quadrature baseband signals is that they have identical spectra, but they are different in the phase response. Both these quadrature baseband signals can be used in the reconstruction of the original baseband signal without any corruption or distortion.

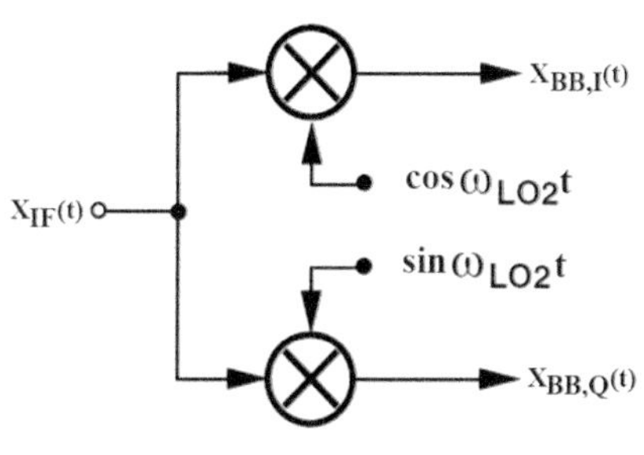

Figure 3.10 I/Q LO-based quadrature down-conversion receiver.

Figure 3.11 shows a symbolic representation of the heterodyne receiver using the quadrature down-conversion method. This architecture has the advantage of avoiding an image-reject filter just before the mixer and enables the LNA not to drive the 50 Ω load and instead a high impedance of the mixer input enables the LNA to be an ultra-low-power design along with improved noise, gain, and linearity. To complement the absence of the image-reject filter and to mitigate the interferers from the image frequency band, the system suggests using a narrowband LNA that will not only suppress the image component but also improve the noise figure performance of the receiver. In this architecture, the LO signal is to be implemented to provide the quadrature signals.

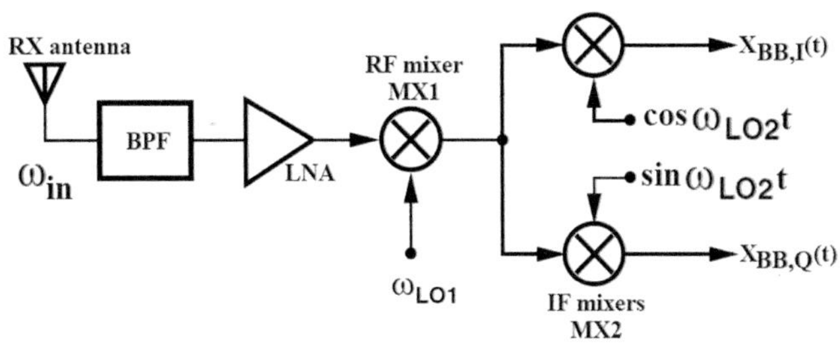

Figure 3.11 *I*/Q LO-based quadrature down-conversion receiver.

3.1.7 Direct-Conversion Receivers

Contrary to the heterodyne receiver architecture, the other possible option is to have only one mixer down-conversion stage from the RF directly to the baseband without any IF stages. Such a receiver architecture is called a direct-conversion receiver (DCR). It is also called the zero-IF architecture. This DCR architecture has recently received more attention because it is a more compact and low-power-consuming system, as the number of building blocks are fewer than that in heterodyne receivers [3–7].

Along with this advantage, the DCR comes with its own specific set of design issues which will be discussed subsequently. The first concern is the asymmetric spectrum due to zero-IF mixer stage and hence it emphasizes the need for the quadrature down-conversion method to prevent the self-corruption of the baseband. The DCR with quadrature down-conversion provides the two baseband quadrature outputs which are shown in Fig. 3.12.

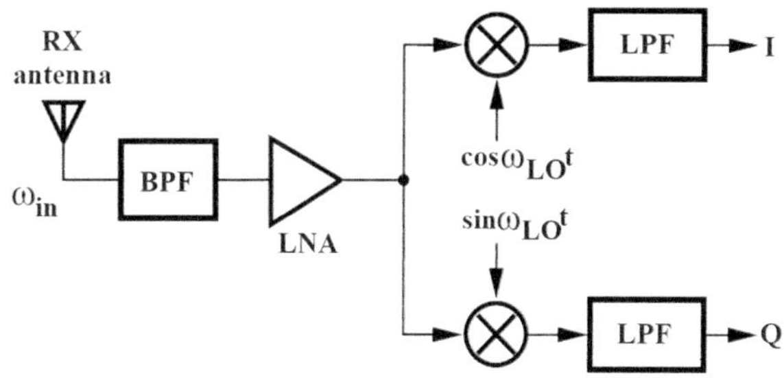

Figure 3.12 Homodyne or direct-conversion receiver (DCR).

The DCR has clearly four advantages over the heterodyne receiver namely,

1. There is no image component and hence there is no requirement for image rejection filters,
2. The channel selection can be achieved just by using on-chip high-order low-pass filters unlike the bulky band-pass filters if the heterodyne receiver,
3. Since there is only one LO frequency, the number of intermodulation and harmonic components are reduced to the bare minimum and hence the spectrum will be cleaner with fewer spurs, and
4. The architecture is simpler with minimum building blocks enabling a compact low-power receiver design.

We now continue to illustrate the system issues specific to the DCRs and preset the various resolving techniques. It is interesting to note that the DCR is part of the heterodyne receiver with zero second IF stage and hence some of these issues will also be present in the latter architecture as well.

3.1.8 LO Leakage

A DCR has a possibility to back-radiate a portion of its internally generated LO power from the receiver antenna. To comprehend

−60 dBm, it produces a −60 dBm + 80 dB = +20 dBm which is a strong down-converted DC signal that yields a very large DC offset in the baseband stages.

An offset like this simply prevents signal identification by overloading the baseband circuitry. Do heterodyne receivers with a zero second IF as shown in Fig. 3.11 have the DC offset problem? Indeed, there are DC offsets in the baseband because of the second LO leaking to the IF mixers' input. The leakage is less than in direct-conversion receivers since the second LO frequency in Fig. 3.11 is less than $f_{in} = \omega_{in}/(2\pi)$, but the DC offset is still significant enough to saturate the baseband stages or at the very least cause significant non-linearity. According to the previous research, baseband I and Q routes for receivers with a final zero-IF need to include a method of offset cancellation. A suitable option would be a high-pass filter (AC coupling), as illustrated in Fig. 3.15(a), wherein R_1 sets the appropriate bias, V_b, for the input of A_1 and C_1 blocks the DC offset. Intersymbol interference (ISI) is introduced, though, because such a network also eliminates a portion of the signal's spectrum at zero frequency, as seen in Fig. 3.15(b). In general, for insignificant ISI, the corner frequency of the high-pass filter, $f_1 = 1/(2\pi R_1 C_1)$, needs to be less than a thousandth of the symbol rate. To find the highest allowable value of f_1 for a particular modulation scheme, meticulous simulations are required. Both the modulation type and the symbol rate affect the viability of on-chip AC coupling. For instance, the 271 kb/s data rate of GSM requires very large capacitors and/or resistors at the corner frequency, which is approximately 20 to 30 Hz [8–11].

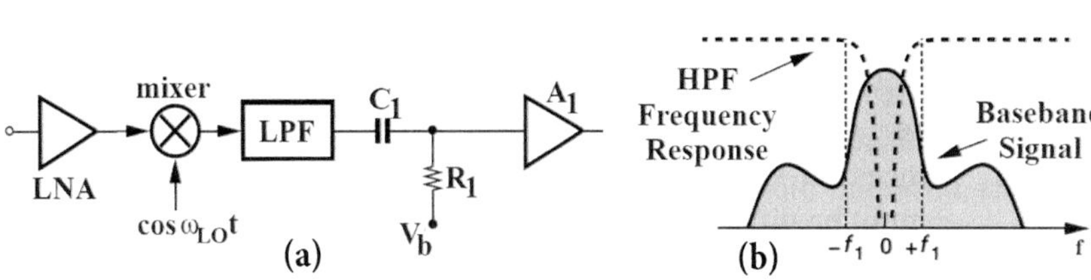

Figure 3.15 Mitigation methods of DC offsets (a) using HPF (b) observed effect on the spectrum.

FSK spectra with (a) small and (b) high-frequency deviations are shown in Fig. 3.16. The delayed responsiveness of AC coupling to transitory inputs is one of its drawbacks. The circuit unavoidably experiences a long time constant with a very low $f_1 = 1/(2\pi R_1 C_1)$, making it incapable of blocking the offset in the event that it changes abruptly. This shift happens when either (a) the LO frequency is changed to a new channel, which modifies the LO leakage, or (b) the LNA's gain is changed to a different value, which modifies the LNA's reverse isolation. (LNA gain switching is required to account for different received signal levels.) These factors combined with the comparatively large size of the capacitors needed make AC coupling uncommon in modern direct-conversion receivers.

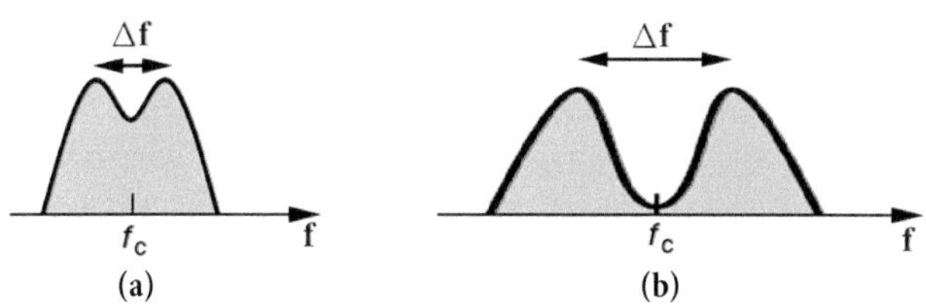

Figure 3.16 Frequency modulation using FSK for (a) small and (b) wide frequency deviation.

First, let us look at the cascade in Fig. 3.17, where *I*1 is taken from node X and indicates that four high-pass networks are needed for the differential outputs of the quadrature mixers. However, 802.11b can function with a high-pass corner frequency of 20 kHz at a maximum bit rate of 20 Mb/s, which is a hardly practical amount for on-chip integration. Reduced energy surrounding the carrier in modulation schemes makes them more suitable for AC coupling in the baseband. Two scenarios for FSK signals are shown in Fig. 3.16: for small modulation indices, the spectrum still has a lot of energy around the carrier frequency, f_c; but, for big modulation indices, the two frequencies produced by ONEs and ZEROs diverge significantly and leave a deep notch at f_c. Baseband can be more easily high-pass filtered if it is down-converted.

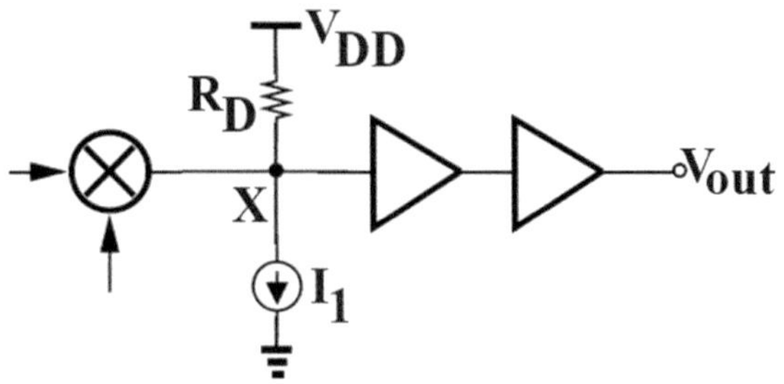

Figure 3.17 DC offset reduction by using current source.

The delayed responsiveness of AC coupling to transitory inputs is one of its drawbacks. The circuit unavoidably experiences a long time constant with a very low $f_1 = 1/(2\pi R_1 C_1)$, making it incapable of blocking the offset if it changes abruptly. This shift happens when either (a) the LO frequency is changed to a new channel, which modifies the LO leakage, or (b) the LNA's gain is changed to a different value, which modifies the LNA's reverse isolation. (LNA gain switching is required to account for different received signal levels.) These factors combined with the comparatively large size of the capacitors needed make AC coupling uncommon in modern direct-conversion receivers.

Digital-to-analog converters (DACs) are used in the most popular method of offset cancellation to draw a correction current. First, let us look at the cascade depicted in Fig. 3.17, in which the value of I_1, which is taken from node X, is changed to push the DC content in V_{out} to zero. $I_1 = \Delta V/R_D$ with appropriate polarity, for instance, if the mixer generates an offset of ΔV at node X and the succeeding stages show no offset. As a result, we are looking for a way to freeze the value of I_1 such that it has no effect on the baseband frequency response. Because of this, I_1 must be under register control and change in discrete steps. I_1 is broken down into units that are switched on or off based on the values kept in the register, as shown in Fig. 3.18. For instance, binary-weighted current sources $4I$, $2I$, and I are managed by the binary word D_3, D_2 and D_1, respectively. A DAC is formed by these current sources.

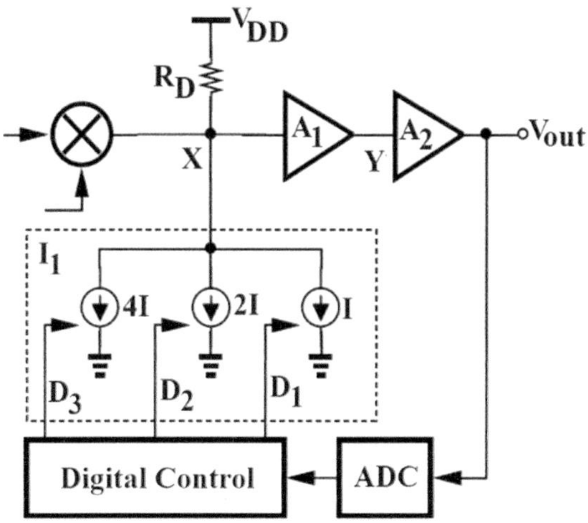

Figure 3.18 Circuit implementation of DC offset reduction by using current sources.

An analog-to-digital converter (ADC) drives the register and digitizes the baseband output (in the absence of signals) when the receiver is turned on. As a result, the whole negative-feedback loop converges to minimize V_{out}. Following that, the values are frozen in the register and kept there while the receiver operates. Despite the somewhat complicated configuration in Fig. 3.18, the DAC and digital controller occupy a significantly lower area than the capacitors in Fig. 3.15(a) due to the scaling of CMOS technology. Furthermore, when receiving signals, the ADC is also employed. Other features can also be accessed using offset's digital storage. For instance, during power-up, the receiver cycles through every possible combination of LO and gain settings, and the necessary values of I_1 are saved in a tiny memory. This is because the offset may change depending on the LO frequency or gain settings before or after the mixer. I_1's value is loaded into the register during reception and retrieved from memory for the specified LO and gain settings. The primary disadvantage of digital storage stems from the limited resolution at which the offset can be eliminated. Higher resolution must be achieved, or several DACs connected to various nodes (e.g., Y and V_{out}) in cascade to restrict the maximum offset, to mitigate this problem.

3.1.10 Even-Order Distortion

The third-order distortion leads to intermodulation and compression through our analysis of non-linearity. Additionally susceptible to even-order non-linearity in the RF circuit are heterodyne designs with a second zero-IF as well as direct-conversion receivers.

Assume that two strong interferers at ω_1 and ω_2 encounter a non-linearity in the LNA of the form $y(t) = \alpha_1 x(t) + \alpha_2 x^2(t)$, as illustrated in Fig. 3.19. A low-frequency beat at $\omega_2 - \omega_1$ is produced by the second-order term, which also yields the product of these two interferers.

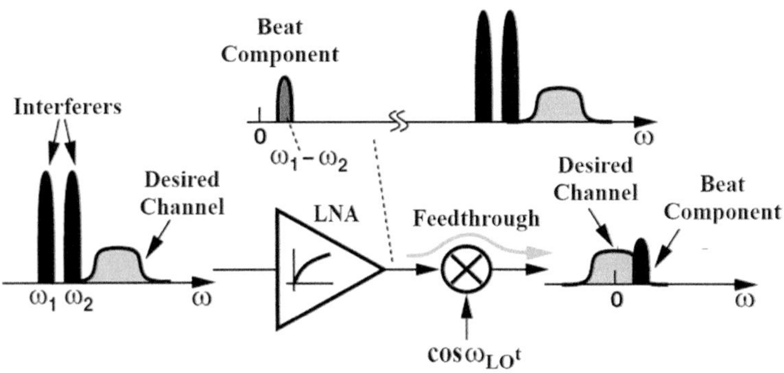

Figure 3.19 DCR issue with even-order distortion.

Such a term becomes insignificant when multiplied by cos $\omega_{LO}t$ in an ideal mixer since it is translated to high frequencies. However, in practice, a portion of the mixer's RF input can appear at the output without frequency translation due to asymmetry in the mixer or the LO waveform. As a result, the down-converted signal becomes distorted because some of the low-frequency beat shows up in the baseband. Naturally, AC coupling can eliminate the beat produced by the LNA, leaving the mixer's input transistor as the main source of even-order distortion.

First, let's look at the circuit depicted in Fig. 3.20 to learn how asymmetries lead to direct feedthrough in a mixer. The output can be expressed as the product of V_{in} and an ideal LO, or as a square wave with a 50% duty cycle that toggles between 0 and 1:

$$V_{out}(t) = V_{in}(t) \cdot S(t) \tag{3.1}$$

$$V_{\text{out}}(t) = V_{\text{in}}(t)\left[S(t) - \frac{1}{2}\right] + V_{\text{in}}(t) \cdot \frac{1}{2} \tag{3.2}$$

We understand that $[S(t) - \frac{1}{2}]$ is a square wave that is DC-free, meaning it only has odd harmonics. Consequently, the product of V_{in} and the odd harmonics of the LO are included in $V_{\text{in}}(t) \times [S(t) - \frac{1}{2}]$. The RF feedthrough to the output is indicated by the second term in Eq. (3.2), $V_{\text{in}}(t) \times \frac{1}{2}$ (with no frequency translation).

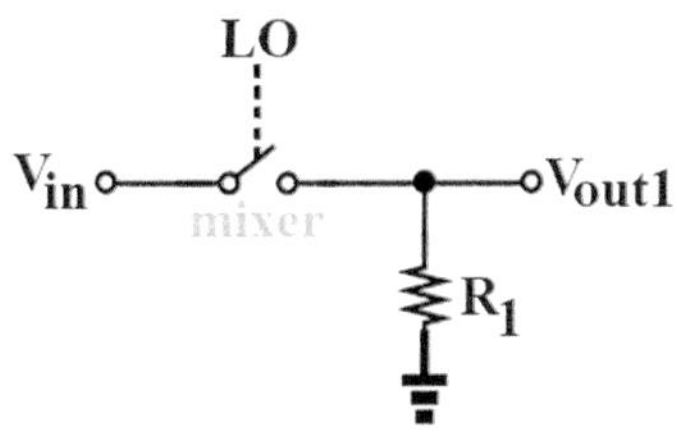

Figure 3.20 Simplified mixer representation with single-ended LO.

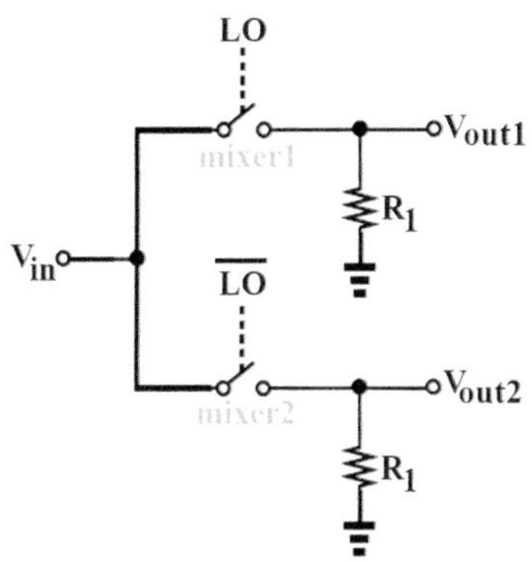

Figure 3.21 Mixer representation with differential LO.

Let us now look at the topology shown in Fig. 3.21, where a second output is produced by a second branch driven by $\overline{\text{LO}}$ (the complement of LO). $\overline{\text{LO}}$ can be expressed as $[1 - S(t)]$, thus,

$$V_{\text{out1}}(t) = V_{\text{in}}(t)S(t) \tag{3.3}$$

$$V_{\text{out2}}(t) = V_{\text{in}}(t)[1 - S(t)]. \tag{3.4}$$

Because $[1 - S(t)]$ displays a DC content of ½, the second output, $V_{out2}(t)$, too has an RF feedthrough equal to $V_{in}(t) \times$ ½. The RF feedthroughs in $V_{out1}(t)$ and $V_{out2}(t)$ are cancelled while the signal components add if the output is sensed differentially. This cancellation is the one that is susceptible to asymmetries; for instance, a net RF feedthrough occurs in the differential output if the switches display an imbalance in their on-resistances.

Even-order distortion is a serious enough issue to warrant measuring quantitatively. Known as the second intercept point (IP_2), this metric is established via a two-tone test $V_{in}(t) = A\cos\omega_1 t + A\cos\omega_2 t$ akin to the one used for IP3, with the exception that the beat component, rather than the intermodulation product, is the output of interest. When the LNA output may be found using

$$V_{out}(t) = \alpha_1 V_{in}(t) + \alpha_2 V_{in}^2(t) \tag{3.5}$$

$$V_{out}(t) = \alpha_1 A(\cos\omega_1 t + \cos\omega_2 t) + \alpha_2 A^2 \cos(\omega_1 + \omega_2)t + \alpha_2 A^2 \cos(\omega_1 - \omega_2)t + \cdots. \tag{3.6}$$

Demonstrating that the beat amplitude increases in proportion to the square of the input tones' magnitude. Thus, on a log scale, the beat amplitude grows with a slope of 2, as seen in Fig. 3.22. It is challenging to formulate the beat amplitude measured in the baseband since it depends on the dimensions and architecture of the device and the net feedthrough of the beat, which is dependent on the mixer and LO asymmetries.

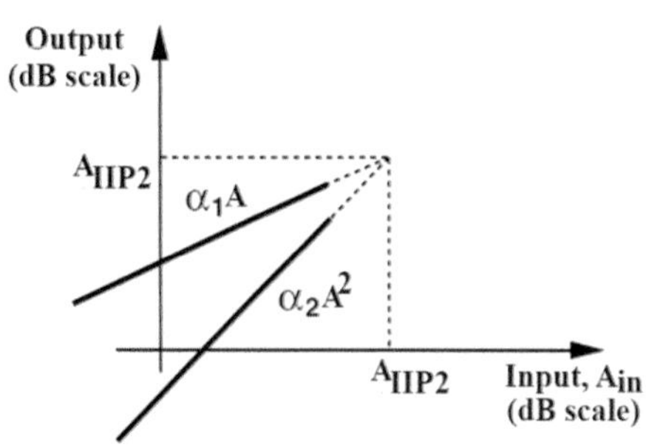

Figure 3.22 Second intermodulation intercept point IP2.

Even when there are no external interferences, even-order distortion can still be observed. Let's say the received signal shows amplitude modulation in addition to frequency and phase signals. Variable-envelope waveforms are produced, for instance, by OFDM, QAM, or basic QPSK with baseband pulse shaping. The signal is represented as $x_{\text{in}}(t) = [A_0 + a(t)] \cos[\omega_c t + \varphi(t)]$, where $a(t)$ is the envelope, which usually fluctuates slowly, indicating that the signal is low-pass. When second-order distortion occurs, the signal looks like

$$\alpha_2 x_{\text{in}}^2(t) = \alpha_2 [A_0^2 + 2A_0 a(t) + a^2(t)] \frac{1 + \cos[2\omega_c t + 2\phi(t)]}{2}. \quad (3.7)$$

Like the beat component in Fig. 3.19, both terms $\alpha_2 A_0 a(t)$ and $\alpha_2 a^2(t)/2$ are low-pass signals that corrupt the down-converted signal as they pass through the mixer with finite attenuation. Since the amplitude information appears as $\alpha_2 A_0 a(t)$, we claim that even-order distortion demodulates AM. The signal may be corrupted by this effect due to either its own envelope or the envelope of a significant interferer. Below, we examine both scenarios. How severe is the aforementioned phenomenon? According to Eq. (3.7), when the envelope variation approaches the input IP_2, the signal-to-noise ratio (SNR) drops to levels that are potentially dangerously low.

The LNA's and maybe the mixer's gain is automatically reduced to considerably lower values at such high input levels to prevent saturation and minimize the self-corruption impact mentioned above. However, the study highlights another, even more challenging circumstance. Large amplitude-modulated interferers that coexist with the intended channel can be demodulated via even-order distortion, allowing the AM component to emerge in the baseband through mixer feedthrough [12–18].

3.1.11 Flicker Noise

As the gain of the LNA/mixer cascade is usually limited to approximately 30 dB due to linearity constraints, the down-converted signal in a direct-conversion receiver is still quite modest, making it vulnerable to noise in the baseband circuits. Furthermore, flicker noise can significantly taint the signal because it is centered around zero frequency. The mixers themselves may also produce flicker noise at their output. Let us assume the down-converted spectrum depicted in Fig. 3.23, where f_{BW} is half of the RF channel bandwidth, to evaluate the influence of flicker noise. $S_{1/f}$ represents the flicker noise, and S_{th} represents the thermal noise at the baseband's end [6, 7].

The corner frequency at which the two-line profiles join is called the cutoff frequency f_c. From this, we can determine the consequence of flicker noise.

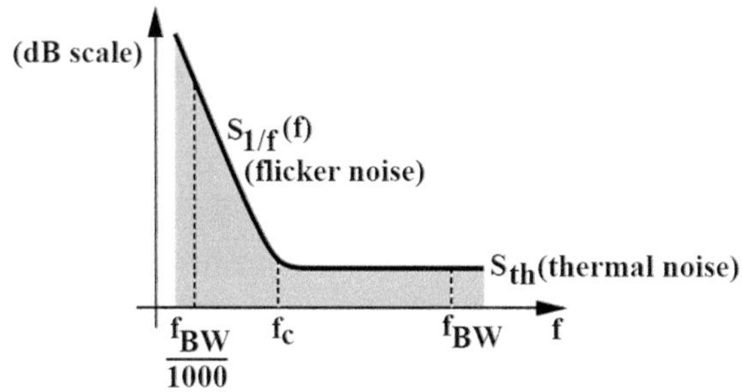

Figure 3.23 Frequency spectrum showing the flicker noise contribution.

The flicker noise issue makes it challenging to use direct conversion for standards with constrained channel bandwidth, as seen by the example above. Under such circumstances, the low-IF design turns out to be a better option.

3.1.12 *I*/*Q* Mismatch

Separation into quadrature phases is necessary for down-conversion of an asymmetrically modulated signal to a zero-IF. This can be achieved by 90°-shifting the LO waveform or the RF signal as shown in Fig. 3.24. As moving the RF signal usually involves significant trade-offs between noise, power, and gain, the method shown in Fig. 3.24(b) is recommended. Either way, as Fig. 3.25 shows, imbalances in the amplitudes and phases of the baseband *I* and *Q* outputs are caused by faults in the 90° phase shift circuit and mismatches between the quadrature mixers.

Significant gain and phase mismatches may also be caused by the baseband stages themselves.

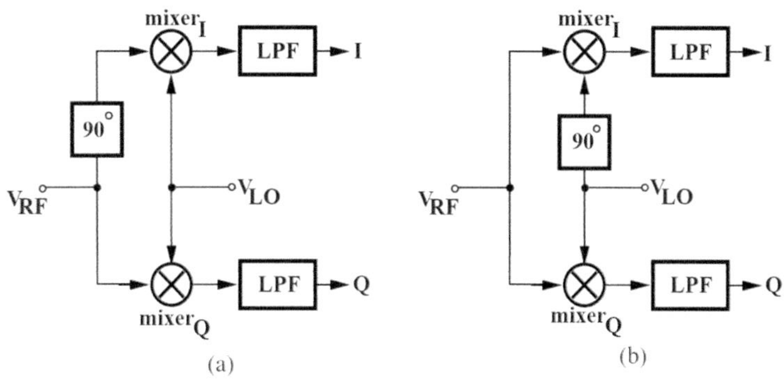

Figure 3.24 Comparison of (a) quadrature RF with fixed LO and (b) quadrature LO with direct RF.

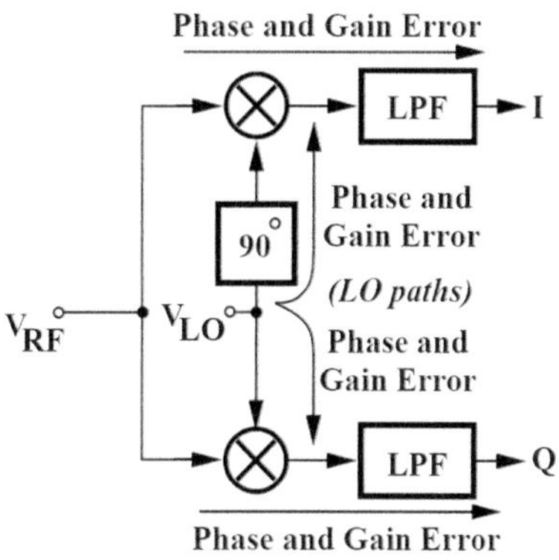

Figure 3.25 Direct RF with quadrature LO DCR with possible sources of *I*/*Q* mismatch.

Compared to heterodyne topologies, direct-conversion receivers typically have bigger quadrature mismatches. For instance, as device dimensions are lowered to get faster speeds, transistor mismatches rise.

Examine the following QPSK signal: $x_{in}(t) = a \cos \omega_c t + b \sin \omega_c t$, where a and b can be either −1 or +1. This will provide an insight into the impact of *I*/*Q* imbalance. Let us now group all the phase and gain mismatches in Fig. 3.25 together with the LO path as shown in Fig. 3.26:

$$x_{\mathrm{LO},I}(t)=2\left(1+\frac{\epsilon}{2}\right)\cos\left(\omega_c t+\frac{\theta}{2}\right) \tag{3.8}$$

$$x_{\mathrm{LO},Q}(t)=2\left(1-\frac{\epsilon}{2}\right)\sin\left(\omega_c t-\frac{\theta}{2}\right), \tag{3.9}$$

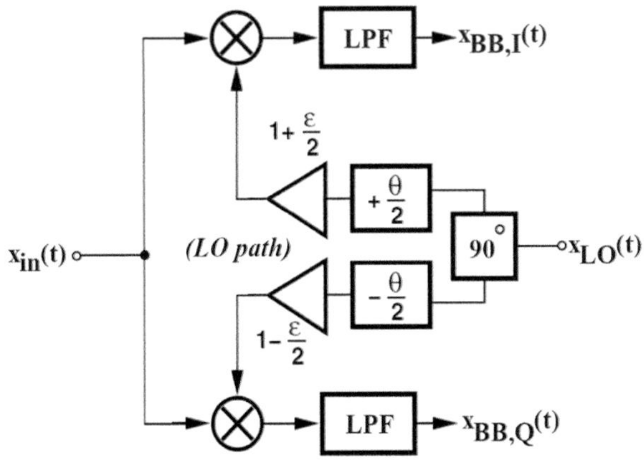

Figure 3.26 LO amplitude I/Q mismatch effects on DCR.

where ε and θ stand for the amplitude and phase mismatches, respectively, and a factor of two is added to simplify the results. The baseband signals that arise from multiplying $x_{\mathrm{in}}(t)$ by the quadrature LO waveforms and low-pass filtering the output are as follows:

$$x_{\mathrm{BB},I}(t)=a\left(1+\frac{\epsilon}{2}\right)\cos\frac{\theta}{2}-b\left(1+\frac{\epsilon}{2}\right)\sin\frac{\theta}{2} \tag{3.10}$$

$$x_{\mathrm{BB},Q}(t)=-a\left(1-\frac{\epsilon}{2}\right)\sin\frac{\theta}{2}+b\left(1-\frac{\epsilon}{2}\right)\cos\frac{\theta}{2} \tag{3.11}$$

The outcomes for the two exceptional situations, $\varepsilon \neq 0, \theta = 0$ and $\varepsilon = 0, \theta \neq 0$, are now examined. $x_{\mathrm{BB},I}(t) = a(1+ \varepsilon/2)$ and $x_{\mathrm{BB},Q}(t) = b(1 - \varepsilon/2)$ in the former scenario, suggesting that the amplitude scales of the quadrature baseband symbols diverge as in Fig. 3.27(a). The constellation's points are shifted, which is more significant shown in Fig. 3.27(b).

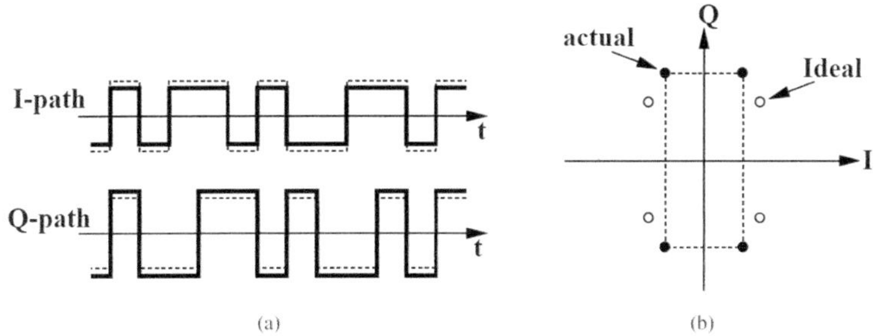

Figure 3.27 RX chain gain mismatch effects between I/Q paths as observed in (a) time domain and (b) on the QPSK modulation scheme constellation.

With $\varepsilon = 0$ and $\theta \neq 0$ a portion of the data symbols in one baseband output taints the other output as observed in Fig. 3.28(a). Furthermore, the constellation shown in Fig. 3.28(b) is stretched along one diagonal and compressed along the other.

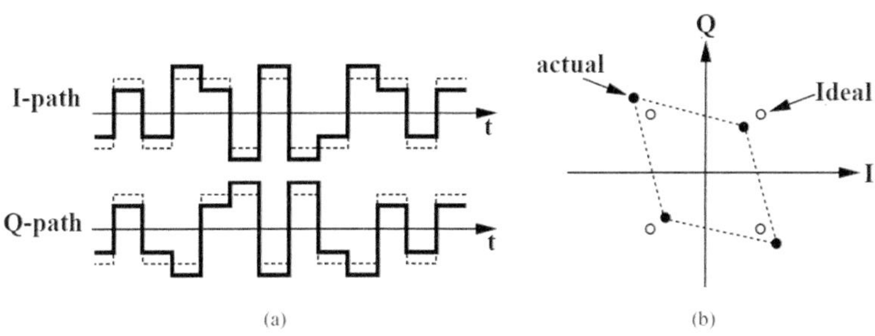

Figure 3.28 RX chain phase mismatch effects between I/Q paths as observed in (a) time domain and (b) on the QPSK modulation scheme constellation.

The maximum allowable I/Q mismatch must be understood while designing an RF receiver to select the appropriate architecture and building blocks. The highest mismatch values that have a negligible impact on performance can be established by simulations for complicated signal waveforms like OFDM with QAM.

As such, the quadrature phase and gain of many high-performance systems need to be calibrated, either upon power-up or on a constant basis. Calibration at power-up can be accomplished by detecting the baseband sinusoids in the analog or digital domain and applying an RF tone to the quadrature mixers' input, as shown in Fig. 3.29 [1]. These sinusoids can be generated at arbitrary low frequencies, making it possible to evaluate the amplitude and phase mismatches with accuracy. The received signal constellation is adjusted prior to detection after the mismatches are identified. An alternative would be to introduce a variable-phase stage, φ, and a variable-gain stage in the LO and baseband routes, respectively, and modify them until the mismatches are small enough, as shown in Fig. 3.30. Keep in mind that the adjustment controls need to be digitally stored for the receiver to function.

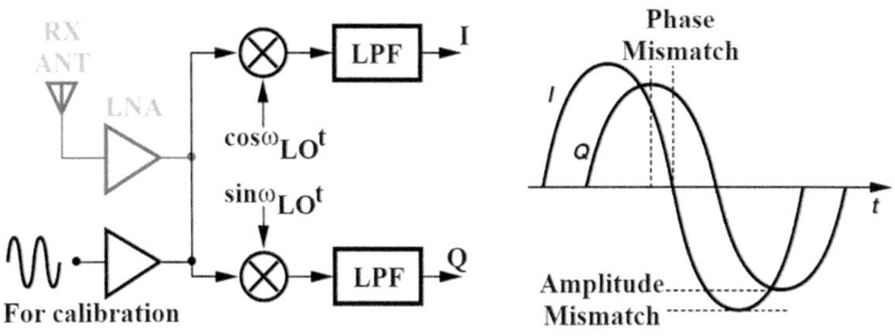

Figure 3.29 DCR I/Q mismatch computation indicating amplitude and phase mismatch.

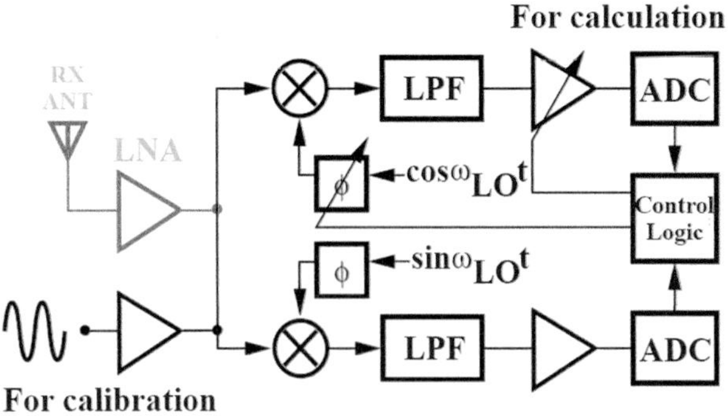

Figure 3.30 DCR I/Q mismatch computation indicating amplitude and phase mismatch and applied corrections.

3.1.13 Mixing Spurs

In contrast to heterodyne devices, direct-conversion receivers are rarely corrupted by spur mixing. This is due to the requirement that $f_1 \approx nf_{\text{LO}}$ for an input frequency (f_1) to fall within the baseband following mixing with nf_{LO}. The selectivity of the antenna, the band-select filter, and the LNA significantly suppress f_1, which is located far from the band of interest because f_{LO} equals the desired channel frequency. If the receiver is intended for a broad frequency range (more than two octaves), the problem of LO harmonics will become apparent.

3.2 Performance Parameters

3.2.1 Noise Figure

The RF receiver performance is limited by the noise to detect the smallest input signals. The noise is a random process and hence it is difficult to analyze the noise as an instantaneous parameter. One of the main contributions towards the noise is the random movement of charge carriers and their collisions with the lattice structure of the semiconductor material due to ambient temperature. Alternatively, by considering the average value the noise power can be determined.

3.2.1.1 Noise as a random process

The unpredictable nature of noise is a problem. The concept of randomness can be challenging for engineers who are accustomed to working with well-defined, deterministic, hard facts. This is especially true if randomness needs to be included quantitatively. We tackle the issue from an intuitive standpoint to get over this fear of randomness. When we say that noise is random, we mean that noise's instantaneous value cannot be predicted. Consider, for illustration, a resistor that is connected to a battery and is generating current. Each electron carrying the current is thermally agitated by the surrounding temperature, which causes it to travel in an erratic manner toward the battery's positive terminal on average. Because of this, the instantaneous current shows various values but the average current stays equal to V_B/R. We look for additional predictable characteristics of noise because it cannot be described in terms of instantaneous voltages or currents. For instance, we are aware that higher ambient temperatures cause electrons to be more thermally agitated, which in turn causes larger current variations. How can we articulate the idea of more substantial random fluctuations for a voltage or current quantity? The average power of the noise,

$$P_n = \lim_{T\to\infty}\frac{1}{T}\int_0^T n^2(t)dt, \tag{3.12}$$

which is defined in parallel with periodic signals, reveals this property. The noise waveform is represented by $n(t)$. This concept basically indicates that we get the average power by computing the area under $n^2(t)$ for a large time T, and normalizing the result with time period T.

The fact that the average power of noise components in circuits is constant is a miracle. For instance, for a resistor operating at constant room temperature, P_n is known and constant. How long should T be in Eq. (3.12)? Noise has a range of frequencies because it is unpredictable. Consequently, T needs to be long enough to hold multiple cycles of the lowest frequency.

For instance, human speech generates noise in a noisy restaurant that ranges from 20 Hz to 20 kHz. To record around 10 cycles of the 20 Hz components, T must be on the order of 0.5 s.

3.2.1.2 Noise spectrum

According to our earlier research, the time-domain perspective of noise only offers a limited amount of information, such as the average power. In contrast, the frequency domain view provides far more knowledge and is more helpful in radio frequency design. It's possible that the reader is already vaguely familiar with the meaning of spectrum. We refer to the range of 20 Hz to 20 kHz as the spectrum of human voice.

This indicates that all the components from 20 Hz to 20 kHz can be observed if we can somehow measure the voice's frequency content. Therefore, how can we determine the frequency content of a signal, such as the component's strength at 10 kHz? The remaining spectrum would need to be filtered away before we could calculate the 10 kHz component's average power.

A constant value is displayed on the power meter when someone talks into the microphone at a steady volume. It is simple to extend it to measure the strength of each frequency component. The average power at each frequency is measured using a bank of 1 Hz band-pass filters centered at f_1 to f_n. The resulting average power of the voice (or noise) carried in a 1 Hz bandwidth at various frequencies is often referred to as the spectrum or the power spectral density (PSD) of $x(t)$ and is indicated by $S_x(f)$. It is interesting to observe that the average power carried is represented by the total area beneath $S_x(f)$.

$$\int_0^\infty S_x(f)df = \lim_{T\to\infty}\frac{1}{T}\int_0^T x^2(t)dt \tag{3.13}$$

Because the spectrum is designed for positive frequencies, it is referred to as one-sided. In certain situations, using a two-sided spectrum makes the analysis easier. The two have equal energies because the latter is an even-symmetric function of the former that has been vertically scaled down by a factor of two.

3.2.1.3 Effect of transfer function on noise

The PSD was defined primarily because it makes it possible to apply many of the frequency domain techniques that are used with deterministic signals to random signals as well. For instance, how can we calculate the PSD at the output of a low-pass filter after applying white noise? It makes sense to us that the output PSD would take on the form of the filter's frequency response, as shown in Fig. 3.31. In actuality, the output spectrum of a linear, time-invariant system with a transfer function $H(s)$ is

$$S_y(f) = S_x(f)|H(f)|^2. \tag{3.14}$$

We note that $|H(f)|$ is squared because $S_x(f)$ is a (voltage or current) squared quantity.

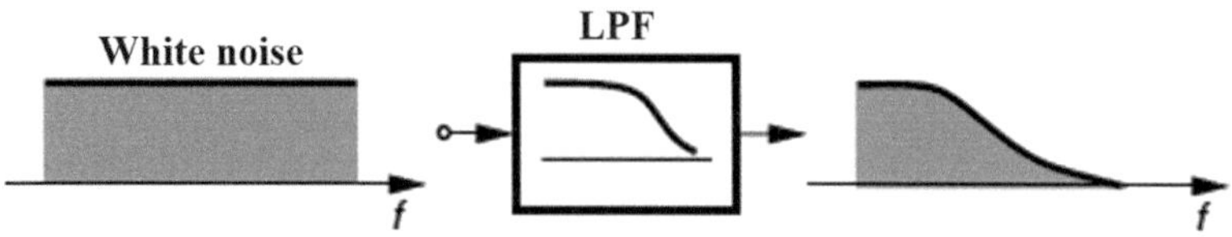

Figure 3.31 Frequency response of white noise through a LPF.

3.2.1.4 Device noise

We want to simulate the noise of a circuit's constituent elements using well-known elements like voltage and current sources to examine the noise performance of circuits. With this kind of representation, conventional circuit analysis methods can be applied.

(a) Thermal noise of resistors

As was previously explained, noise is caused by the random agitation of charge carriers in resistors by ambient thermal energy.

A parallel current source with a PSD $\overline{I_n^2} = \overline{V_n^2} / R_1 = 4kT / R_1$ of which is Norton equivalent shown $\overline{V_n^2}$ = $4kTR_1$, in Fig. 3.32(b) or a series voltage source with PSD of the Thevenin equivalent seen in Fig. 3.32(a), can be used to represent the noise. Sometimes the analysis is made simpler by the model selection. The sources' polarity is irrelevant, but it needs to remain constant for the duration of a circuit's computations.

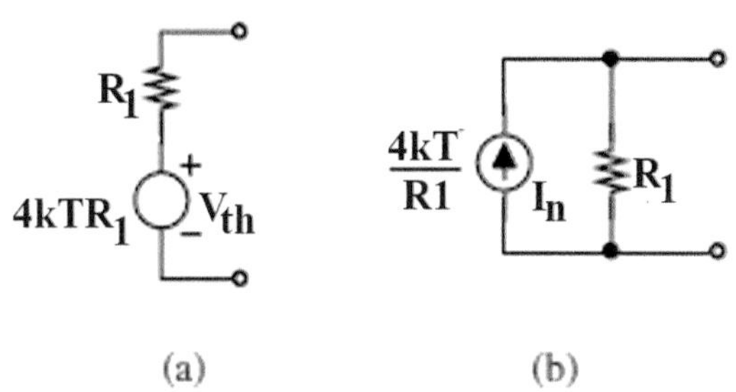

Figure 3.32 Receiver thermal noise equivalent (a) Thevenin and (b) Norton's models.

Can we extract energy from a resistor if it transforms ambient heat into a noise voltage or current? Specifically, does R_2 receive energy from the configuration depicted in Fig. 3.33? Interestingly, R_2 likewise generates a noise PSD of 4 kTR_2, so even if R_1 and R_2 live at the same temperature, no net energy is transmitted between them. But let's say R_2 is maintained at T = 0 K. Subsequently, R_1 keeps absorbing thermal energy from its surroundings, transforms it into noise, and transfers the energy to R_2. The average amount of power sent to R_2 is equivalent to

$$P_{R2} = \frac{\overline{V_{out}^2}}{R_2} = \overline{V_n^2}\left(\frac{R_2}{R_1 + R_2}\right)^2 \frac{1}{R_2} = 4kT\frac{R_1R_2}{(R_1 + R_2)^2}. \tag{3.15}$$

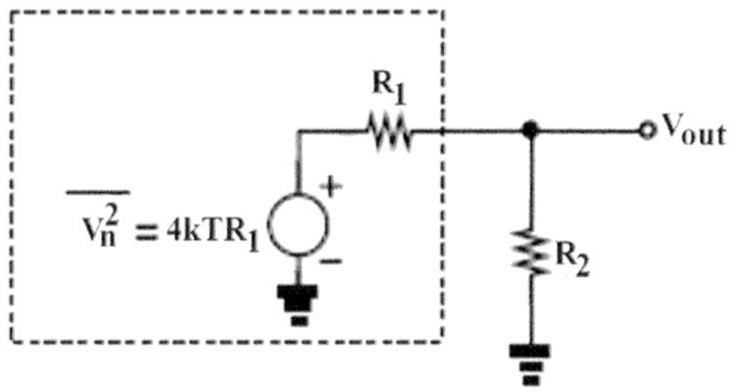

Figure 3.33 Noise transfer between resistors.

The maximum value of this amount is reached if $R_2 = R_1$:

$$P_{R2,max} = kT \tag{3.16}$$

Known as the available noise power, kT has a power per unit bandwidth dimension and is independent of the resistor value. It is demonstrable to the reader that at T = 300 K, the value of kT = −173.8 dBm/Hz. It is not necessary for a circuit to have an explicit resistor of value R_1 in order for it to display a thermal noise density of $\overline{V_n^2}$ = $4kTR_1$. After all, according to Eq. (3.14), the surrounding circuit may change the noise density of a resistor to a higher or lower value.

It should be noted that a passive circuit that emits heat must have a physical resistance and consequently causes thermal noise. Lossy circuits are noisy, as we informally put it.

As a receiving element shown in Fig. 3.34(b), the antenna produces a thermal noise PSD for a transmitting antenna that wastes energy by radiation in accordance with the equation

$$\overline{V_{n,\text{ant}}^2} = 4kTR_{\text{rad}}, \tag{3.17}$$

where R_{rad} is the radiation resistance.

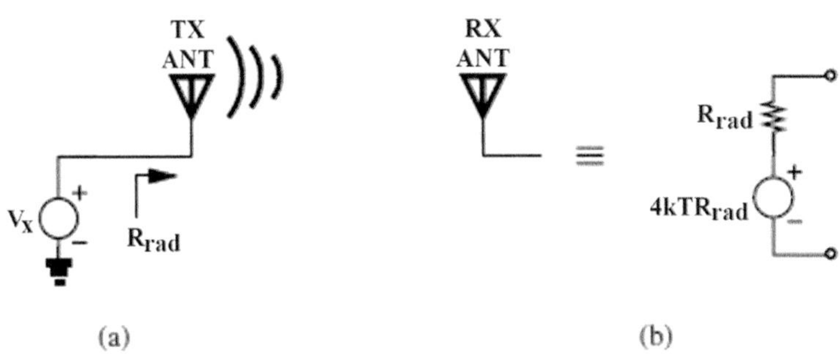

Figure 3.34 (a) TX antenna, (b) RX antenna generating thermal noise.

(b) Noise in MOSFETs

A current source connected between the source and drain terminals of MOS transistors operating in the saturation region can be used to estimate the thermal noise of the transistors in Fig. 3.35(a):

$$\overline{I_n^2} = 4kT\gamma g_m, \tag{3.18}$$

where γ is the excess noise coefficient and gm is the transconductance. For long-channel transistors, the value of γ is 2/3, while in short-channel devices, it can even reach a value of 2 [1]. The true value of γ depends on additional factors [1] and is typically measured for every CMOS technology generation.

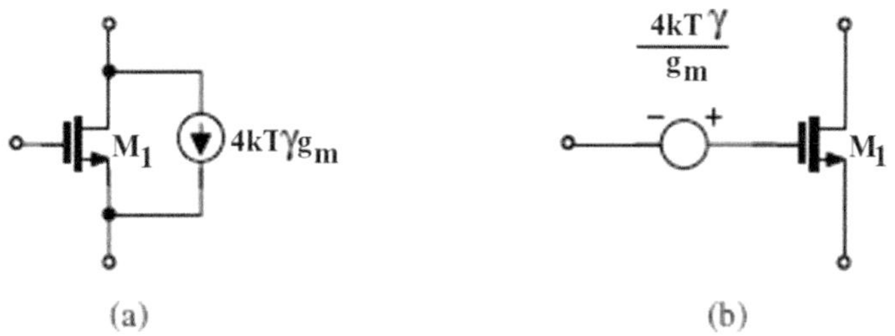

Figure 3.35 MOSFET channel's thermal noise modelled as (a) current and (b) voltage sources.

As an alternative, the noise $\overline{V_n^2} = 4kT\gamma/g_m$ can be represented by a voltage source connected in series with the gate as illustrated in Fig. 3.35(b). The gate resistance of MOSFETs contributes to thermal noise as well; this impact gets more significant as the gate length is reduced. The resistance of a device with width W and length L shown in Fig. 3.36(a) is represented as

$$R_G = \frac{W}{L}R, \tag{3.19}$$

where R_G stands for polysilicon gate's sheet resistance.

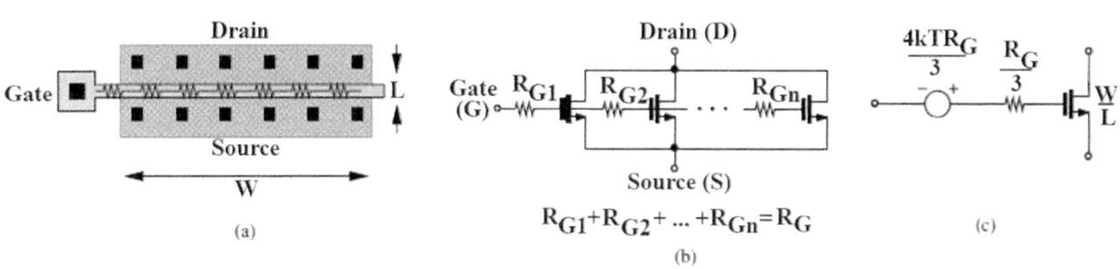

Figure 3.36 (a) MOSFET polysilicon gate resistance, (b) distributed noise equivalent circuit, and (c) lumped noise equivalent circuit model.

This noise must be far lower in a good design than the channel's noise as

$$4kT\frac{R_G}{3} \ll \frac{4kT\gamma}{g_m}. \tag{3.20}$$

Physical resistances are also present at the drain and gate terminals; however, they can be reduced by using several fingers. A gate-induced noise current [1] is produced at very high frequencies when thermal noise current passing through the channel couples capacitively to the gate as illustrated in Fig. 3.37.

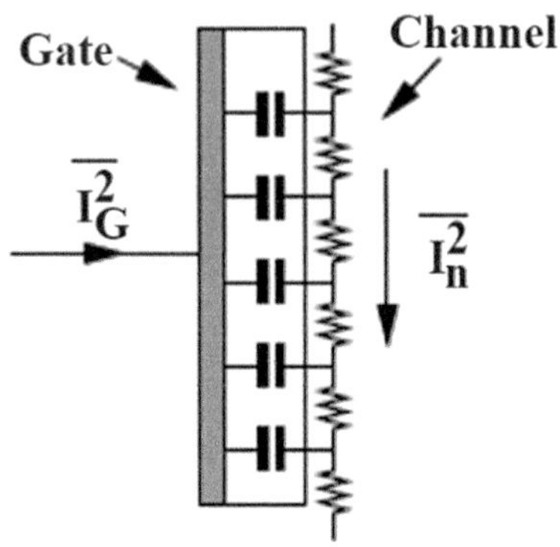

Figure 3.37 MOSFET gate-induced channel noise, $\overline{I_G^2}$.

This effect is not modelled in typical circuit simulators, but its significance has remained unclear. In this chapter, we neglect the gate-induced noise current. MOS devices also suffer from flicker or $1/f$ noise. Modelled by a voltage source in series with the gate, this noise exhibits the following:

$$\overline{V_n^2} = \frac{K}{WLC_{ox}} \frac{1}{f}, \tag{3.21}$$

where K is a process-dependent constant. In most CMOS technologies, K is lower for PMOS devices than for NMOS transistors because the former carry charge well below the silicon-oxide interface and hence suffer less from surface states of the dangling bonds [1]. Slowly varying noise components take on a large amplitude due to the $1/f$ dependence. The intended signal's spectrum or the time scale of interest determines which frequency in the noise integration to choose as the lowest [1]. The $1/f$ noise PSD intercepts the thermal noise PSD at a frequency known as the $1/f$ noise corner frequency, or f_c, for a given device size and bias current. According to the previous example, f_c can be derived by converting the flicker noise voltage to current and equating the result to the thermal noise current, as shown in Fig. 3.38:

$$\frac{K}{WLC_{ox}}\frac{1}{f_c}g_m^2 = 4kTG\gamma g_m \tag{3.22}$$

Consequently,

$$f_c = \frac{K}{WLC_{ox}}\frac{g_m}{4kT\gamma}. \tag{3.23}$$

In today's MOS technology, the corner frequency is in tens or even hundreds of megahertz range.

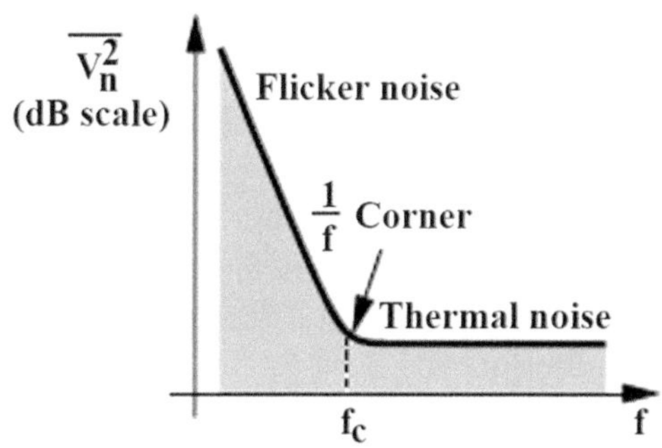

Figure 3.38 Frequency spectrum showing flicker noise and thermal noise.

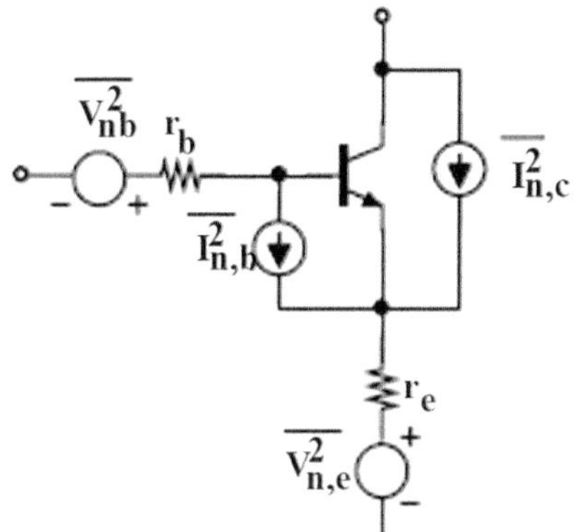

Figure 3.39 Bipolar transistor's noise sources.

3.2.1.5 Representation of noise in circuits

Now that the noise of devices has been defined, our goal is to create metrics that indicate the noise performance of circuits, or how noisy a particular circuit is. Noise referred by input, how is it possible to see circuit noise in a lab setting? Since we can only access the output, we are limited to measuring the output noise. Regretfully, the output noise makes it impossible to compare circuits fairly: A circuit with high gain may have high output noise rather than high noise. We refer the noise to the input as a result.

A series voltage source and a parallel current source are used in analog design to simulate input-referred noise as in Fig. 3.40 [1]. The former is derived by dividing the output noise by the voltage gain, or, equivalently, by shorting the input port of models A and B and equal their output noises. In a similar vein, the latter is calculated by equating the output noises (that is, by dividing the output noise by the transimpedance gain) while the input ports remain open. It could seem from the example above that M1's noise is counted twice. It can be demonstrated that [1] the two input-referred noise sources are frequently correlated but also necessary and sufficient.

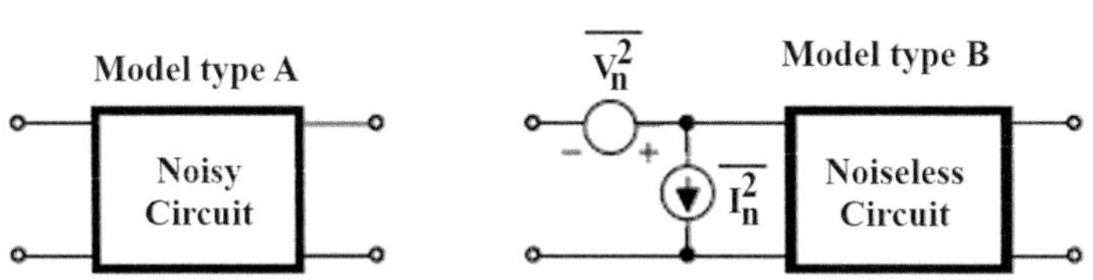

Figure 3.40 Determination of input-referred noise.

At high frequencies, the computation and application of input-referred noise sources become challenging. For instance, measuring the transimpedance gain of an RF stage is very difficult. Because of this, RF designers use the idea of noise figure, which is another measurement-friendly metric for noise performance.

(a) Noise figure

The SNR, which is calculated by dividing the signal power by the noise power, is a crucial factor in circuit and system design. Therefore, it is useful to inquire as to how the SNR decreases while the signal passes through a certain circuit. Even when the circuit serves as an attenuator, the output SNR is identical to the input SNR if there is no noise in the circuit. We define the circuit's noise figure (NF) as

$$NF = \frac{SNR_{in}}{SNR_{out}}, \tag{3.24}$$

where NF = 1 for a noiseless stage, in order to quantify the circuit's level of noise. Given that every variable in this ratio has a power (or voltage squared) dimension, we may describe NF in decibels as follows:

$$NF|_{dB} = 10 \cdot \log_{10}\left(\frac{SNR_{in}}{SNR_{out}}\right) \tag{3.25}$$

It should be noted that most texts refer to Eq. (3.24) as the noise factor and Eq. (3.25) as the noise figure. The definition of NF in Eq. (3.24) may seem difficult when compared to input-referred noise since it depends on both the noise of the circuit under examination and the SNR supplied by the stage before it. In reality, even if the circuit may have finite internal noise, $SNR_{in} = \infty$ and $NF = \infty$ if the input signal is noise-free. In this scenario, only the input-referred noise can be given and NF is not a useful parameter.

In general, the noise figure can be calculated more easily than what Eq. (3.24) might imply. Let's say, for illustration purposes, that an antenna detects the signal received by a low-noise amplifier as shown in Fig. 3.41(a). The model depicted in Fig. 3.41(b) results from the antenna radiation resistance, or R_S, producing thermal noise as expected by Eq. (3.17). In this case, $\overline{V_{n\,RS}^2}$ stands for the antenna's thermal noise and $\overline{V_n^2}$ for the LNA's output noise. SNR_{in} at the LNA's input and SNR_{out} at its output need to be calculated.

When V_{in} and V_{RS} appear at the input of the LNA, they are attenuated by $\alpha = Z_{in}/(Z_{in} + R_S)$ if the LNA has an input impedance of Z_{in}. That is,

$$\mathrm{SNR}_{\mathrm{in}} = \frac{|\alpha|^2 V_{\mathrm{in}}^2}{|\alpha|^2 \overline{V_{\mathrm{RS}}^2}}, \tag{3.26}$$

where V_{in} is the antenna's received signal's root mean square value. Assuming a voltage gain of A_v from the LNA input to the output, we may calculate $\mathrm{SNR}_{\mathrm{out}}$ by recognizing that the output signal power equals $V_n^2 |\alpha|^2 A_v^2$.

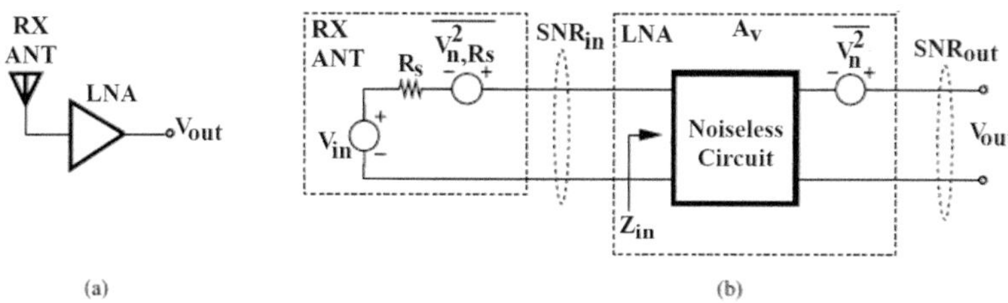

Figure 3.41 (a) RX antenna feeding the LNA, (b) the equivalent noise circuit.

There are two parts to the output noise: (a) the antenna noise that the LNA, $\overline{V_{\mathrm{RS}}^2}\, |\alpha|^2 A_v^2$, amplified, and (b) the LNA, $\overline{V_n^2}$, output noise. Since there is no correlation between these two components, we just combine the PSDs and write $\mathrm{SNR}_{\mathrm{out}}$:

$$\mathrm{SNR}_{\mathrm{out}} = \frac{V_{\mathrm{in}}^2 |\alpha|^2 A_v^2}{\overline{V_{\mathrm{RS}}^2} |\alpha|^2 A_v^2 + \overline{V_n^2}} \tag{3.27}$$

$$\mathrm{NF} = \frac{V_{\mathrm{in}}^2}{4kTR_S} \cdot \frac{\overline{V_{\mathrm{RS}}^2} |\alpha|^2 A_v^2 + \overline{V_n^2}}{V_{\mathrm{in}}^2 |\alpha|^2 A_v^2} = \frac{1}{\overline{V_{\mathrm{RS}}^2}} \cdot \frac{\overline{V_{\mathrm{RS}}^2} |\alpha|^2 A_v^2 + \overline{V_n^2}}{|\alpha|^2 A_v^2} = 1 + \frac{\overline{V_n^2}}{|\alpha|^2 A_v^2} \cdot \frac{1}{\overline{V_{\mathrm{RS}}^2}} \tag{3.28}$$

The NF can also be defined as the total noise at the output divided by the noise at the output caused by the source impedance. The NF is frequently referred to as the spot noise figure to highlight the narrow bandwidth because it is typically stated for a 1 Hz bandwidth at a certain frequency.

According to Eq. (3.30), the source impedance affects the NF both through $\overline{V_n^2}$ and $\overline{V_{RS}^2}$. The input noise current, $\overline{I_{n,\text{in}}^2}$, partially flows through R_S if we model the noise using input-referred sources. This causes a source-dependent noise voltage of $\overline{I_{n,\text{in}}^2} R_S^2$ at the input, and consequently a proportional noise at the output. As a result, the NF needs to be defined in relation to a source impedance, which is usually 50 Ω. Equation (3.30) can be simplified for manual analysis and simulations by observing that the numerator is the total noise measured at the output:

$$\text{NF}=\frac{1}{4kTR_S}\cdot\frac{\overline{V_{n,\text{out}}^2}}{A_0^2}, \tag{3.29}$$

where the voltage gain from V_{in} to V_{out} (rather than the gain from the LNA input to its output) is represented by $A_0 = |\alpha|A_v$, and $\overline{V_{n,\text{out}}^2}$ comprises both the source impedance noise and the LNA noise. We state, in general, that we simply divide the total output noise by the gain in order to calculate the NF by normalizing the outcome to match the noise of R_S.

It is significant to remember that the aforementioned derivations still hold true in the event that there is no real power transmission from the antenna to the LNA or from the LNA to a load. Since derivations in Fig. 3.41(b) are based on voltage (squared) numbers rather than power quantities, they all hold true even if Z_{in} goes to infinity and no power is transmitted to the LNA. In other words, no inconsistency occurs in the presence of impedance mismatches or even infinite input impedances if the derivations take noise and signal voltages into account. There is a significant distinction between conventional microwave design and contemporary RF design.

3.2.1.6 Level of noise in cascaded stages

Given that a receiver chain consists of multiple stages, it is preferable to calculate the NF of the entire cascade relative to each level. Examine the cascade shown in Fig. 3.42(a), where A_{v1} and

A_{v2} represent the two stages' unloaded voltage gains. Additionally displayed are the two stages' input and output impedances as well as their output noise voltages. First, we use a direct way to determine the NF of the cascade Eq. (3.30). To do this, we just need to calculate the total noise at the output caused by the two stages, divide the result by $(V_{out}/V_{in})^2$, normalize it to 4 kTR_S, and add one. Considering the loadings, we express the total voltage gain as

$$A_0 = \frac{V_{out}}{V_{in}} = \frac{R_{in1}}{R_{in1} + R_S} A_{v1} \frac{R_{in2}}{R_{in2} + R_{out1}} A_{v2}. \quad (3.30)$$

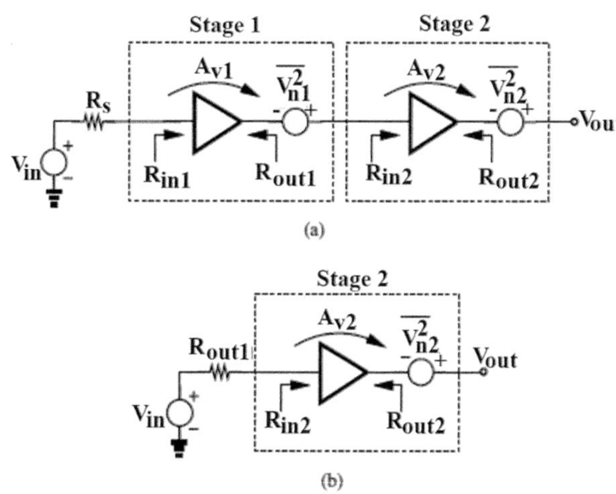

Figure 3.42 (a) Noise determination of cascaded stages (b) equivalent reduced representation.

The two stages' combined output noise, represented by $\overline{V^2_{n,out}}$, is made up of two parts: (a) $\overline{V^2_{n2}}$, and (b) $\overline{V^2_{n1}}$ amplified by the second stage. As it arrives at the second stage's input, V_{n1} is scaled by a factor of $R_{in2}/(R_{in2} + R_{out1})$ because it perceives an impedance of R_{in2} to its right and R_{out1} to its left. Consequently,

$$\overline{V^2_{n,out}} = \overline{V^2_{n2}} + \overline{V^2_{n1}} \frac{R^2_{in2}}{(R_{in2} + R_{out1})} A^2_{v2}. \quad (3.31)$$

The overall NF is therefore expressed as

$$\mathrm{NF}_{\mathrm{tot}} = 1 + \frac{\overline{V^2_{n,\mathrm{out}}}}{A_0^2} \cdot \frac{1}{4kTR_S} \tag{3.32}$$

$$\mathrm{NF}_{\mathrm{tot}} = 1 + \frac{\overline{V^2_{n1}}}{\left(\dfrac{R_{\mathrm{in1}}}{R_{\mathrm{in1}} + R_S}\right)^2 A_{v1}^2} \cdot \frac{1}{4kTR_S} + \frac{\overline{V^2_{n2}}}{\left(\dfrac{R_{\mathrm{in1}}}{R_{\mathrm{in1}} + R_S}\right)^2 A_{v1}^2 \left(\dfrac{R_{\mathrm{in2}}}{R_{\mathrm{in2}} + R_{\mathrm{out1}}}\right)^2 A_{v2}^2} \cdot \frac{1}{4kTR_S}. \tag{3.33}$$

Regarding a source impedance of R_S, the first two terms make up the NF of the first stage, NF_1. The second stage's noise is represented by the third term, but how can it be stated in terms of this stage's noise figure? Now, let us examine the second stage independently and calculate its noise figure in relation to R_{out1}'s source impedance shown in Fig. 3.42(b). Returning to Eq. (3.30), we have

$$\mathrm{NF}_2 = 1 + \frac{\overline{V^2_{n2}}}{\dfrac{R^2_{\mathrm{in2}}}{(R_{\mathrm{in2}} + R_{\mathrm{out1}})^2} A_{v2}^2} \cdot \frac{1}{4kTR_{\mathrm{out1}}}. \tag{3.34}$$

It follows from Eqs. (3.23) and (3.24) that

$$\mathrm{NF}_{\mathrm{tot}} = \mathrm{NF}_1 + \frac{\mathrm{NF}_2 - 1}{\dfrac{R^2_{\mathrm{in1}}}{(R_{\mathrm{in1}} + R_S)^2} A_{v1}^2 \dfrac{R_S}{R_{\mathrm{out1}}}}. \tag{3.35}$$

What is represented by the denominator? This number is actually the first stage's available power gain, which is calculated by dividing the available power at the output ($P_{\mathrm{out,av}}$), or the power it would supply to a matched load, by the available source power ($P_{S,\mathrm{av}}$), or the power the source would supply to a matched load. This can be confirmed with Eq. (3.36) using Stage 1 of the circuit in Fig. 3.42(a) with a load equivalent of R_{out1}:

$$P_{\text{out,av}} = V_{\text{in}}^2 \frac{R_{\text{in1}}^2}{(R_S + R_{\text{in1}})^2} A_{v1}^2 \cdot \frac{1}{4R_{\text{out1}}} \tag{3.36}$$

Likewise, V_{in}'s power delivery to a load of R_S is determined by

$$P_{S,\text{av}} = \frac{V_{\text{in}}^2}{4R_S}. \tag{3.37}$$

It is true that the denominator in Eq. (3.35) is equal to the ratio of Eqs. (3.36) and (3.37). Using these insights, we compose

$$\text{NF}_{\text{tot}} = \text{NF}_1 + \frac{\text{NF}_2 - 1}{A_{P1}}, \tag{3.38}$$

where A_{P1} stands for the first stage's available power gain. Remembering that NF_2 is calculated in relation to the first stage's output impedance is crucial. For m cascaded stages,

$$\text{NF}_{\text{tot}} = 1 + (\text{NF}_1 - 1) + \frac{\text{NF}_2 - 1}{A_{P1}} + \cdots + \frac{\text{NF}_m - 1}{A_{P1} \cdots A_{P(m-1)}}. \tag{3.39}$$

This finding, known as Friis' equation [1], implies that the first few stages of a cascade are the most crucial since the noise supplied by each level reduces as the overall gain preceding that stage grows. On the other hand, when the input of a stage experiences attenuation (loss), the NF of the subsequent circuits is amplified.

The example above illustrates a common scenario in contemporary RF design: there is no attempt to ensure impedance matching between the two stages, and the interface between the two stages lacks a 50 Ω impedance. In these situations, direct computation of the NF is more appealing because Friis' equation gets complicated. Although the second stage in the example above has an infinite input impedance, Eq. (3.34) allows direct technique to be extended to more practical scenarios.

Equation (3.34) states that even with complex input and output impedances, the following calculations must be made: (1) $\overline{V_{n1}^2}$ must be divided by the unloaded gain from V_{in} to the first stage's output; (2) $\overline{V_{n2}^2}$, the second stage's output noise, must be calculated with this stage driven by the first stage's output impedance; and (3) $\overline{V_{n2}^2}$ must be divided by total voltage gain from V_{in} to V_{out}.

Notably, the noise of the cascaded stages after the LNA is divided by the gain of the LNA. As shown in Fig. 3.43 for a LNA with mixer cascade stages, the input noise currents are ignored, and the input-referred noise voltages of the LNA and mixer are indicated by the variables $\overline{V_{n,\mathrm{LNA}}^2}$ and $\overline{V_{n,\mathrm{mix}}^2}$ respectively. For convenience, we will assume that the mixer has a unity voltage gain, and we will write the entire output noise as

$$A_{v1}^2\left(\overline{V_{n,\mathrm{LNA}}^2}+4kTR_S\right)+\overline{V_{n,\mathrm{mix}}^2}. \tag{3.40}$$

Therefore, the total noise figure is given as

$$\mathrm{NF}_{\mathrm{tot}} \frac{A_{v1}^2\left(\overline{V_{n,\mathrm{LNA}}^2}+4kTR_S\right)+\overline{V_{n,\mathrm{mix}}^2}}{A_{v1}^2}\frac{1}{4kTR_S} \tag{3.41}$$

$$\mathrm{NF}_{\mathrm{tot}}=\mathrm{NF}_{\mathrm{LNA}}+\frac{\overline{V_{n,\mathrm{mix}}^2}}{A_{v1}^2}\cdot\frac{1}{4kTR_S}. \tag{3.42}$$

Hence, alternatively, in the NF equation (3.42), the noise of the second stage is the mixer's noise is divided by the total gain from the input voltage source to the LNA output.

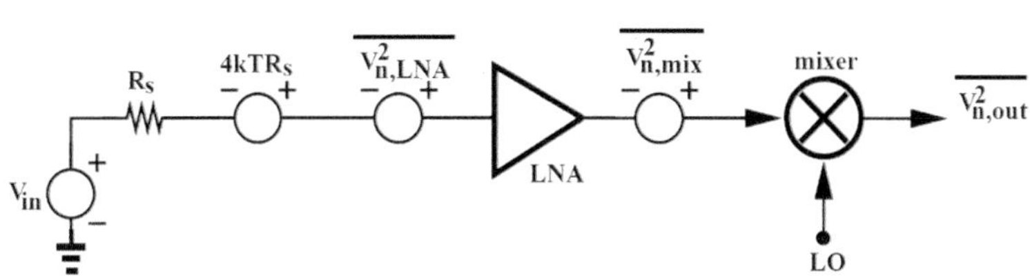

Figure 3.43 Noise model of receiver chain with LNA followed by mixer.

3.2.1.7 Lossy circuits noise figure

At the front end of RF transceivers are passive circuits like filters, and their loss can have serious consequences. The circuit's undesired resistive components, which turn input power into heat and reduce signal power at the output, are the source of the loss. Moreover, keep in mind that resistive components also produce thermal noise. In other words, passive lossy circuits cause noise in addition to signal attenuation.

Our goal is to demonstrate that the power loss of a passive (reciprocal) circuit, which is represented as $L = P_{in}/P_{out}$ (where P_{in} is the available source power and P_{out} is the available output power), equals the noise figure of the circuit. The available power is the power that a specific source or circuit would supply to a conjugate-matched load, as stated in the derivation of Friis' equation. If the input and output match, the proof is simple.

Here, we examine a more general scenario. Let's take a look at the configuration in Fig. 3.44(a), where a load impedance of R_L is pushed while a source impedance of R_S drives the lossy circuit. According to Eq. (3.38), the source power can be derived as Pin = $V_{in}^2/(4R_S)$. We build the Thevenin equivalent depicted in the Fig. 3.44(b) to ascertain the available output power, yielding $P_{out} = V_{They}^2/(4R_{out})$. As a result, the loss is given as

$$L = \frac{V_{in}^2}{V_{Thev}^2}\frac{R_{out}}{R_S}. \tag{3.43}$$

We use an analogous circuit depicted in Fig. 3.44(c) to calculate the noise figure as

$$\overline{V_{n,out}^2} = 4kTR_{out}\frac{R_L^2}{(R_L + R_{out})^2}. \tag{3.44}$$

If R_L is noiseless, the only thing that can be ascertained is the noise figure of the lossy circuit. While the circuit generates an output voltage of $V_{out} = V_{Thev}R_L/R_L(R_L + R_{out})$ in response to V_{in}, one can determine the voltage gain from V_{in} to V_{out} as indicated in Fig. 3.44(b).

$$A_0 = \frac{V_{Thev}}{V_{in}}\frac{R_L}{R_L + R_{out}} \tag{3.45}$$

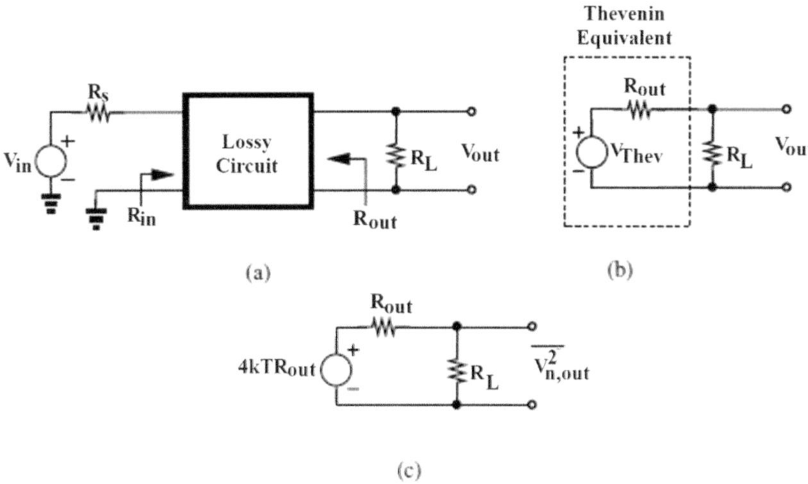

Figure 3.44 Noise determination of the (a) passive lossy network (b) voltage source based on Thevenin's equivalent (c) reduced equivalent representation.

In other words, Eq. (3.44) is divided by the square of Eq. (3.45) and normalized to $4kTR_S$, the NF is equal to

$$\text{NF}=4kTR_{\text{out}}\frac{V_{\text{in}}^2}{V_{\text{Thev}}^2}\frac{1}{4kTR_S} \tag{3.46}$$

$$\text{NF} = L. \tag{3.47}$$

From this, we can observe that the noise figure of a lossy passive circuit is equivalent to its loss.

3.2.2 Linearity

A typical linear system with a gain of α will have a transfer characteristic as

$$y(t) = \alpha x(t). \tag{3.48}$$

When the input is large, then the system will be pushed towards its non-linearity and hence the system will generate higher order terms of input resulting in a complex transfer characteristic as

$$y(t) = \alpha_0 + \alpha_1 x(t) + \alpha_2 x^2(t) + \alpha_3 x^3(t) + \cdots, \tag{3.49}$$

where α_i are the coefficients of the higher-order terms that characterize the output.

One measure of receiver non-linearity is the third intercept point (IIP_3), which can be determined by using a two-tone intermodulation characteristics. In this method, two tones of equal amplitude and separated in frequency by an offset within the system's bandwidth is input to the system and the output intermodulation components are measured. The main challenge in defining the relative IM of a circuit is that it becomes significant only when the value of A is provided. Practically speaking, we would rather have a single measurement that accurately depicts the circuit's intermodulation behavior without requiring knowledge of the input level used for the two-tone test. Thankfully, there is a measure known as the third intercept point (IP_3).

The idea behind IP_3 stems from our previous finding that the output IM products increase more sharply ($\propto A^3$) as the amplitude (A) of each tone increases. Therefore, if we keep raising A, the amplitude of the IM products will finally match the fundamental tones' amplitude at the output. The input third intercept point (IIP_3) is the input level at which this happens, as shown on a log-log scale in Fig. 3.45. In a similar vein, OIP_3 represents the corresponding output. We refer to the input amplitude in the following derivations as A_{IIP3}.

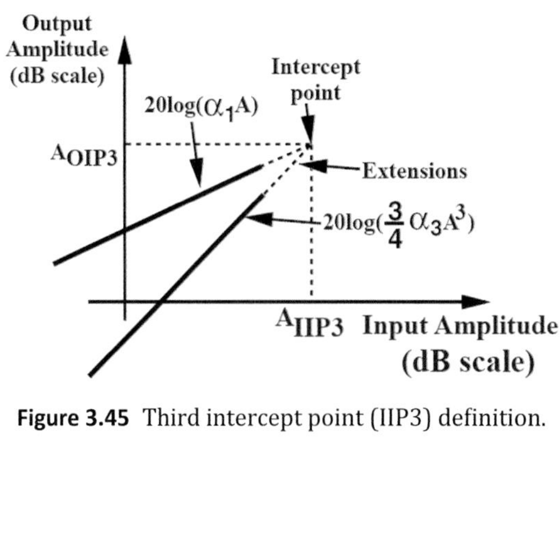

Figure 3.45 Third intercept point (IIP3) definition.

To produce the IP_3, we boost input A_{in}, plot the amplitudes of the IM products and the basics on a log-log scale, and extrapolate these plots based on their slopes as three and one, respectively, as shown in Fig. 3.46. In simulations or experiments, extrapolation can be very difficult, therefore we usually take a shortcut that yields a good first approximation.

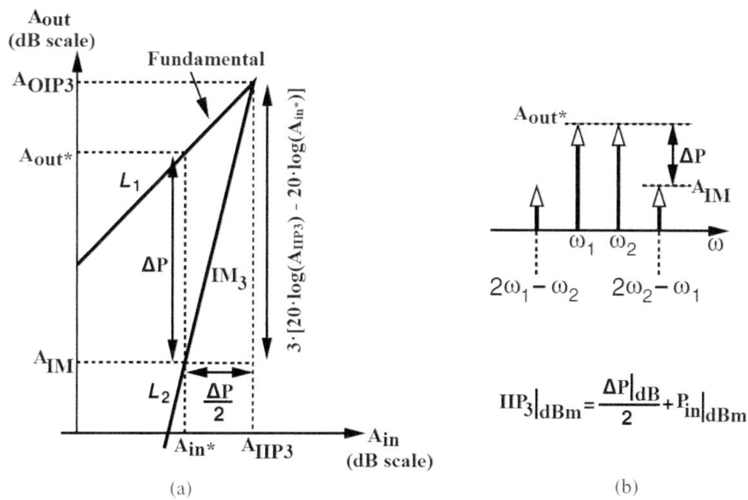

Figure 3.46 Derivation of the third intercept point (IIP3) based on two-tone test.

In Fig. 3.46(a), if we assume that the input is equal to A_{IIP3}, then the (extrapolated) fundamental tones and the (extrapolated) output IM products are of the same size. The input has now been decreased to a level A_{in^*}. In other words, the input change is equivalent to $20{\cdot}\log(A_{IIP3})$ – $20{\cdot}\log(A_{in^*})$. The IM products have a slope of three on a log-log scale, while the fundamentals have a slope of unity. As a result, the slope of the difference between the two plots increases to 2. We indicate ΔP as the difference of $20{\cdot}\log(A_{out^*}) - 20{\cdot}\log(A_{IM})$ as shown in Fig. 2.46(b).

Hence, we can deduce the following important derivation of the third intercept point as

$$\mathrm{IIP}_3(\mathrm{dBm}) = \frac{\Delta P(\mathrm{dB})}{2} + P_{in}(\mathrm{dBm}). \tag{3.50}$$

Alternatively, the IIP_3 can be found for a specific input level by adding the result to the input level, where all the values are stated as logarithmic (dB) quantities, and half the difference between the output fundamental and IM levels. This rule is represented by the simplified notation shown in Fig. 3.46(b). The fact that the IP_3 is measured without extrapolation is crucial in this scenario.

3.2.2.1 Non-linearity of cascaded stages

Since cascaded stages as shown in Fig. 3.47 are used in RF systems to process signals, it is critical to understand how each stage's non-linearity relates to the cascade's input. In this case, we ascertain a cascade's IP_3. Unless otherwise specified, we will henceforth refer to the input IP_3 as A_{IP3} for the sake of conciseness.

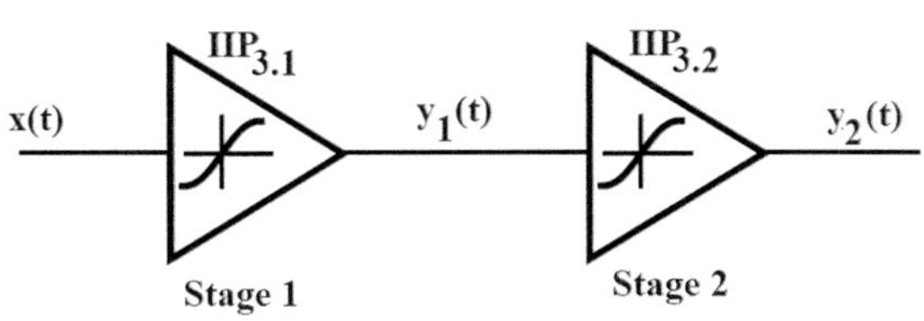

Figure 3.47 Non-linearity of cascaded stages.

The overall non-linearity of the cascaded stages is determined by

$$\frac{1}{A_{IP3}^2} \approx \frac{1}{A_{IP3,1}^2} + \frac{\alpha_1^2}{A_{IP3,2}^2}, \tag{3.51}$$

where α_1 is the gain of stage 1.

Because the IP_3 of each stage in a cascade chain is correspondingly scaled down by the total gain preceding that level, if each stage has a gain larger than unity, the non-linearity of the later stages becomes more vital.

3.2.3 Selectivity

The wireless communication channel is in an environment which is considered harsh on the transceiver design. The most significant limitation may come from the restricted channel bandwidth allotted to every user. The application of complex baseband processing methods like coding, compression, and bandwidth-efficient modulation is required since, according to Shannon's theorem, this corresponds to a finite data rate. According to Shannon's theorem, a communication channel's attainable maximum data rate is equal to $B \cdot \log_2(1 + \text{SNR})$, where SNR is the signal-to-noise ratio and B is the bandwidth.

The transceiver's RF front-end architecture is also impacted by the narrow channel bandwidth. To prevent leaking to neighboring channels, the transmitter must use narrowband modulation and amplification, as shown in Fig. 3.48, and the receiver must be able to process the target channel while rejecting powerful in-band and out-of-band interferers to a sufficient degree.

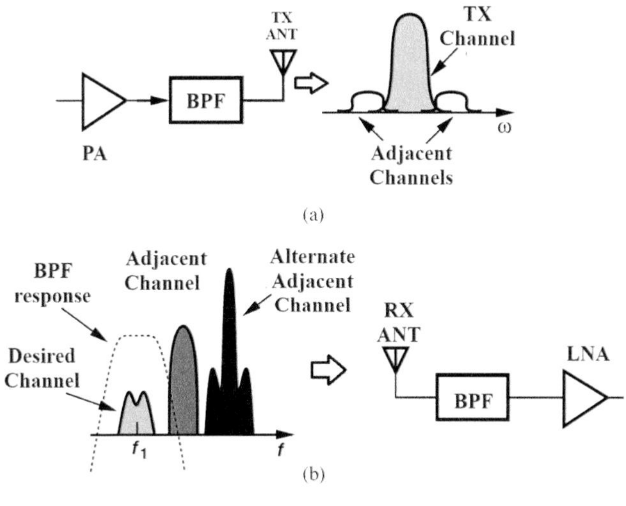

Figure 3.48 (a) TX front-end and (b) RX front-end environments.

Specifically, we realized that a receiver's linearity had to be good enough to handle interference without suffering from compression or severe intermodulation. We may then ask if the receiver linearity criteria can be loosened by just filtering out the interference. Unfortunately, there are two problems with this. First, the filter needs to have a very high selectivity (high Q), since an interferer can be only one or two channels away from the target signal as shown in Fig. 3.49. When the degree of interference is, let's say, 55 to 65 dB higher than the intended signal level, the needed value of Q rises to unreasonably huge numbers, like 10^6. Second, such a filter would require a variable yet precise center frequency, a highly challenging property to accomplish, since a different carrier frequency may be provided to the user at various times.

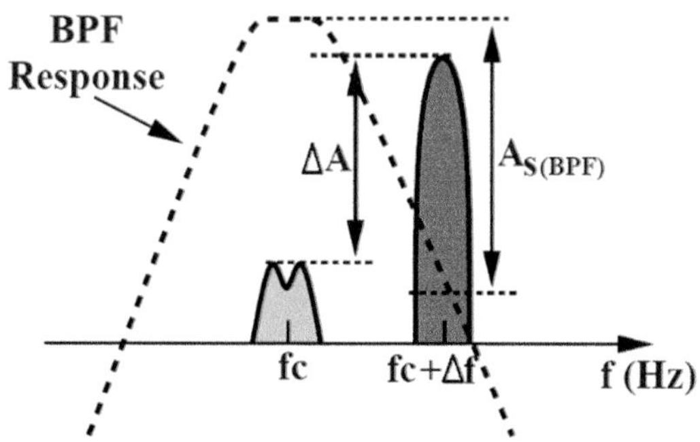

Figure 3.49 Band-pass filter stopband rejection of an interferer.

3.2.3.1 Band and channel selection

The filtering method discussed above is referred to as channel-selection filtering because it rejects the interference in the other channels while selecting the desired signal channel. Here, we note two important things: To prevent compression or excessive intermodulation, all receiver chain stages prior to channel-selection filtering must be sufficiently linear. Secondly, since channel-selection filtering is very challenging at the input carrier frequency, it must be postponed to a later point where the desired channel's center frequency is significantly lower and the necessary filter Q's are therefore more reasonable.

As is well known, the center frequency divided by the 3 dB bandwidth can be used to approximate the Q of a band-pass filter. However, most receiver front ends do provide a band-select filter as shown in Fig. 3.50, which suppresses components that might be generated by users who do not fit the standard of interest by selecting the full receive band and rejecting out-of-band interferers. As a result, we can distinguish between in-band interferers, which are usually eliminated close to the end of the receiver chain, and out-of-band interferers. The trade-off between selectivity and in-band loss affects the front-end band-select (passive) filter since sharpening the band-pass frequency response's edges requires increasing the filter's order, or the number of cascaded sections inside the filter. The front-end loss significantly increases the receiver's overall NF, which is quite undesired. As a result, the filter is made to have a modest loss (0.5 dB to 1 dB) and some frequency selectivity.

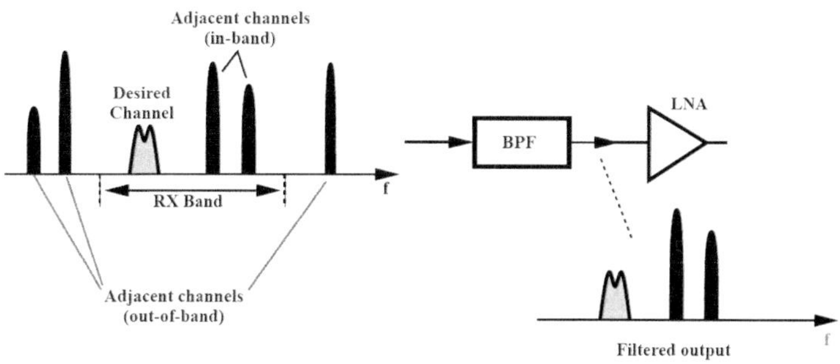

Figure 3.50 BPF filter-based band selection.

The trade-off between selectivity and in-band loss affects the front-end band-select (passive) filter since sharpening the band-pass frequency response's edges requires increasing the filter's order, or the number of cascaded sections inside the filter. The front-end loss significantly increases the receiver's overall NF, which is quite undesired. As a result, the filter is made to have a modest loss (around 1 dB) and some frequency selectivity.

Furthermore, the significance of regulated spectrum re-generation is achieved by carefully selecting the PA topology and modulation strategy. The front-end BPF is unable to reduce the out-of-channel signals generated by the PA. Thus, the PA is to be designed with better linearity performance such that the spectral regeneration can be minimized in the transmitter output.

3.2.4 Sensitivity

Numerous factors determine an RF receiver's performance. We address two of these in this section—sensitivity and dynamic range.

detects the strength of the signal and modifies the receiver gain. With the upper end being relatively easily achieved, this compression-based DR can exceed 100 dB, excluding the interferers. The second category, referred to as the spurious-free dynamic range (SFDR), shows constraints resulting from interference as well as noise. The higher end is defined as the greatest input level in a two-tone test for which the third-order IM products do not exceed the integrated noise of the receiver; the lower end is still equal to the sensitivity.

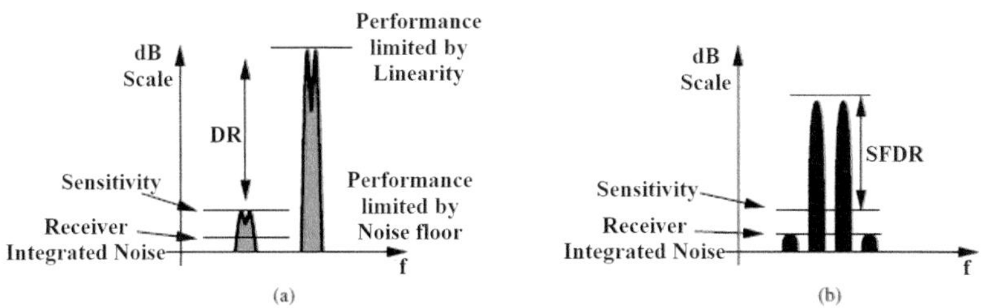

Figure 3.51 (a) Dynamic range and (b) spurious-free dynamic range.

Two tones (modulated or unmodulated) with similar amplitudes are applied to the receiver input, and their level is increased until the IM products exceed the integrated noise floor, as illustrated in Fig. 3.51(b). The SFDR is obtained by dividing each tone's power by its sensitivity. The highest relative level of interference that a receiver can handle and still generate a signal with a sufficient quality from a low input level is represented by the SFDR.

Is it appropriate to measure the different levels shown in Fig. 3.51(b) at the circuit's output or input? The output port is a more logical choice for this kind of measurement because the IM components are only visible at the output. In this instance, the sensitivity, which is typically an input-referred number, needs to be scaled by the circuit's gain in order to be referenced to the output. Alternatively, the IM magnitudes of the output can be referenced to the input by dividing them by the gain. In our SFDR computations, we employ the latter strategy.

We modify Eq. (3.56) as follows to get the upper bound of the SFDR:

$$P_{\mathrm{IIP3}} = P_{\mathrm{in}} + \frac{P_{\mathrm{out}} - P_{\mathrm{IM,out}}}{2}, \tag{3.58}$$

wherein no actual power may be delivered at the input or output ports; however, for the sake of brevity, we have designated P_x = 20·log A_x. Additionally, $P_{\mathrm{IM,out}}$ shows the quantity of IM products at the output. By writing $P_{\mathrm{IM,in}} = P_{\mathrm{IM,out}} - G$, we can indicate the IM level to the input if the circuit displays a gain of G (in dB). In a similar way, $P_{\mathrm{in}} = P_{\mathrm{out}} - G$ provides the input level for each tone. As a result, Eq. (3.58) becomes

$$P_{\mathrm{IIP3}} = P_{\mathrm{in}} + \frac{P_{\mathrm{in}} - P_{\mathrm{IM,in}}}{2} \tag{3.59}$$

$$P_{\mathrm{IIP3}} = \frac{3P_{\mathrm{in}} - P_{\mathrm{IM,in}}}{2} \tag{3.60}$$

$$P_{\mathrm{in}} = \frac{2P_{\mathrm{IIP3}} + P_{\mathrm{IM,in}}}{3}. \tag{3.61}$$

The value of P_{in} that causes P_{IM} to match the receiver's integrated noise is the upper bound of SFDR:

$$P_{\mathrm{in,max}} = \frac{2P_{\mathrm{IIP3}} + (-174\mathrm{dBm} + \mathrm{NF} + 10\log B)}{3} \tag{3.62}$$

The difference (in dB) between $P_{\mathrm{in,max}}$, and sensitivity is known as the SFDR:

$$\mathrm{SFDR} = P_{\mathrm{in,\,max}} - (-174\mathrm{dBm} + \mathrm{NF} + 10\log B + \mathrm{SNR}_{\mathrm{min}}) \tag{3.63}$$

$$\mathrm{SFDR} = \frac{2(P_{\mathrm{IIP3}} + 174\mathrm{dBm\text{-}NF\text{-}}10\log B)}{3} - \mathrm{SNR}_{\mathrm{min}} \tag{3.64}$$

3.3 Receiver Sub-System Design

The design of a low-power 2.4 GHz band receiver as shown in Fig. 3.52. It is used to directly decode QPSK modulated signals at a maximum data rate of 11 Mbps even with very low input power level, i.e., less than –85 dBm. The receiver design is based on DCR or homodyne architecture with external inductors for input matching to offset based on the bond-wire inductances. The receiver chain design includes a single-ended LNA design, a 25% duty cycle LO-driven active mixer, a passive low-pass filter to attenuate high-frequency components and spurious components and differential programmable-gain amplifier (PGA) topology for baseband signals *I* and *Q*.

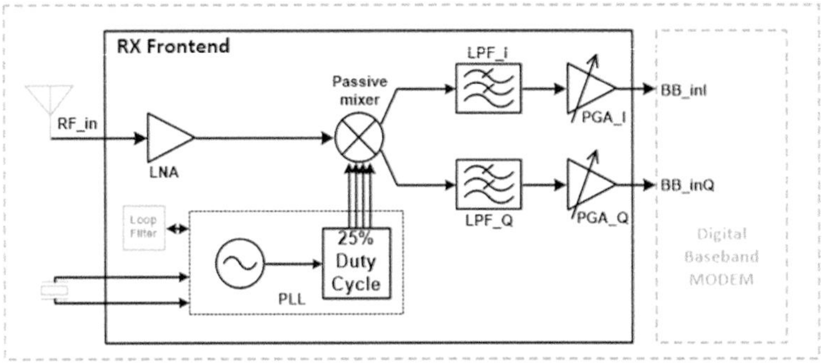

Figure 3.52 Overall receiver front-end architecture with internal LO.

The RF input is given to the gate of the source-degenerated transistor of the LNA. Its gain and center frequency (f_c) are tunable via digital pins and voltage-controlled capacitors, respectively. An RF PMOS transistor at the output acts as a load. The low-power differential PGA has a dB-linear tunable gain range of –30 to +60 dB using a 5-bit digital control along with a bandwidth tunability feature is available for larger capacitive loads.

3.3.1 LNA Design

The LNA is one of the most important design components for a receiver in the RF front end (RFFE). To give input RF signals enough gain with the least amount of noise contribution in the RF front end, a two-stage LNA is constructed. A better noise figure and a respectable gain are what the first stage of the LNA is meant to give. Improving the overall receiver front-end gain is the aim of the second stage. With an LC tank serving as the load, the first stage is developed utilizing a conventional source degraded common-source cascaded LNA in accordance with these specifications. An inverter-based self-biased LNA topology was used in the second stage to achieve a greater gain. Two further aspects of the proposed LNA are its frequency tunability and gain tunability. In the first stage of the LNA, gain tunability is accomplished by pulling a known current from the main branch. A varactor is employed in the circuit of the LC tank, which serves as the load for the first stage of the LNA, to achieve frequency tunability. While LNA gain tuning is accomplished with a gain tuning voltage generating circuit and 2-bit digital control, frequency tuning is accomplished with analog voltage control. External inductors manufactured on the PCB are used to construct the input-matching network for the LNA. The external inductors are located at the first stage's load terminal, source degeneration terminal, and LNA input. Figure 3.53 shows an illustration of the suggested LNA topology.

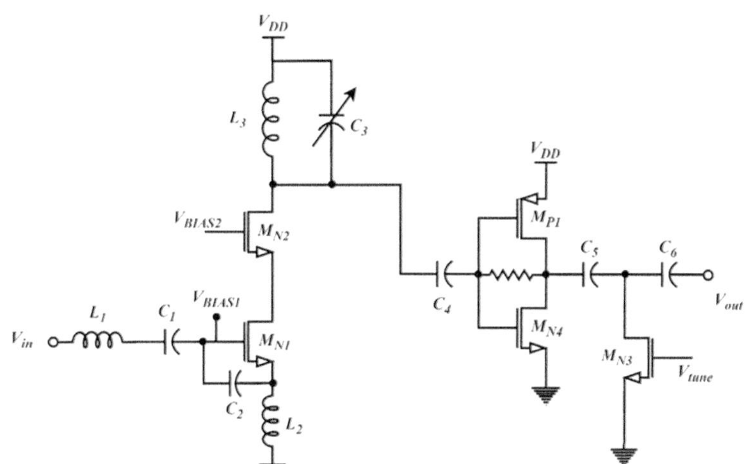

Figure 3.53 Two-stage LNA schematic.

The LNA is built to withstand fluctuations in PVT. Figure 3.54 shows the post-layout simulation results for the LNA voltage gain, noise figure, and input matching for a 10% supply voltage and temperature fluctuation over the range of –25°C to 85°C.

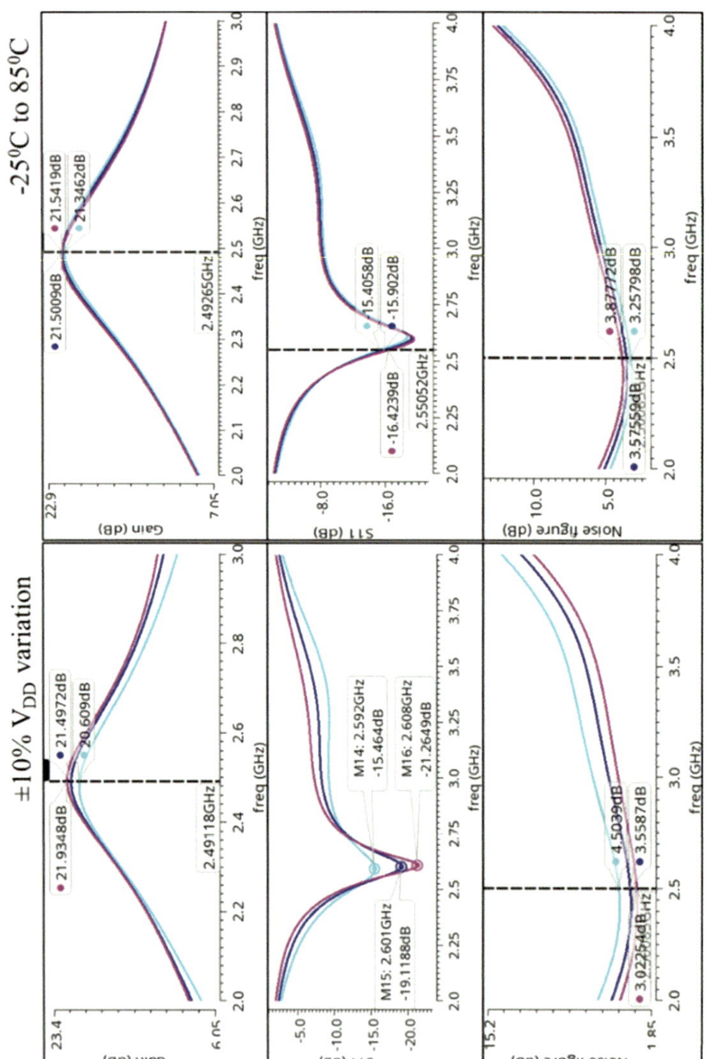

Figure 3.54 Two-stage LNA post-layout simulation results.

At 3.45 mW of power consumption, the LNA produced a noise figure of 3.6 dB and a voltage gain of 21.5 dB. As seen in Fig. 3.54, the designed LNA additionally has the capability of gain adjusting via the voltage V_{tune}. Two digital bits and a matching circuit that converts the bits to analog voltage are used to create the tuning voltage. Figure 3.55 shows the gain performance for various bit settings. As will be covered later, a passive mixer is used to combine LO signal with an amplified RF signal.

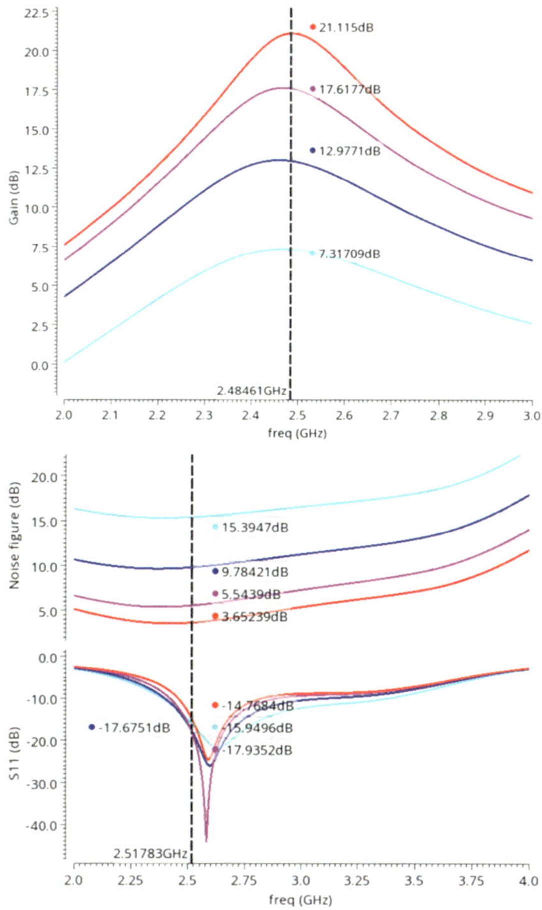

Figure 3.55 Two-stage LNA post-layout simulation results with different gain settings.

8. M. Valla, G. Montagna, R. Castello, R. Tonietto, and I. Bietti, A 72 mW CMOS 802.11a direct conversion front-end with 3.5 dB NF and 200 kHz 1/*f* noise corner, *IEEE Journal of Solid-State Circuits*, vol. 40, no. 4, pp. 970–977, Apr. 2005.
9. R. Bagheri, A. Mirzaei, S. Chehrazi, E. Heidari, M. Lee, M. Mikhemar, W. Tang, and A. Abidi, An 800 MHz–6 GHz software-defined wireless receiver in 90 nm CMOS, *IEEE Journal of Solid-State Circuits*, vol. 41, no. 12, pp. 2860–2876, Dec. 2006.
10. T. Nguyen, N. Oh, V. Le, and S. Lee, A low-power CMOS direct conversion receiver with 3 dB NF and 30 kHz flicker-noise corner for 915 MHz band IEEE 802.15.4 ZigBee standard, *IEEE Transactions on Microwave Theory and Techniques*, vol. 54, no. 2, pp. 735–741, Feb. 2006.
11. N. Poobuapheun, C. Wei-Hung, Z. Boos, and A. Niknejad, A 1.5 V 0.7-2.5 GHz CMOS quadrature demodulator for multiband direct-conversion receivers, *IEEE Journal of Solid-State Circuits,* vol. 42, no. 8, pp. 1669–1677, Aug. 2007.
12. B. Tenbroekl, J. Strange, D. Nalbantis, C. Jones, P. Fowers, S. Brett, C. Beghein, and F. Beffa, Single-chip tri-band WCDMA/HSDPA transceiver without external SAW filters and with integrated TX power control, *IEEE International Solid-State Circuits Conference*, 2008, pp. 202–203.
13. B. R. Carlton, J. S. Duster, S. S. Taylor, and J. C. Zhan, A 2.2 dB NF, 4.9-6 GHz direct conversion multi-standard RF receiver front-end in 90 nm CMOS, *IEEE Radio Frequency Integrated Circuits Symposium*, 2008, pp. 617–620.
14. A. Mirzaei, X. Chen, A. Yazdi, J. Chiu, J. Leete, and H. Darabi, A frequency translation technique for SAW-less 3G receivers, *IEEE Symposium on VLSI Circuits*, 2009, pp. 280–281.
15. A. Mirzaei, H. Darabi, J. Leete, X. Chen, K. Juan, and A. Yazdi, Analysis and optimization of current-driven passive mixers in narrowband direct-conversion receivers, *IEEE Journal of Solid-State Circuits*, vol. 44, no. 10, pp. 2678–2688, Oct. 2009.
16. D. L. Kaczman et al., A single-chip tri-band (2100, 1900, 850/800 MHz) WCDMA/HSDPA cellular transceiver, *IEEE Journal of Solid-State Circuits*, vol. 41, no. 5, pp. 1122–1132, May 2006.
17. A. Mirzaei, H. Darabi, J. Leete, and Y. Chang, Analysis and optimization of direct-conversion receivers with 25% duty-cycle current-driven passive mixers, *IEEE Transactions on Circuits and Systems I: Regular Papers*, vol. 57, no. 9, pp. 2353–2366, Sep. 2010.

18. M. Mikhemar, A. Mirzaei, A. Hadji-Abdolhamid, and J. Chiu, A 13.5 mA sub-2.5 dB NF multi-band receiver, *IEEE Symposium on VLSI Circuits*, 2012, pp. 82–83.

Chapter 4

Transmitter Sub-System

4.1 Transmitter Architectures

To support the higher data rate of the next-generation wireless communication with limited bandwidth, the recent-day RF transmitters are based on digital modulation termed shift keying such as amplitude shift keying (ASK), which is also called on-off keying (OOK), BPSK, QPSK. QAM16, etc. The modern-day transmitter works as a modulator based on the digital baseband for basic modulation schemes such as OOK, BPSK, GFSK, QPSK, etc. and as a frequency up-converter for complex modulation schemes such as QAM16, orthogonal frequency division multiplexing (OFDM), etc. using the carrier signal called local oscillators (LO). Hence, the RF transmitter functions as a modulator, up-converter, and power amplifier (PA) [1–7].

In the next section, we will study the various transmitter architectures, their advantages and drawbacks to get a better understanding of the recent RF transmitters.

4.1.1 Direct-Conversion Transmitters

Most transmitter architectures resemble the receiver topologies discussed in Chapter 3, but they execute their activities in a

CMOS RF and mm-Wave Transceivers and Synthesizers
Bharatha Kumar Thangarasu, Nagarajan Mahalingam, Kaixue Ma, and Kiat Seng Yeo

ISBN 978-981-5129-80-9 (Hardcover), 978-1-003-67356-9 (eBook)
www.jennystanford.com

different sequence. A TX cascade of quadrature up-conversion mixers and a PA, for instance, is suggested by an RX cascade of an LNA and quadrature down-conversion mixers. Any narrowband modulated signal can be expressed using the generalized GMSK waveform expression as

$$\begin{aligned} x(t) &= A(t)\cos[\omega_c t + \phi(t)] \\ &= A(t)\cos\omega_c t\cos[\phi(t)] - A(t)\sin\omega_c t\sin[\phi(t)]. \end{aligned} \tag{4.1}$$

Thus, the quadrature baseband signals are defined by us as

$$x_{BB,I}(t) = A(t)\cos[\phi(t)] \tag{4.2}$$

$$x_{BB,Q}(t) = A(t)\sin[\phi(t)]. \tag{4.3}$$

The transmitter can be built in the manner depicted in Fig. 4.1. This topology, known as a "direct-conversion" transmitter, uses a "quadrature up-converter" to transfer the baseband spectrum straight to the RF carrier. The purpose of the PA and matching network, which come after the up-converter, is to supply the most power possible to the antenna and filter out-of-band components that arise from the non-linearity of the PA (Chapter 2).

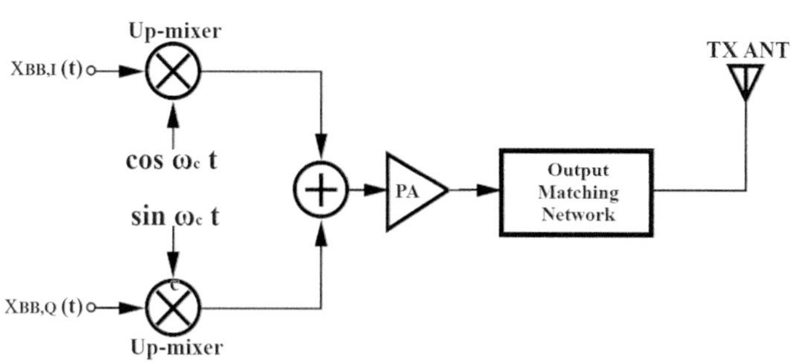

Figure 4.1 Block diagram of a direct-conversion transmitter.

It should be noted that the noise of the mixers is far less important here than it is in receivers since the baseband signal is generated in the transmitter and has a suitably big amplitude (several hundred millivolts). The PA is where the TX design starts. This circuit is designed to provide the antenna with the necessary power (and, if applicable, sufficient linearity). Because the PA uses big transistors to handle high currents, it has a considerable input capacitance. In order to act as a buffer between the up-converter and the PA, a pre-driver is usually placed in between. This chapter examines several TX topologies, and the direct-conversion strategy offers the most compact solution and a comparatively "clean" output. In other words, the output spectrum is comparable to that of direct-conversion receivers in that it only includes the intended signal surrounding the carrier frequency (and its harmonics) and lacks any spurious components. However, there are additional problems with straight up-conversion that need to be addressed.

4.1.1.1 *I*/*Q* mismatch

Since conventional double sideband (DSB) modulation consists of a carrier signal and two identical sidebands as shown in Fig. 4.2, it may be deemed wasteful of bandwidth and power. On the other hand, single sideband (SSB) modulation produces the final signal using just one sideband, as the name suggests and as shown in Fig. 4.3. Stated differently, SSB modulation offers a significantly more effective means of communication than DSB modulation. Both in terms of the radio spectrum utilized and the power required to broadcast the signal, it is significantly more efficient. Because of its benefits, SSB modulation has been utilized extensively for a long time to provide efficient communications.

Let's use the following mathematical example to demonstrate DSB:

$$\cos(A)\cdot\cos(B)=\frac{1}{2}\cdot[\cos(A+B)+\cos(A-B)], \tag{4.4}$$

where baseband (data) frequency is B, and carrier frequency is A. As a result, two sidebands are provided by DSB modulation, as the name suggests: $(A + B)$ and $(A - B)$.

As for SSB:

$$\cos(A)\cdot\cos(B)+\sin(A)\cdot\sin(B)=\cos(A-B) \tag{4.5}$$

or

$$\cos(A)\cdot\cos(B)-\sin(A)\cdot\sin(B)=\cos(A+B) \tag{4.6}$$

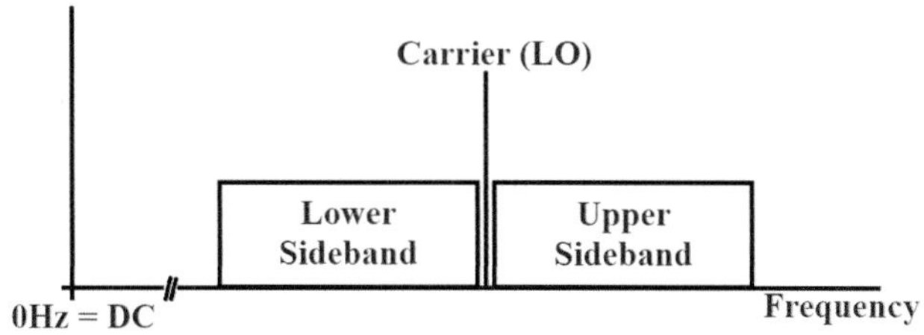

Figure 4.2 Double sideband output spectrum.

Lower sideband (LSB) is represented by (A – B), and upper sideband (USB) by (A + B).

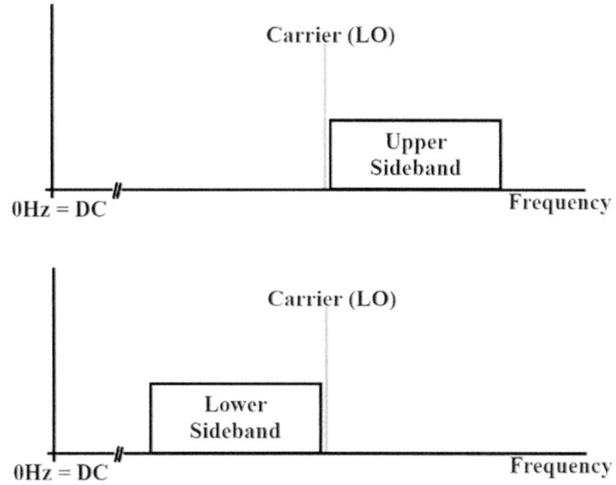

Figure 4.3 SSB modulation with USB as desired and LSB as desired output.

Furthermore, the sine and cosine have an exact 90° phase offset, as shown in Fig. 4.4.

Figure 4.5 shows a definition of the sine and the cosine in terms of a unit circle:

As shown in Fig. 4.6, cosine is in phase, hence the term "I" signal. Moreover, sine is the quadrature phase, thus its symbol "Q."

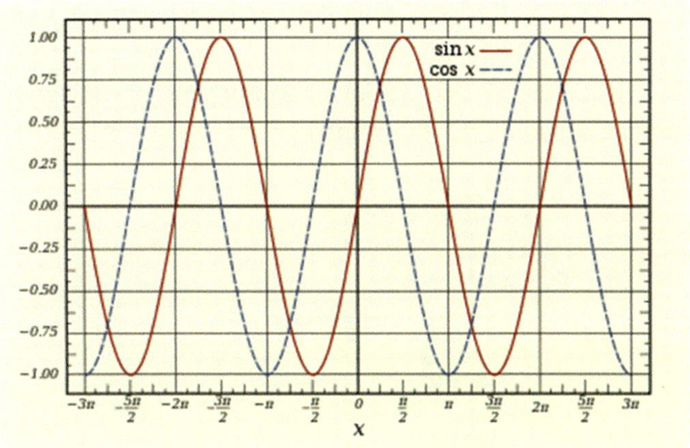

Figure 4.4 I/Q signals in time domain.

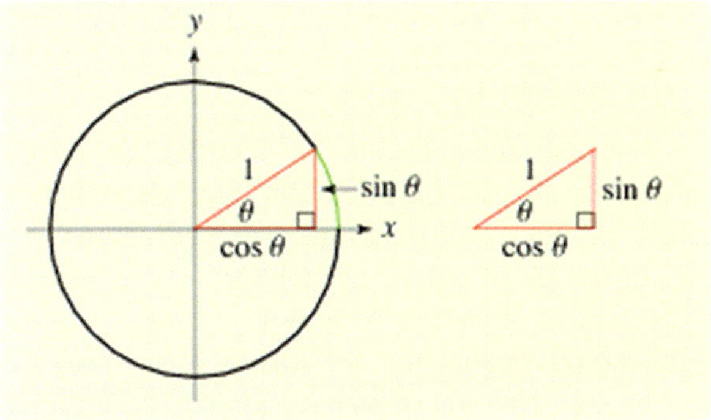

Figure 4.5 Unit circle: trigonometric relations.

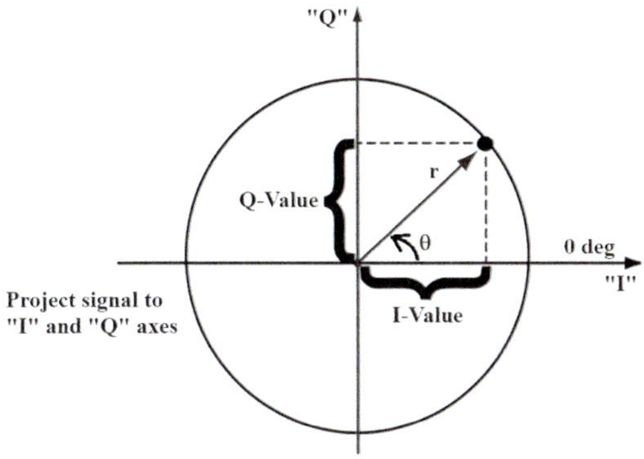

Figure 4.6 Polar-to-rectangular conversion.

From the polar-to-rectangular conversion, we can find $I = r \cdot \cos(\theta)$ and $Q = r \cdot \sin(\theta)$.

As a result, the block diagram for creating an $(A - B)$ signal using SSB modulation looks as shown in Fig. 4.7.

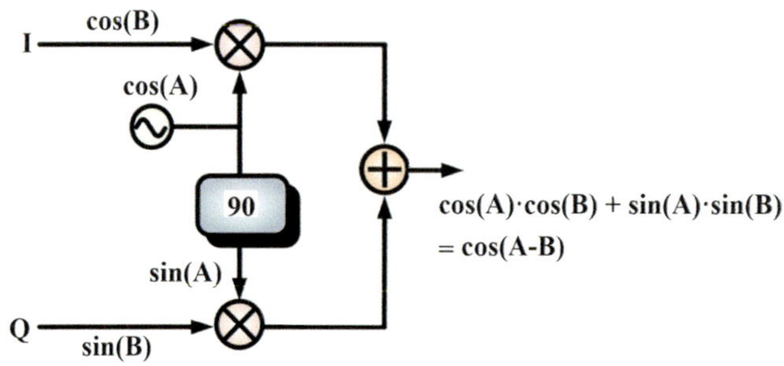

Figure 4.7 SSB signal generation receiver block diagram.

4.1.1.2 *I/Q* imbalance

As was previously established, in theory, the spectrum for SSB modulation should only contain one desirable sideband. In actuality, though, the spectrum will contain at least three tones. One of the three tones—referred to as the "image"—is the unwanted sideband, as shown in Fig. 4.8.

The unwanted sideband, or "image," that arises from an imbalance in *I/Q*. The *I/Q* imbalance among the differential *I/Q* input channels, becomes more severe in the event of broadband operation as compared to narrowband operation. This leads to the image product, which exacerbates the system performance.

The inherent benefits of a direct up-conversion (DUC) transmitter are its high integration level and conceptual simplicity. As a result, it has gained popularity recently, particularly for handheld devices like cell phones. The DUC transforms baseband to RF directly, as its name suggests, as shown in Fig. 4.1. The ideal *IQ* modulator would have completely symmetrical in-phase and quadrature arms to provide optimum signal quality. In other words, the *I* and *Q* channels ought to be precisely 90° out of phase and have gains that are equal.

The LO's high frequency makes it impossible to digitally integrate the *IQ* modulator for DUC transmitters. However, even though designers aim for a symmetrical *IQ* modulator circuit, small discrepancies arise between the in-phase and quadrature circuits on the same device due to changes in the manufacturing process. In addition, there are phase and gain imbalances between the two branches of an analog *IQ* modulator.

Put differently, the *I* and *Q* channels of a real-world DUC quadrature modulator might have varying gains, and the LO signals might not be precisely 90°; out of phase. *I*/*Q* amplitude imbalance, also known as *I*/*Q* gain imbalance, is a sign that the gains on the *I* and *Q* channels diverge. *I*/*Q* phase imbalance is another sign that the *I* and *Q* channels are not precisely 90°; out of phase.

I/*Q* imbalance is the combined term for phase imbalance and *I*/*Q* gain, as seen in Fig. 4.9, in terms of constellation.

Naturally, *I*/*Q* imbalance leads to error vector magnitude (EVM) degradation and deteriorates modulation accuracy because of constellation distortion. Furthermore, as previously stated, an imbalance in *I*/*Q* results in an unwanted sideband, or image. Sideband suppression is the amplitude difference between the signal and the image as shown in Fig. 4.8. A plot that can be used to link sideband suppression to quadrature imbalance and *I*/*Q* gain imbalance is depicted in Fig. 4.8. In this case, it is noteworthy that unless the gain imbalance is likewise corrected, enhancing the quadrature phase imbalance has no bearing on sideband suppression.

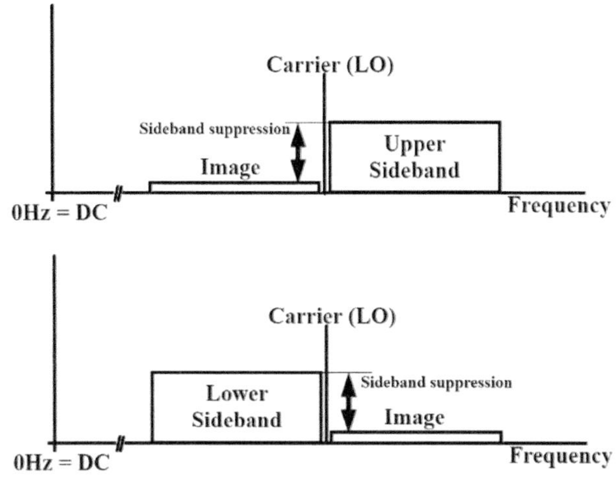

Figure 4.8 SSB signal generation receiver block diagram.

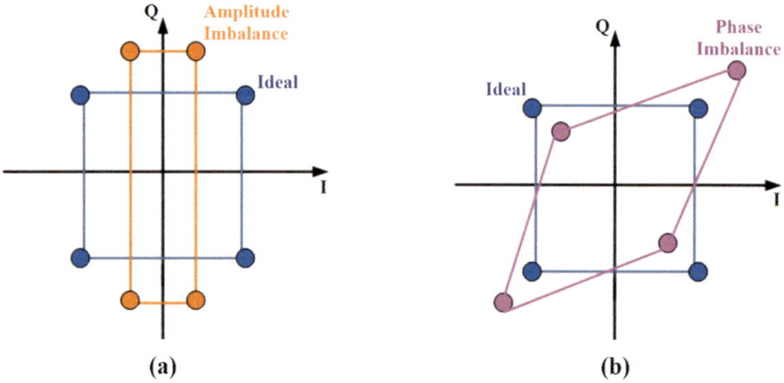

Figure 4.9 Constellation with I/Q (a) amplitude imbalance and (b) phase imbalance.

Stated differently, I/Q phase imbalance has less of an impact on sideband suppression than I/Q gain imbalance normally does. Even though the I/Q imbalance is unavoidable, we can minimize it as much as we can. We can take advantage of the fact that we can optimize the sideband suppression by varying the amplitude and phase offsets between the I and Q channels.

The gain delta between I and Q is changed in the first pass, as seen in the above figure. For a gain differential of roughly −0.1 dB, the sweep produces a sideband suppression of about −57 dBc. Next, adjust the I–Q skew. This results in a phase adjustment of −0.05° and a further sideband suppression down to −60 dBc. In this instance, the phase sweep only marginally improves the sideband suppression that the first-pass gain adjustment produces.

The phenomena demonstrate that, as previously noted, I/Q phase imbalance has less of an impact on sideband suppression than I/Q gain imbalance. Thus, to limit the unwanted sideband leakage, gain and phase must be adjusted successively in many steps.

With modification and corrections, the sideband suppression does get better in the frequency domain as shown in Fig. 4.10.

Additionally, in the RF scenario, care should be taken throughout the PCB layout process to ensure that the differential I/Q channels have an identical physical architecture to mitigate the performance reduction brought on by the image product.

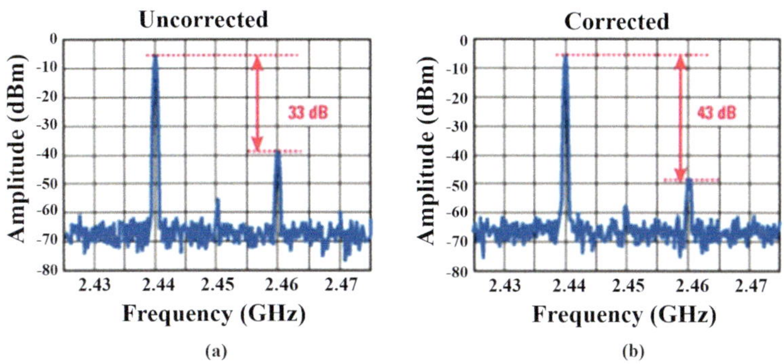

Figure 4.10 Transmitter spectrum output with I/Q imbalance (a) uncorrected and (b) corrected.

Both *Tx* I/Q and *Rx* I/Q signals take on a differential shape to prevent interference from the outside world from affecting their precision during modulation and demodulation.

Four I/Q signals namely $I+$, $I-$, $Q+$, and $Q-$ will be present. Additionally, the phase relationship is depicted in Fig. 4.11.

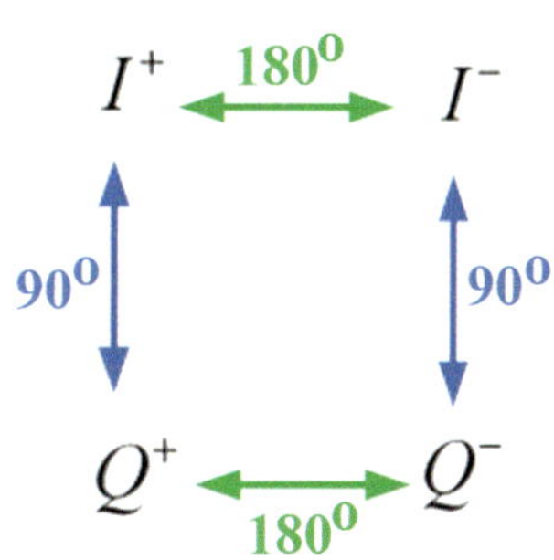

Figure 4.11 I/Q signal phase relationship.

The four traces on the *IQ* signal route, which go from the DAC output to the modulator input, should ideally be symmetrical on both the positive and negative sides of each channel as well as between the I and Q channels. In reality, trace lengths are not precisely matched because of differences in PCB design specifications and manufacturing constraints. *IQ* phase errors are the outcome of mismatches, which skew the signal in one channel from the other. As seen in Fig. 4.12, there are two common forms of trace length mismatches.

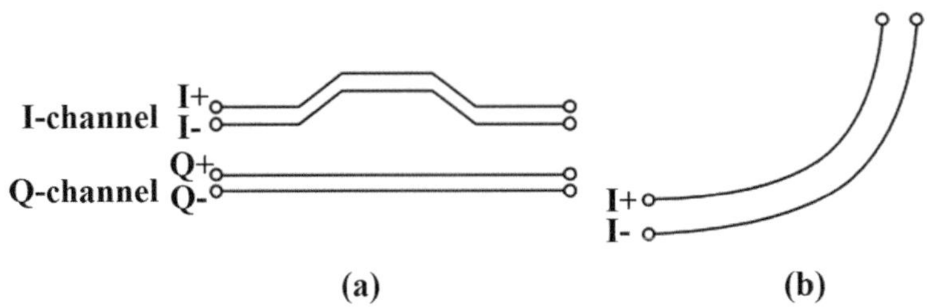

Figure 4.12 *I/Q* traces imbalance (a) *I* and *Q* channel mismatch (b) differential lines mismatch.

The *IQ* phase imbalance is mitigated by trace discrepancies between the *I* and *Q* channels. A differential signal is distorted by mismatches between the positive and negative sides of a channel, which skew the two sides away from 180° out of phase. This results in phase imbalance as well as gain. In a differential pair, the traces are usually arranged quite near to one another. The likelihood of a mismatch is minimal.

On the other hand, if the differential pair is lengthy, the exterior trace adds somewhat more length to the total length than the internal one each time it turns on the PCB. Up to a point where the mismatch begins to affect sideband suppression, it can accumulate. As a result, we must employ a few techniques to reduce the mismatch brought on by PCB turns.

As was already noted, there will always be an *I/Q* imbalance. Regarding PCB, our greatest effort should be focused on ensuring that the four *I/Q* signals—*I*+, *I*–, *Q*+, and *Q*– —have the same lengths. Naturally, to lessen the possibility of a mismatch, try to minimize the PCB trace lengths of the four *I/Q* signals. In the absence of such, sideband suppression will worsen.

In addition, we must take the temperature into account. In general, sideband suppression increases with decreasing temperature. The vulnerability of higher-order modulation schemes, like 64-QAM, to *IQ* gain imbalance is significantly greater. A constellation diagram with several orders of modulation is a simple method to see this effect.

For this reason, while building an RFIC that supports complicated modulation schemes, it is crucial to reduce gain or phase imbalance. Several constellation plots with rising modulation orders and a persistent gain imbalance are shown in Fig. 4.13.

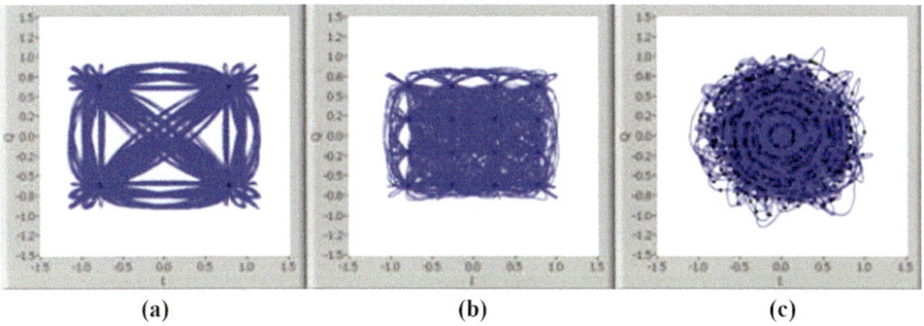

Figure 4.13 I/Q gain imbalance in (a) QPSK (b) 16-QAM and (c) 64-QAM.

4.1.1.3 Carrier leakage

As was already mentioned, the spectrum for SSB modulation will really contain at least three tones. Three types of frequency components are identified: the desired sideband, or signal; the undesired sideband, or image; and the carrier leakage, or LO leakage [1].

The two main causes of carrier leakage, sometimes referred to as carrier feedthrough and I/Q origin offset, are as follows:

a. LO leakage

b. DC Offset of I/Q channels.

A passive mixer usually requires a high LO power to achieve low conversion loss. The LO signal may leak via the RF port, which could lead to a considerable amount of LO leakage, because of the finite mixer port-to-port isolation and strong LO power.

Furthermore, high amounts of carrier leakage are also caused by excessive DC offsets in I/Q channels. Baseband in DUC transmitters instantly transforms to RF with LO. Therefore, in I/Q channels with LO, DC offsets also directly convert to LO leakage if there exist DC offsets. Both the baseband port of each up-conversion mixer and the analog baseband circuitry generating the quadrature signals in the transmitter of Fig. 4.14 show DC offsets.

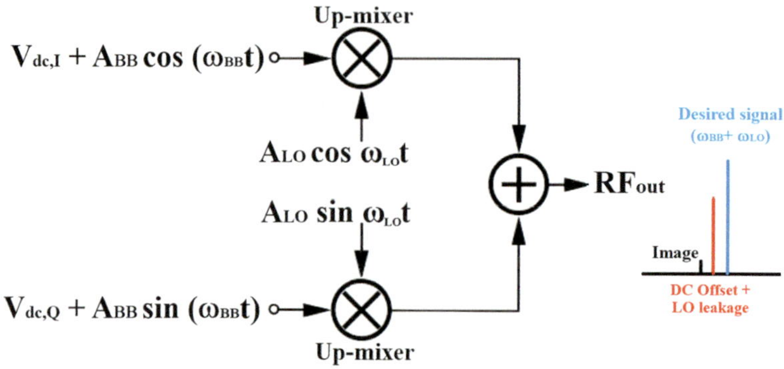

Figure 4.14 Contributors for carrier leakage in a transmitter.

Hence, the output signal is expressed as

$$V_{out}(t)=[A(t)\cos\phi+V_{OS1}]\cos\omega_c t-[A(t)\sin\phi+V_{OS2}]\sin\omega_c t, \quad (4.7)$$

where V_{OS1} and V_{OS2} stand for the total DC offsets related to the mixers' input port. Consequently, a portion of the unmodulated carrier is present in the up-converter output:

$$V_{out}(t)=A(t)\cos(\omega_c t+\phi)+V_{OS1}\cos\omega_c t-V_{OS2}\sin\omega_c t \quad (4.8)$$

Known as "carrier leakage," relative carrier leakage is defined as

$$\text{Relative Carrier Leakage}=\frac{\sqrt{V_{OS1}^2+V_{OS2}^2}}{\sqrt{\overline{A^2(t)}}}. \quad (4.9)$$

There are two negative outcomes of this phenomenon. It first causes the EVM at the TX output to increase by distorting the signal constellation. For instance, if a QPSK signal is represented by $V_{out}(t)$,

$$V_{out}(t)=\alpha_1(V_0+V_{OS1})\cos\omega_c t-\alpha_2(V_0+V_{OS2})\sin\omega_c t. \quad (4.10)$$

When this equation is applied to a perfect direct-conversion receiver, DC offsets, or changes in the constellation both horizontally and vertically, affect the baseband quadrature outputs (Fig. 4.15). If the transmitter's output power needs to be adjusted over a large range by altering the baseband signals' amplitude, the second impact becomes apparent. To prevent the near-far effect, for instance, one must reduce the broadcast power as one approaches the base station. A conceptual diagram of the power control loop is shown in Fig. 4.15.

The mobile device is asked to modify its broadcast power by the base station, which detects the power level it receives from the device. When there is little distance between the two, the mobile output power must be lowered to extremely low levels, which results in the spectrum depicted in Fig. 4.16 when carrier leakage is present. It is challenging to determine the true signal strength in this situation since the carrier power predominates. The constellation is shown below:

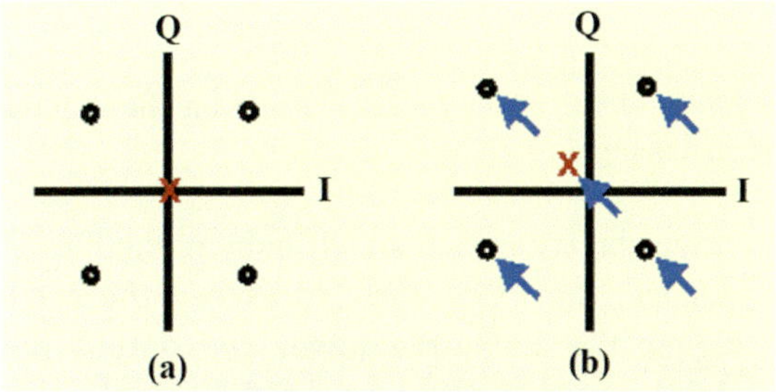

Figure 4.15 Constellation diagram (a) without *IQ* offset (b) with IQ offset.

Naturally, DC offsets in I/Q channels lead to EVM degradation and deteriorate modulation accuracy because of constellation distortion. As previously stated, the I/Q imbalance is more severe in the case of broadband operation as opposed to narrowband operation. Similarly, when broadband operation is included, carrier leakage gets worse.

The carrier leakage and signal overlap in the case of broadband operation and DUC transmitter are depicted in Fig. 4.16.

In the absence of countermeasures, the signal is diminished but the carrier leakage remains constant. Consequently, carrier leakage is much more than signal in low-power mode, as illustrated in Fig. 4.16, which lowers SNR. EVM varies inversely with SNR, as shown below:

$$\text{EVM} = \sqrt{\frac{1}{\text{SNR}}} \tag{4.11}$$

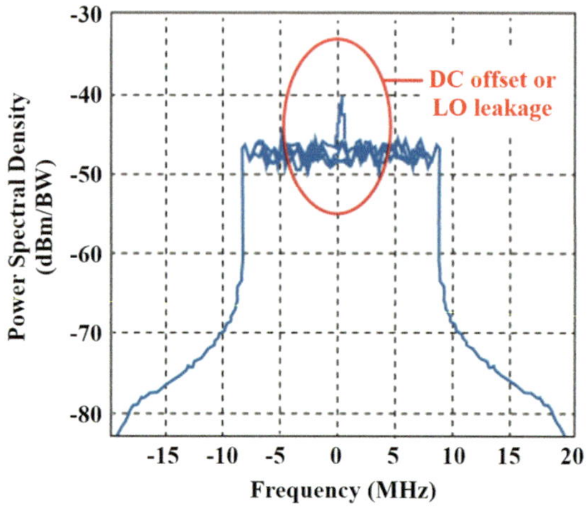

Figure 4.16 Effect of DC offset or LO leakage on the modulation signal power spectral density.

In other words, when too much gain reduction is used due to carrier leakage, the EVM starts to exceed the predetermined limit; the higher the EVM, the lower the output power. As a result, the LO leakage criteria in wireless applications change with output power. However, due to increased EVM, the lower output power sets a limit due to carrier leakage. Hence, the carrier leakage will reduce the dynamic range.

Of course, in an ideal world, carrier leakage would not exist at all in the absence of DC offsets in I/Q channels, as indicated by the following formula:

$$\text{LO leakage} = \left| \frac{DC_I + DC_Q}{\sqrt{\left(I^2 + Q^2\right)}} \right| \tag{4.12}$$

Hence, with a DC-blocking high-pass filter (HPF), DC offsets can be minimized. Nevertheless, the carrier leakage is inevitable, and can't be rejected by means of DC block HPF as shown in Fig. 4.17.

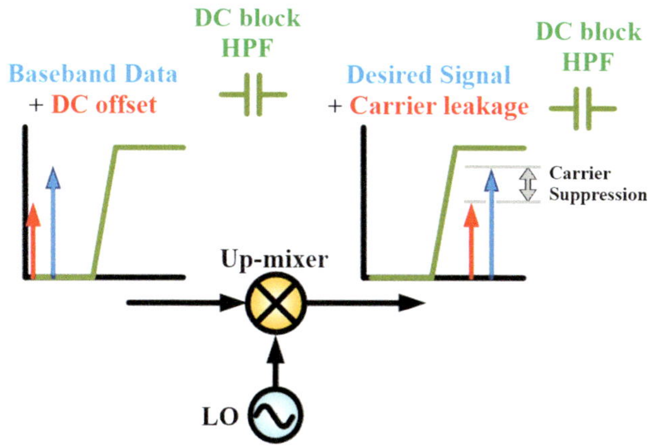

Figure 4.17 Effect of DC-blocking HPF on the DC offset and carrier leakage.

As previously mentioned, the DC block functions as an HPF. The DC offset and baseband data are rejected prior to the mixer. On the other hand, the baseband data and the DC offset pass the DC block following the mixer. DC blocking is therefore not the effective way to stop carrier leakage.

A calibration circuit is integrated into some transceivers to address the problem of carrier leakage. Carrier suppression is defined as the amplitude difference between the desired signal and the carrier leakage as shown in Fig. 4.17.

After calibration, the carrier suppression does improve in the frequency domain. Individual parts of an RF transmitter implementation, such as the digital-to-analog converter, can have small errors in DC offset and gain. Consequently, it is crucial to apply gain or DC offset changes to the baseband I or Q signals when thinking about a DAC, or direct quadrature modulator. As an example, the carrier suppression does indeed get better with successive adjustments to the DC offset of the I/Q differential input. Hence, the intrinsic faults of DUC transmitters are unwanted sideband and the carrier leakage.

4.1.1.4 TX linearity

The criteria for spectrum regrowth (adjacent-channel power) and/or the acceptable distortion of the signal to be transmitted must be taken into consideration while selecting the linearity of transmitters. When it comes to variable-envelope modulation methods, both effects become crucial. The compression that a variable-envelope signal undergoes is usually what defines its distortion. It is established how close the signal's level can approach the 1 dB compression point when the bit error rate or constellation deterioration is insignificant, as seen in Fig. 4.18, where the signal is exposed to a non-linear characteristic in simulations. For instance, the 64-QAM OFDM signal's average power needs to stay roughly 8 dB below the circuit's P1dB [1, 3].

We state that the circuit needs to run at "8 dB back-off." Stated otherwise, if the circuit is at the 1 dB compression point with a peak swing of V_0, then the average signal swing cannot be greater than $V_0/2.5$. Compression of the signal can occur at any point in a TX chain. In a well-designed system, the stages that come before it must stay well below compression as the PA output gets closer to P1dB since the biggest voltage swing happens at the PA's output, which controls the TX's compression. In order to do this, we need to limit the output swing of the pre-driver and the stages that come before it, as well as increase the PA's gain. The PA design is further burdened by this necessity.

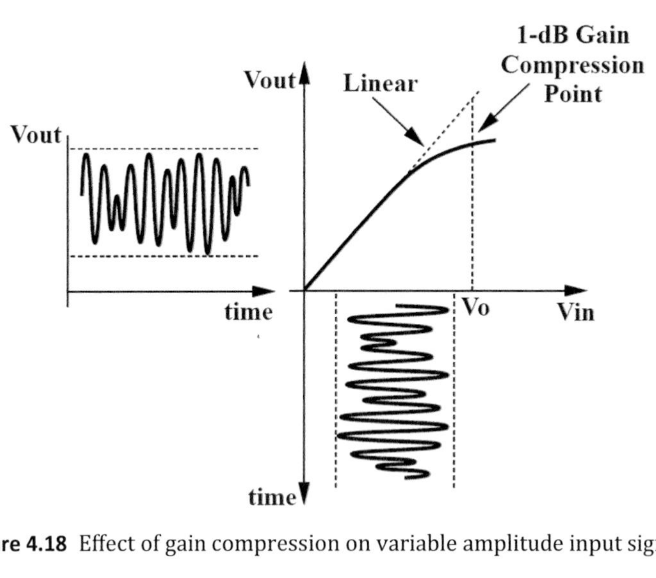

Figure 4.18 Effect of gain compression on variable amplitude input signal.

4.1.1.5 Oscillator pulling

While the above-mentioned concerns are relevant to most transmitter layouts, one in particular becomes crucial in direct-conversion topologies. The PA output shows very significant amplitude swings (20 V_{pp} for 1 W given to a 50 Ω load), as shown in Fig. 4.19(a). These large swings relate to different portions of the system via the printed-circuit board traces, silicon substrate, and package parasitics. Consequently, a sizable portion of the PA output most likely couples to the LO. The PA driver may still draw the LO even while the PA is off chip. Take note that in direct-conversion transmitters, the center frequency of the PA output spectrum equals ω_{LO}.

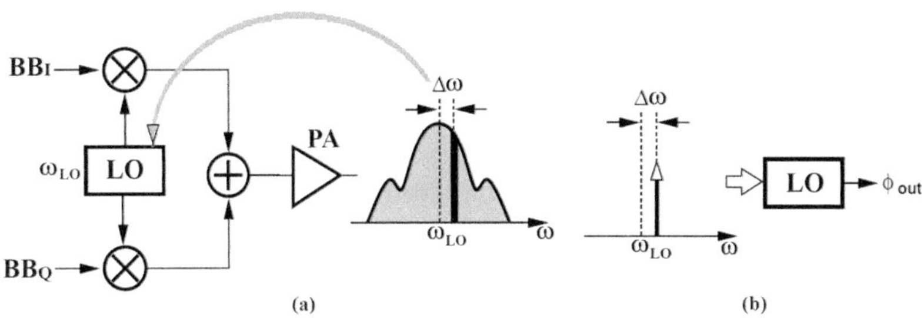

Figure 4.19 Direct-conversion TX (a) effect of LO pulling by PA output (b) injection concept on LO.

Let's model a slice of the output spectrum that is centered at $\omega_{LO} + \Delta\omega$ using an impulse with the same energy (Fig. 4.19(b)). We thus ask what happens when an oscillator running at a frequency of ω_{LO} is "injected" with a sinusoid at a frequency of $\omega_1 = \omega_{LO} + \Delta\omega$ such that $\omega_{LO} >> \Delta\omega$. This phenomenon, which goes by the name "injection pulling," has been well investigated and examined [1, 7]. Under these circumstances, there occurs periodic modulation of the oscillator's output phase, φ_{out}. As you can see in Fig. 4.20(a), φ_{out} really spends a portion of the period at 90° (relative to the input phase) before quickly rotating 360°. As a result, the input and output waveforms look like those in Fig. 4.20(b). The output spectrum is shown to be highly asymmetric (Fig. 4.20(c)), with most of the impulses being positioned at a distance from the input frequency, ω_{inj} (= ω_{LO} + $\Delta\omega$). Observe that the interval between the impulses is not equal to $\Delta\omega$, but rather equals the phase variation's frequency in Fig. 4.20(a).

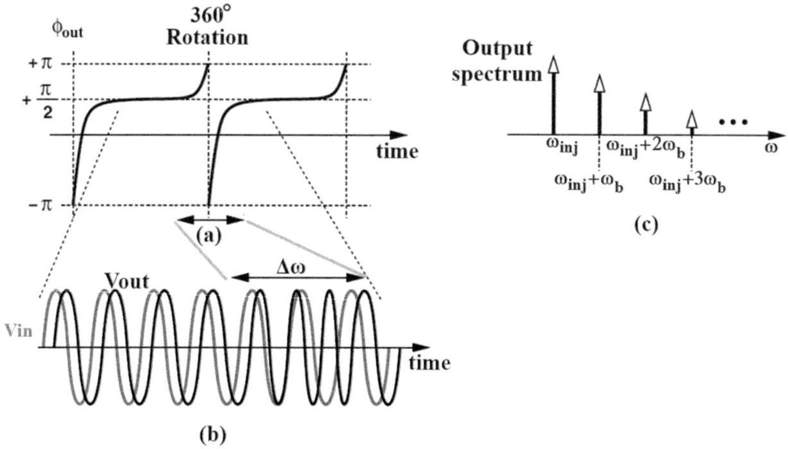

Figure 4.20 (a) LO phase variation caused by injection pulling; (b) variation of output frequency in time domain; (c) corresponding output spectrum.

What output power level is necessary for injection pulling to be noticeable? The response is contingent on multiple factors: (1) the larger internal voltage and current swings of the LO have lesser effect of pulling; (2) the Q of the oscillator tank employed; (3) if the PA output is differential, its coupling to the LO is 30–40 dB lower than single-ended PA output; (4) The degree to which the synthesizer's feedback loop counteracts the pulling [1]; (5) The packaging type and layout symmetry. However, for standard designs, injection pulling could become problematic when the PA output rises above 0 dBm. The PA output frequency and oscillator frequency need to be sufficiently divergent (e.g., by more than 20%) to prevent injection pulling, but this is not feasible in the architecture shown in Fig. 4.19. The transmitter architectures and frequency plans that resulted from this idea are listed in Section 4.3.3.

4.1.1.6 Modern direct-conversion transmitters

Most transmitters used in direct conversion today steer clear of oscillators that match the PA output frequency. We refer to the former as the LO frequency, ω_{LO}, and the latter as the carrier frequency, ω_c, to prevent confusion. The objective is completed by selecting ω_{LO} that is sufficiently distant from ω_c and using techniques like frequency division and mixing to get ω_c from ω_{LO}. In Fig. 4.21, $\omega_{LO} = 2\omega_c$ is shown in a typical case. The LO is followed by a divide-by-2 circuit, which produces $\omega_{LO}/2$ with quadrature phases. Two factors make this architecture popular: (1) injection pulling is much decreased, and (2) the divider easily delivers the carrier's quadrature phases, an otherwise challenging process.

However, injection pulling is still not totally eliminated by the architecture shown in Fig. 4.21. The LO may still be pulled since the PA non-linearity generates a finite amount of power at the carrier's second harmonic. Nevertheless, this effect can be mitigated with the right arrangement and isolation methods. The main problem with the construction shown in Fig. 4.21 is the divider's speed requirement. The divider could end up being the transceiver's main speed restriction if it operates at twice the carrier frequency. However, as will be demonstrated in the discussion that follows, there are more significant shortcomings with other transmitter topologies. Therefore, it makes perfect sense to invest a significant amount of time and resources in divider design in order to make this architecture possible.

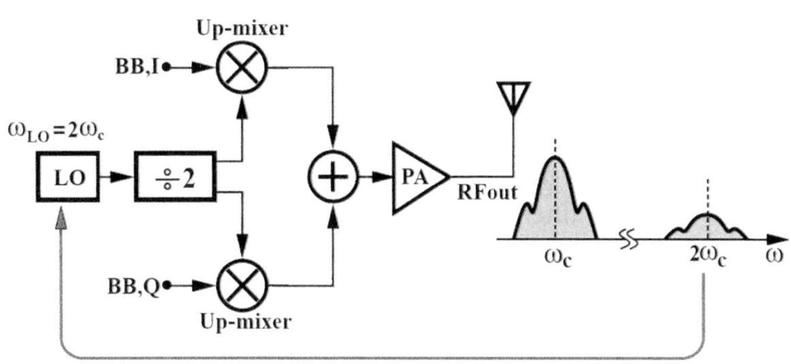

Figure 4.21 LO at twice the carrier frequency reduction of LO pulling effect.

A different method of determining LO frequencies is by mixing. For instance, $\omega_1 + \omega_2$ or $\omega_1 - \omega_2$ can be produced by combining the outputs of two oscillators operating at ω_1 and ω_2.

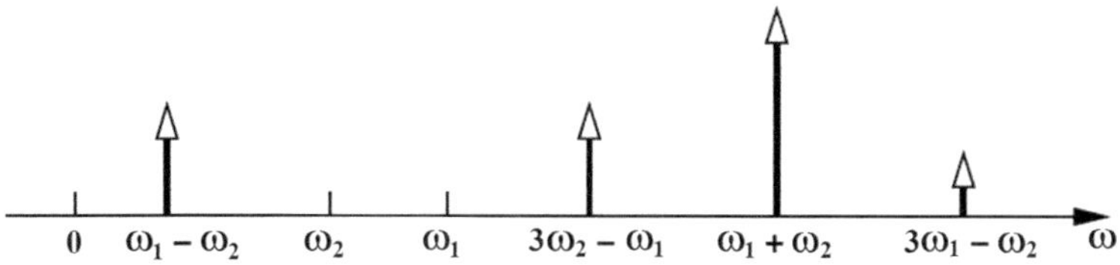

Figure 4.24 Non-linearity and mismatch effects on DUC with SSB mixer output spectrum.

In the mixer example shown in Fig. 4.25, the port sensing driven by V_{in2} is non-linear, whereas that generated by V_{in1} is linear. V_{in1} is multiplied by a square wave that alternates between 0 and 1 in the mixing circuit. That is, the powerful spurs in Fig. 4.24 are caused by V_{in2}'s third harmonic, which is only one-third of its fundamental.

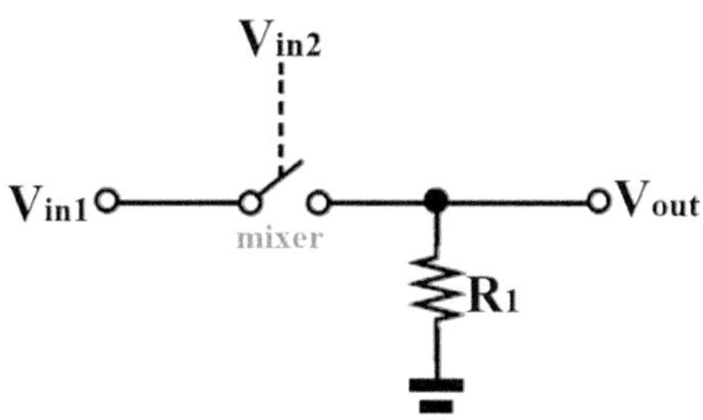

Figure 4.25 Switch-based depiction of mixer.

How serious are the spurs up above? Only one port in a typical mixer design can be linearized, keeping a small third harmonic in that port. The other is extremely non-linear to maintain a fair profit (or loss). Thus, we deduce that, of the two spurs at $3\omega_1 - \omega_2$ and $3\omega_2 - \omega_1$, only one may be suppressed to levels that are acceptable, while the other one stays only 10 dB (a factor of one-third) below the component that is required. For instance, if $\omega_2 = \omega_1/2$, then $3\omega_1 - \omega_2 = \omega_1/2$ and $3\omega_2 - \omega_1 = \omega_1/2$; we can suppress the latter by linearizing the port sensing ω_2, but the former still needs significant filtering.

The quadrature phases of the carrier must be provided by the SSB mixer to be used in a direct-conversion TX. This is achieved by copying the SSB mixer as illustrated in Fig. 4.26 and observing that $\sin(\omega_1)t \cos(\omega_2)t + \cos(\omega_1)t \sin(\omega_2)t = \sin(\omega_1 + \omega_2)t$.

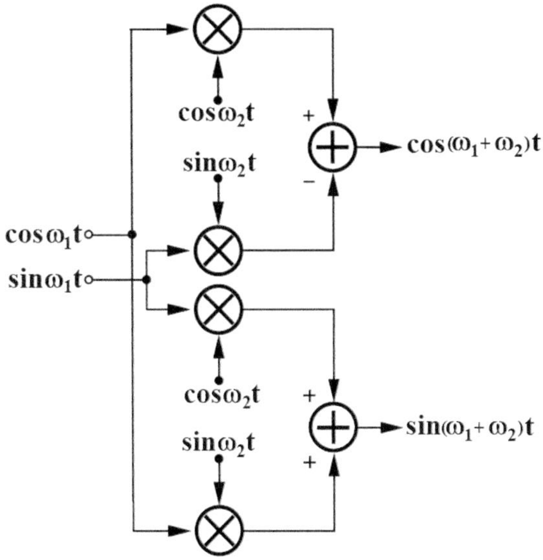

Figure 4.26 I/Q SSB mixer structure.

A direct-conversion TX with SSB mixing for internal carrier generation is seen in Fig. 4.27. This architecture is immune to injection pulling because the carrier and LO frequencies differ enough. This architecture suppresses the carrier sideband at $\omega_1/2$, but has two disadvantages: (1) it is difficult to deal with the spurs at $5\omega_1/2$ and other harmonic-related frequencies, and (2) the LO needs to give quadrature phases.

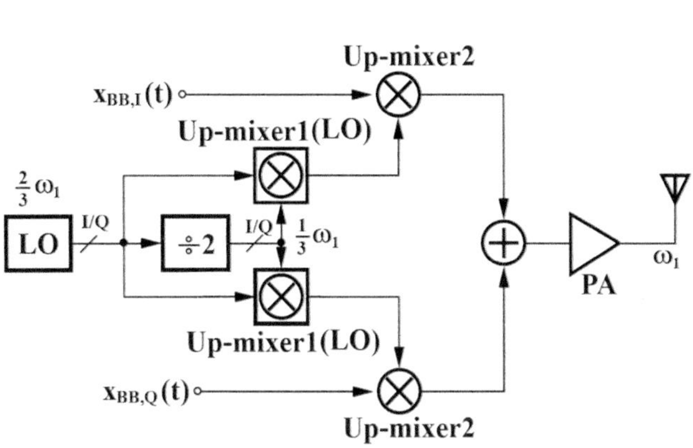

Figure 4.27 DUC TX with SSB mixer structure with LO path.

4.1.1.7 Heterodyne transmitters

Another way to prevent injection pulling is to up-convert the signal in two stages, keeping the LO frequency away from the PA output spectrum. Such a TX topology is depicted in Fig. 4.28. In this instance, the baseband I and Q signals are up-converted to an IF of ω_1, combined with ω_2 to translate to a carrier frequency of $\omega_1 + \omega_2$. This stage is followed by a band-pass filter since the second mixer also generates an output at $\omega_1 - \omega_2$. This architecture has the same benefit as its receiver counterpart in that the I/Q up-conversion happens at a frequency much lower than the carrier, resulting in fewer gain and phase mismatches. The mismatch effect quantification equations are the same as the direct-conversion TX equations derived in Section 4.3.2.

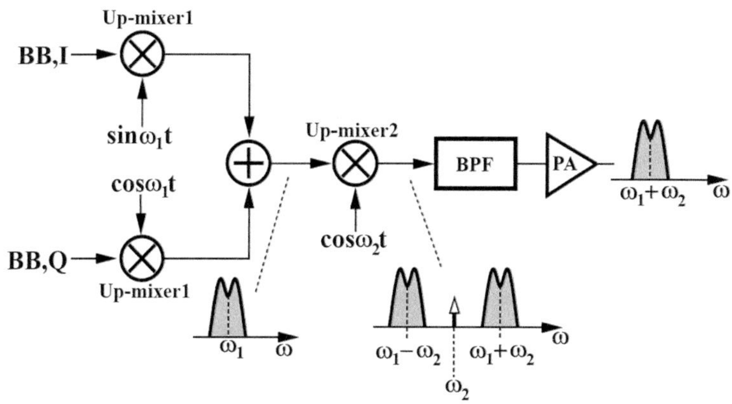

Figure 4.28 Heterodyne two-stage TX structure.

We remove the first oscillator from the above TX and extract the necessary phases from the second oscillator (Fig. 4.29). Therefore, $3\omega_1/2$ is the carrier frequency. Let's examine how non-idealities affect this architecture. The first and second LOs are the names we give to the LO waveforms at $\omega_1/2$ and ω_1, respectively.

At the output of the quadrature up-converter, the DC offsets in the baseband yield a component at $\omega_1/2$, whereas at the RF mixer's input, the DC offset yields another component at ω_1 (Fig. 4.30). The former can be reduced as previously said. The latter needs to be filtered out, along with the bottom sideband at $\omega_1/2$. Though significantly smaller than the lower sideband, the leakage at $\omega 1$ is closer to the upper sideband than the lower sideband. As a result, the filter that comes after the RF mixer needs to be built to attenuate both to levels that are acceptable.

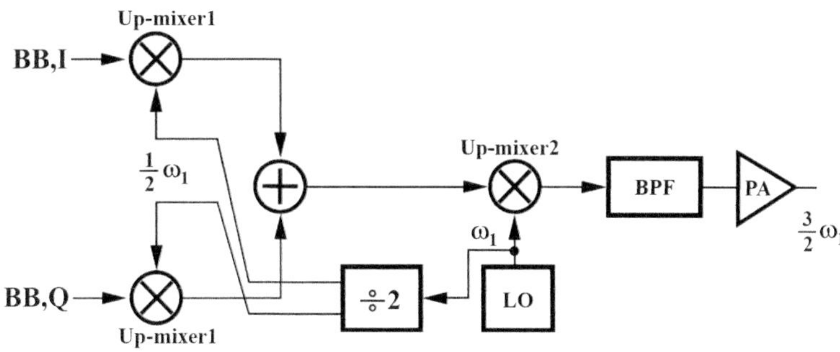

Figure 4.29 Two-stage heterodyne sliding-IF TX.

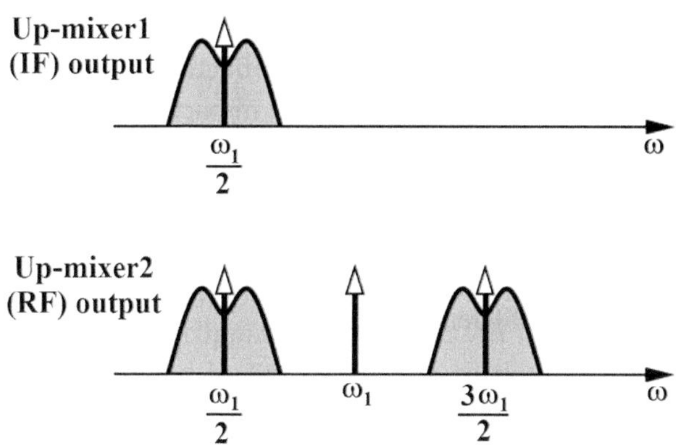

Figure 4.30 Heterodyne TX showing carrier leakage.

4.2 Transmitter Performance Parameters

4.2.1 Efficiency

As the cost of electricity rises for both direct supply and battery charging from storage, most wireless communication sources, such as base stations and mobile devices, struggle to maintain low operating costs. This situation has led to a rise in the need for energy-efficient transmitters during the last few decades.

4.2.2.3 Total harmonic distortion

It is also known as THD, and its definition is

$$\mathrm{THD} = \sqrt{\frac{\text{power in distortion}}{\text{power in fundamental}}}$$

Consequently, it may be concluded that

$$\mathrm{THD} = \sqrt{\mathrm{HD2}^2 + \mathrm{HD3}^2 + \cdots}. \tag{4.25}$$

4.2.2.4 Intermodulation distortion

Since there are often multiple tone inputs in real-world circuits, examining the circuit with a two-tone input reveals several fascinating characteristics pertaining to the non-linearity of the designs.

$$\begin{aligned} v_i &= V_1 \cos\omega_1 t + V_2 \cos\omega_2 t \\ v_o &= c_0 + c_1 v_i + c_2 v_i^2 + c_3 v_i^3 + \cdots \end{aligned} \tag{4.26}$$

4.2.2.5 Second-order intermodulation

Considering the second-order term as $c_2 v_i^2$ will yield two terms as $\cos(\omega_1 + \omega_2)t$, or the second-order intermodulation (IM2) product. This implies that the two input tones will self-modulate because of second-order non-linearity. The two tones' identical amplitude ratio is known as the normalized IM2 distortion.

$$\mathrm{IM2} = \frac{\text{amplitude of the intermod}}{\text{amplitude of the fundamental}} = \frac{c_2}{c_1} V_i \tag{4.27}$$

It is noteworthy that IM2 = 2×HD2. This example illustrates the effects of the second-order non-linearity term IM2 on the system response, using ω_1 = 800 MHz and ω_2 = 900 MHz.

This indicates that four terms are formed, two of which are IM2 and the other two of which are HD2 terms that have the potential to impair system performance, in addition to a DC term as shown in Fig. 4.31.

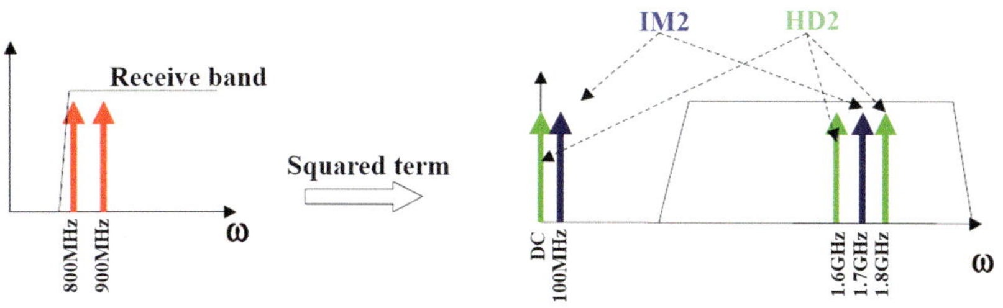

Figure 4.31 Consequences of the IM2.

4.2.2.6 Third-order intermodulation

Then, another intriguing and extremely important non-linearity distortion is seen if we consider the cubic term $c_3 v_i^3 = c_3 \left[V_1 \cos\omega_1 t + V_2 \cos\omega_2 t \right]^3$ that was extracted from the two-tone input signal. Two frequency terms the $2\omega_1 - \omega_2$ and $2\omega_2 - \omega_1$ become more noticeable among the numerous terms that result from extending the cosine relation.

In contrast to the 2nd-order non-linearity, which only produces in-band IM2 for wideband systems, the cubic non-linearity formed by IM3 can produce in-band IM components for narrowband systems, as demonstrated in Fig. 4.32. As a result, the IM3 terms are more serious and cannot be eliminated. Furthermore, the IM3 = 3×HD3 is observed using the cosine trigonometry.

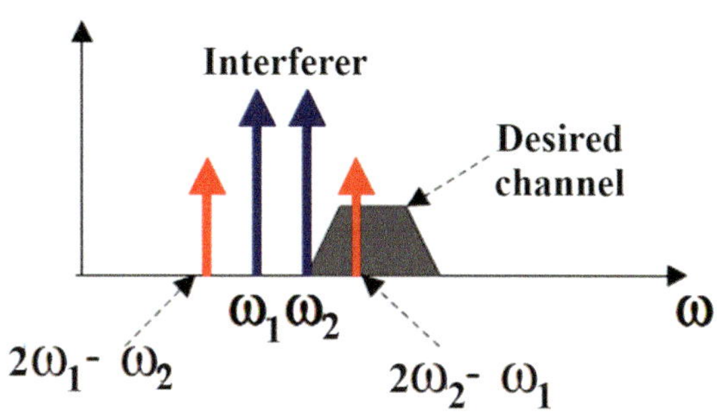

Figure 4.32 Consequences of the IM3.

4.2.2.7 Third-order intercept point

The third-order IM3 term increases with a slope of three in the dB scale, but the fundamental term always increases with a unit slope as shown in Fig. 4.33. As a result, the IM3 component grows more quickly even though it starts later and at greater input powers than the fundamental component. However, the circuit's supply voltage will limit how far both the fundamental and the IM3 may saturate. Therefore, an intercept point known as the IP3, a virtual point that offers a metric of the circuit's non-linearity, will be found via extrapolation. The amplitude of the IM3 component and the fundamental are equal at the IP3 intercept point. Because of this, we can determine via back-calculating that, for any input power on the dBm scale, we do not actually need to sweep the input power all the way to the IIP3 point. A detailed illustration of the third intermodulation point equations is provided in Chapter 3, Section 3.2.2.

$$P_{\text{in}}\ (\text{dBm}) = \text{IIP3}\ (\text{dBm}) + 0.5 \times \text{IM3}\ (\text{dB}) \tag{4.28}$$

and we can also find the equivalent output power as

$$P_{\text{out}}\ (\text{dBm}) = \text{OIP3}\ (\text{dBm}) + 0.5 \times \text{IM3}\ (\text{dB}). \tag{4.29}$$

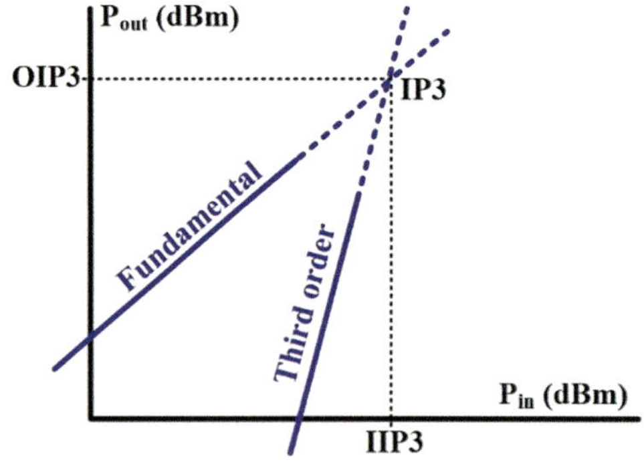

Figure 4.33 Third intercept point (IIP3) depiction.

4.2.2.8 Gain compression

Using trigonometric simplification and a single-tone input signal to the non-linear device, it is seen that the fundamental frequency component has two amplitude terms. If the amplitude of the third-order non-linear fundamental component is similar to that of the linear component, the amplitude reduction known as gain compression will occur as depicted in Fig. 4.34. The first term is the fundamental linear component, and the second term is caused by the third-order non-linearity, which is typically a negative value.

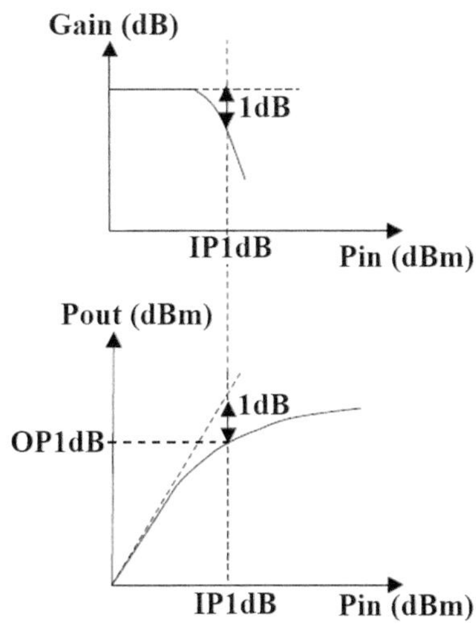

Figure 4.34 Gain compression point depiction.

P1dB = IP3 – 9.6 dB is the relationship between the 1 dB compression point and the IP3 point, according to one computation. Assuming that there is only third-order non-linearity, this relationship can be established.

4.2.3 Spectral Purity

Due to harmonic distortion and intermodulation brought on by the non-linearity terms in the device, several undesirable frequency components that were absent from the input signal are created in the output. These distortion components, also known as spectrum regrowth, interfere with the required output signal in addition to producing unwanted emissions.

The transmitter chain circuits, which include the PA, will function in the linear region for small input signals. As a result, the output signal will be a perfect duplicate of the input signal in both the temporal and frequency domain spectrums. As a result, the output signal will likewise be band limited if the baseband input signal is.

On the other hand, increasing the transmitter power will cause the transmitter chain circuits to become non-linear to achieve a greater communication distance. Both the higher-order function of the input and a linear function of the input will be the output in this non-linear region. The output will contain the input's fundamental and harmonics if the input is a single-tone sinusoid. demonstrating that gain compression will occur at the output. We can supply a two-tone sinusoidal signal, as illustrated in Fig. 4.35, to further examine the impact of the non-linearity on the output frequency response. The result will be

$$y(t) = \alpha_1(A_1 \cos\omega_1 t + A_2 \cos\omega_2 t) + \alpha_2(A_1 \cos\omega_1 t + A_2 \cos\omega_2 t)^2 + \alpha_3(A_1 \cos\omega_1 t + A_2 \cos\omega_2 t)^3. \quad (4.30)$$

Numerous frequency components, such as the harmonics and the intermodulation product terms like $[N\omega_1 \pm M\omega_2]$, where N and M are integers, can be obtained by solving the trigonometric equation. As a result, the intermodulation distortion and harmonic distortion covered in the preceding sub-section are produced.

1. Based on the output spectrum plot, the terms resulting from second-order intermodulation (IM2) can be easily filtered at the transmitter front end since they are distant from the input tones and will not be important.
2. Likewise, it is possible to filter out and disregard the third-order intermodulation sum terms, namely $[2\omega_1 + \omega_2]$ and $[\omega_1 + 2\omega_2]$.
3. The only terms that are close to the primary linear two-tones are the third intermodulation difference terms, $[2\omega_1 - \omega_2]$ and $[\omega_1 - 2\omega_2]$, making it challenging to filter out these two terms. In order to give the transmitter a spurious free transmission spectrum, the designer must make sure that the third-order intercept point is towards larger input powers. These two terms of IM3 will cause significant non-linear distortions.

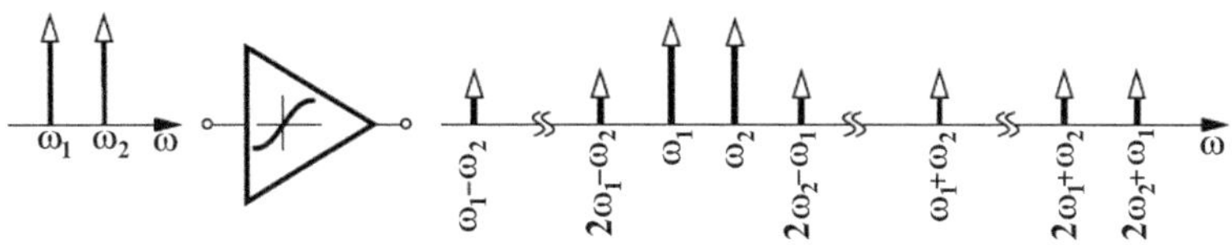

Figure 4.35 Two-tone test to analyze the non-linearity based on inter-modulation components.

4.2.4 Peak-to-Average Power Ratio

A major design issue with OFDM is the enormous dynamic range between a signal's peak and average power. Peak-to-average power ratios (PAPR) lead to sub-optimal transmission performance. This is accurate since the OFDM waveform is made up of several sinusoidal signals, which can behave in both constructive and destructive ways. As a result, there will eventually come a time when the peak signal intensity to average ratio increases significantly. A signal with a high PAPR will cause a PA to operate in its non-linear region [10–12].

A high PAPR causes two problems. Out-of-band distortion is the first cause of adjacent-channel interference and spectral emission mask (SEM) violation. Second, in-band distortion adversely affects throughput performance. To address these problems, the PA of a radio must operate below its 1 dB compression point (P1dB). Such backing off promotes linearity, albeit at the cost of reduced PA efficiency. We must use a PAPR reduction strategy to maintain acceptable linearity while enabling the PA to operate at high efficiency.

Furthermore, the amplitude of the sinusoids is constant. The amplitudes of the sinusoids will, however, vary dynamically. This amplitude volatility combined with the destructive interference results in a low average output envelope. The OFDM waveform has a high PAPR as a result.

When a PA receives a multi-carrier signal, like an OFDM signal, with a high PAPR and a very wide dynamic demand, Fig. 4.36 shows how the PA behaves in its non-linear zone. This results in out-of-band distortion, which disrupts the SEM, interferes with nearby channels, and emits undesired radiation. This leads to poor cellular system coverage, dropped calls, and poor service quality.

Reducing power to avoid PA compression will negatively impact cell coverage and power efficiency.

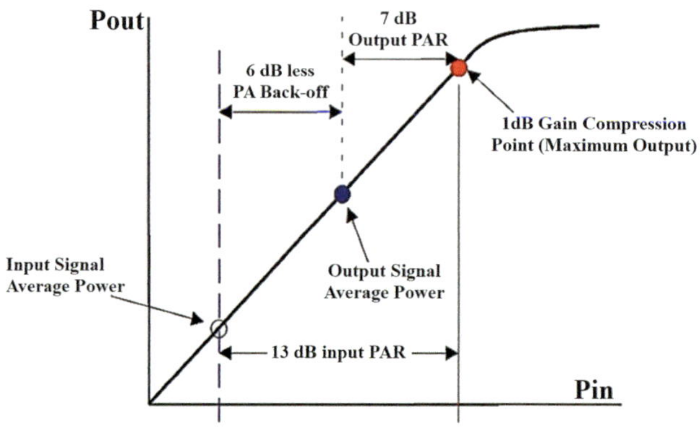

Figure 4.36 PAPR reduction techniques using PA output back-off.

One wise first step would be to reduce PAPR. In Fig. 4.36, the input signal's PAPR is 13 dB. By employing a PAPR reduction strategy, this can be lowered to 7 dB, allowing for 6 dB less back-off and greatly boosting PA efficiency. High PAPR is one of the primary practical problems with multi-carrier transmissions, such as an OFDM signal.

4.2.4.1 Methods of PAPR reduction

Up to now, a lot of strategies have been proposed to reduce high PAPR in communication networks. The two main categories of these techniques are distortion-based and distortionless. The former group will exhibit some distortion that deviates from the SEM. Techniques like peak windowing, peak cancellation, noise shaping, and clipping fall into this category. Some distortion is acceptable because the system's design attempts to enhance efficiency [1].

Techniques for scrambling, sometimes referred to as distortionless techniques, increase system complexity by necessitating the delivery of extra side information to the receiving device. Moreover, this protocol for user equipment has not been described by the 3GPP. This category contains techniques such as Partial Transmit Sequence (PTS) and Selected Mapping (SLM).

However, using similar techniques in 4G and 5G mobile networks is not feasible. As a result, techniques based on distortion are more desirable [1].

4.3 Transmitter Sub-System Design

The design of a high-efficiency 2.4 GHz band transmitter is shown in Fig. 4.37. It is used to directly modulate the QPSK signals at a maximum data rate of 11 Mbps for transmission. The transmitter design is based on single-stage I/Q DUC or homodyne architecture with completely on-chip solution.

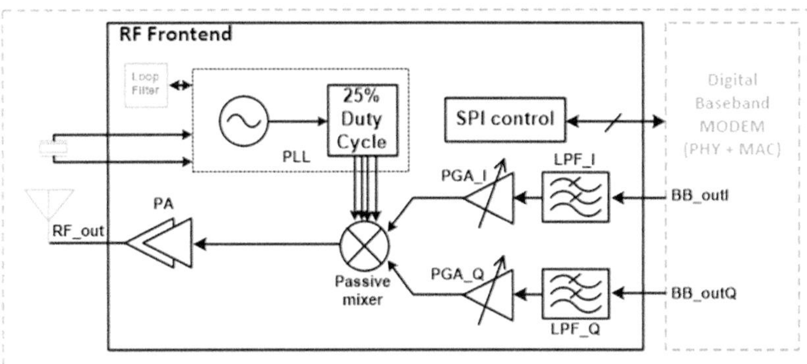

Figure 4.37 Transmitter front-end architecture with internal LO.

The transmitter chain includes a differential to single-ended power amplifier, a quadrature differential passive up-conversion mixer and 5-bit programmable gain amplifier (PGA) with baseband low pass filter. The mixer accepts a 25% duty cycled quadrature differential LO signal to directly up-convert the quadrature differential baseband input to the 2.4 GHz RF band. The transmitter chain is designed to improve the spectral purity of the transmitter output by minimizing all the spurious components such as the LO leakage, image component, higher harmonic distortion components. The typical testbench of the transmitter chain is as shown below with all the corresponding building blocks schematic along with the layouts.

4.3.1 PA Design

The PA is a single-stage differential amplifier that converts a differential input to single-ended output using an on-chip transformer-based balun. The transistors are stabilized using stability resistors at the input side to enable unconditional stability of the PA design over varying input and output impedance checked over a wide frequency range. The PA design uses a differential symmetric structure to ensure better matching and cancellation of even-order harmonics. The output matching network is based on the transformer-coupled load which is designed using the load-pull analysis for improved efficiency. The input matching network is built using two L-section LC network. Based on this the overall PA has minimum components to achieve the desired performance. By adjusting the PA biasing an output power control can be achieved in addition to the PGA. Figure 4.38 shows an illustration of the suggested PA topology.

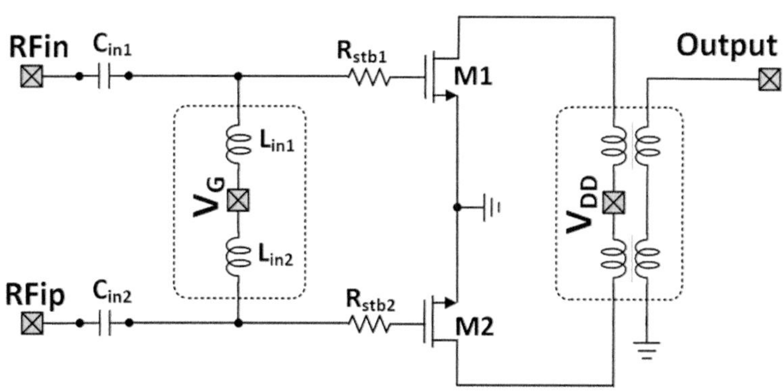

Figure 4.38 Differential one-stage PA schematic.

The PA is built to withstand PVT variations. It achieves a gain better than 10 dB. with improved input and output return losses as shown in Fig. 4.39. The design consumes only 21 mW from a power supply of 1.1 V. The stability factors namely the $Kf > 1$ and $B1f > 0$ are achieved over a large frequency range from DC to 100 GHz. Since the design is implemented with minimum components it has overall improved efficiency will minimal loss and compact design. The PA can additionally have variable output power control by adjusting its gate bias voltage which is shown later in the overall transmitter design performance. With such a minimum supply voltage and DC current, the overall PA design has improved efficiency and acceptable linearity performance.

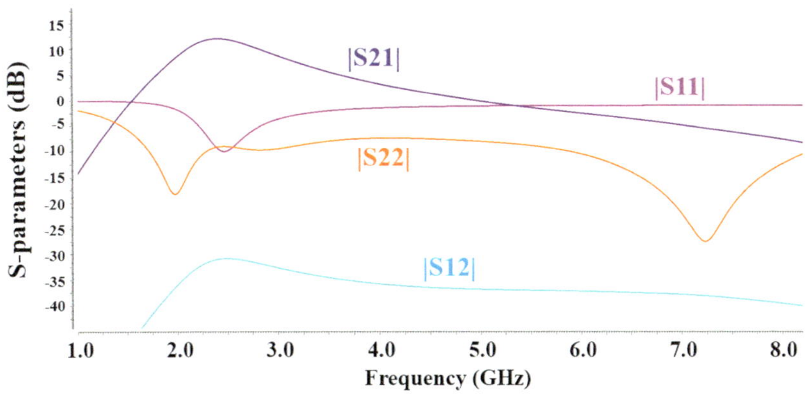

Figure 4.39 PA post layout S-parameter simulation results.

4.3.2 Passive Mixer

The internal *I/Q* differential LO signals with a 25% duty cycle drive the passive mixer as shown in Fig. 4.40. The passive mixer is built using four RF NMOS transistors to enable minimal insertion loss and no DC power consumption. The circuit schematic shown in Fig. 4.40 suggests the correct polarity and phase LO and baseband *I/Q* signals driving the passive mixer to enable single side band (SSB) modulation that can significantly suppress the image sideband and the carrier signals. This is verified by the overall TX chain simulation results.

4.3.3 Programmable Gain Amplifier

The same precisely scaled five-transistor operational transconductance amplifier (5T-OTA) with gain tunability via variable load is the PGA unit cell as used in the receiver (Section 3.3.3) is used in the TX chain as shown in Fig. 4.41. The main difference is the number of bits which for the TX chain is 5-bit gain control

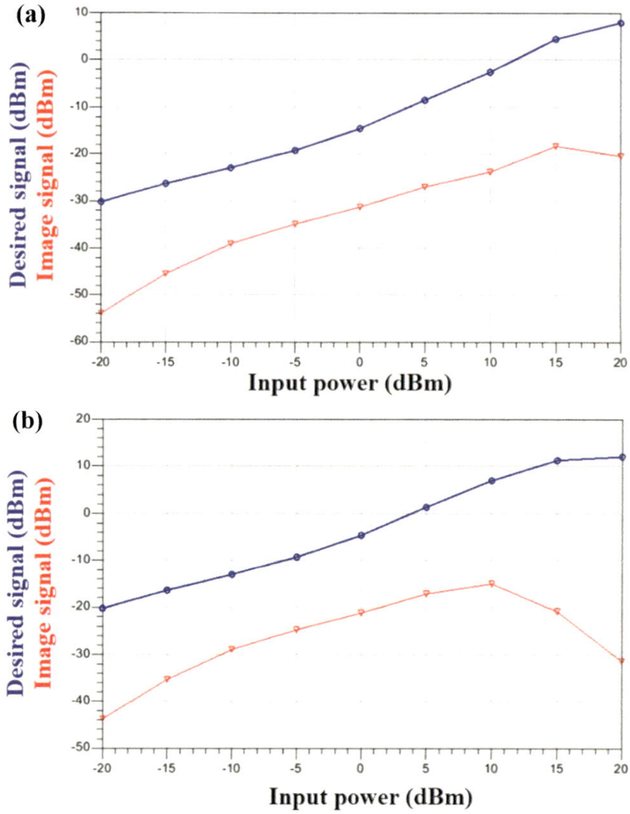

Figure 4.44 Simulated spur rejection at (a) mixer output and (b) PA output.

4.4 Conclusion

In this chapter, an overview of the RF transmitter sub-system design and development is given. The transmitter sub-system's various topology options and their benefits with the accompanied design issues are discussed. The transmitter design's critical and complementary performance parameters, namely the efficiency, and linearity are explained in detail. Furthermore, the extension of these two parameters on the transmitter system performance such as the spectral purity and the PAPR are illustrated in this chapter. Consequently, an example RF transmitter sub-system design and evaluation is presented.

References

1. B. Razavi, *RF Microelectronics*, 2nd ed., Upper Saddle River, NJ: Prentice Hall, 1998.
2. A. A. Abidi, Direct-conversion radio transceivers for digital communications, *IEEE Journal of Solid-State Circuits*, vol. 30, pp. 1399–1410, Dec. 1995.
3. S. Lerstaveesin and B. S. Song, A complex image rejection circuit with sign detection only, *IEEE J. Solid-State Circuits*, vol. 41, pp. 2693–2702, Dec. 2006.
4. K. Feher, *Wireless Digital Communications*, New Jersey: Prentice-Hall, 1995.
5. R. Steele, ed., *Mobile Radio Communications*, New Jersey: IEEE Press, 1992.
6. B. Razavi, A 900 MHz/1.8 GHz CMOS transmitter for dual-band applications, *IEEE Journal of Solid-State Circuits*, vol. 34, pp. 573–579, May 1999.
7. R. Adler, A study of locking phenomena in oscillators, *Proceeding of the IEEE*, vol. 61, no. 10, pp. 1380–1385, Oct. 1973.
8. B. Razavi, A study of injection locking and pulling in oscillators, *IEEE Journal of Solid-State Circuits*, vol. 39, pp. 1415–1424, Sep. 2004.
9. M. Zargari et al., A 5 GHz CMOS transceiver for IEEE 802.11a wireless LAN systems, *IEEE Journal of Solid-State Circuits*, vol. 37, pp. 1688–1694, Dec. 2002.
10. S. A. Sanielevici et al., A 900 MHz transceiver chipset for two-way paging applications, *IEEE Journal of Solid-State Circuits*, vol. 33, pp. 2160–2168, Dec. 1998.
11. B. Razavi, A 5.2 GHz CMOS receiver with 62 dB image rejection, *IEEE Journal of Solid-State Circuits*, vol. 36, pp. 810–815, May 2001.
12. A. Parsa and B. Razavi, A new transceiver architecture for the 60 GHz Band, *IEEE Journal of Solid-State Circuits*, vol. 44, pp. 751–762, Mar. 2009.

Missing is not lost; it can be recovered.
Mistakes are not failures; they can be road to success.
Job loss is not a dereliction of duty; it can be a new beginning.
A gaffe is not a breach of trust; it can be corrected.

失掉不是失去。还可以找回。
失误不是失败，还可以成功。
失业不是失职，还可以从来。
失言不是失信，还可以纠正。

—Kiat Seng Yeo

Chapter 5

Transceiver System Integration

Generally, for the complete integration as system-on-chip (SoC) of the radio frequency (RF) transceiver system, we need to consider the internal interfaces between the three sub-systems namely the receiver, the transmitter, and the frequency synthesizer. Besides internal interfaces, external interfaces are also very critical to the overall architectural design. In this chapter, we will study both these types of interfaces that eventually form the complete RF transceiver SoC. The internal interfaces are determined in Section 5.1.1 and the remaining sections of the chapter describe the external interfaces and the post-fabrication processing involved in building and developing the final prototype.

5.1 Integration

The RF transceiver's sub-systems are internally interconnected such that the local oscillator (LO) signal from the frequency synthesizer will be provided to the up-conversion mixer in the transmitter chain and to the down-conversion mixer in the receiver chain. The transmitter chain and the receiver chain have the RF front-end interface at one end and on the other end has the baseband interface.

CMOS RF and mm-Wave Transceivers and Synthesizers
Bharatha Kumar Thangarasu, Nagarajan Mahalingam, Kaixue Ma, and Kiat Seng Yeo

ISBN 978-981-5129-80-9 (Hardcover), 978-1-003-67356-9 (eBook)
www.jennystanford.com

For the baseband interface, there are two possible options as shown in Fig. 5.1. One option is to support the direct digital signals from the receiver and to the transmitter chain. The second option is to have analog signal at the RF transceiver baseband section such that in the receiver chain an analog-to-digital signal converter (ADC) is used to interface with the digital baseband and a digital-to-analog converter (DAC) in the transmitter chain. Hence, we can achieve the RF to digital conversion to support various digital modulation schemes.

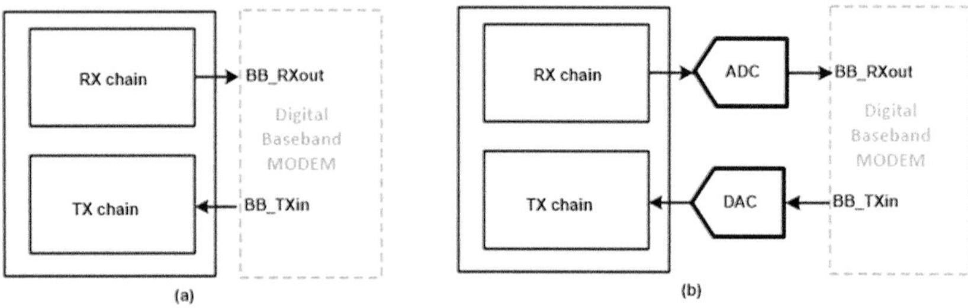

Figure 5.1 RF transceiver baseband interface (a) direct digital and (b) analog-to-digital conversion.

For the RF front-end interface, there are three possibilities shown in Fig. 5.2 as follows:

(a) Dedicated antennas for the transmitter and receiver chains which is generally a 50 Ω impedance.

(b) A single antenna interfaced to both the transmitter output and the receiver input using a transmit/receiver (T/R) switch to support time domain duplexing (TDD).

(c) Alternatively, a single antenna interfaced to the transmitter and receiver chains using the duplexer to support the frequency domain duplexing (FDD).

To support reconfiguration, which is a post-fabrication RF transceiver function modification process in which by using digital switches within the RF transceiver are controlled using digital settings. The digital setting is established using a digital controller with an external interface such as a serial peripheral interface (SPI). The details of SPI are provided in subsequent sub-sections.

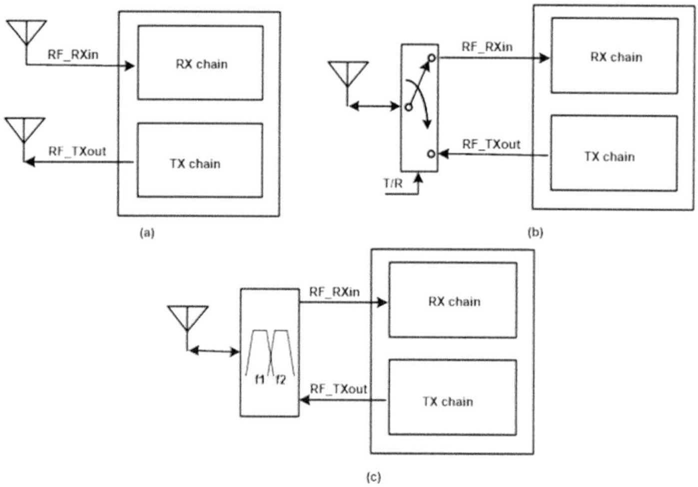

Figure 5.2 RF transceiver front-end interface (a) direct (b) T/R switch-based TDD and (c) duplexer-based FDD.

5.1.1 Sub-Systems

The RF transceivers are built using integration of the sub-systems namely the receiver, the transmitter, and the frequency synthesizer to achieve its basic operation. The details of these three sub-systems are explained in Chapters 3, 4 and 7, respectively, in this book. The interfaces between the sub-systems are shown in Fig. 5.3.

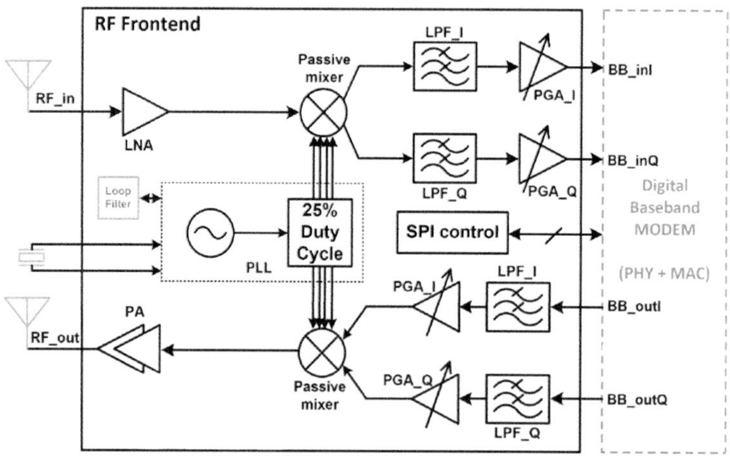

Figure 5.3 RF transceiver block diagram.

Additionally, the system requires supplementary sub-systems namely the SPI, and the electrostatic discharge (ESD) protection input/output pad ring to enable the RF transceiver to be utilized for various applications. The details for the SPI are provided subsequently and the ESD protection is explained in Section 5.1.2.

5.1.1.1 Serial peripheral interface

A serial peripheral interface is referred to as SPI [1]. It is a synchronous serial communication protocol. It serves as a means of wired communication for microcontrollers and other peripheral devices, such as input and output devices. High-speed data transfer is permitted. It is well known in embedded systems and digital communication applications. Sensors, ADCs, DACs, shift registers, SRAM, and other devices are among the devices that frequently use SPI to simultaneously send and receive data.

A synchronous, full-duplex interface based on master-slave is possible using SPI. On either the rising or falling clock edge, the data from the master and slave are synchronized. Data can be transmitted simultaneously by the master and slaves without any waiting delays. Three to four wires can be used for the SPI interface protocol.

SPI operates in a little different way. As a "synchronous" data bus, it uses distinct lines for data and a "clock" to maintain precise synchronization between the two sides. The receiver is instructed by the oscillating clock signal on precisely when to sample the bits on the data line. This could be the clock signal's rising (low to high) or falling (high to low) edge. The receiver looks at the data line to read the next bit as soon as it detects that clock edge. Devices will have a maximum speed at which they can function, but since the clock is transmitted with the data, it is not necessary to define the speed.

SPI master device components: The master device is a simple device that manages the data translation process on the SPI bus. It produces the clock signal and manages the data flow. The microcontroller or specialized SPI controller serves as the master device in many applications. Peripheral devices that are linked to the SPI bus and managed by master devices are referred to as slave devices. Each slave device has a unique slave chip select (CS) line that enables the master to choose which slave device to connect to.

SPI bus: SPI bus refers to a physical link that facilitates data transfer between the slave and the master devices. As seen below, SPI devices have four signal lines:

(a) *SPI serial clock* (SCLK): In a serial clock, the slave and master devices use this clock signal to synchronize the timings of data transfers.

(b) *Slave chip select* (CS): Every slave device has a specific CS pin when using slave chip select and when the master will need to speak with that particular slave. The same SCLK, master in slave out (MISO), and MOSI lines can be used by several slave devices to interface with the master, but the CS lines need to be separated.

(c) *Master out slave in* (MOSI): This feature allows MOSI to communicate with other slave devices by sharing data or information from the master.

(d) *Master in slave out* (MISO): This type of sharing allows the master to access data or information from the slave device.

One master and one or more slaves are possible for SPI interfaces. The SPI link between the master and slaves is depicted in Fig. 5.4.

Data transfer protocol: For straightforward data transfers, synchronous serial communication (SPI) is utilized. Full-duplex mode data transport and reception occur simultaneously. The master-slave will initiate the data transfer by producing the clock pulses. Every clock cycle, one bit of data will be sent from the master to the slave as well as from the slave to the master.

Data rate: Depending on the length of the transmission line and the slave devices' master capabilities, the SPI bus can handle various data transfer rates. The bit rate is expressed in megahertz (MHz) or bits per second (bps).

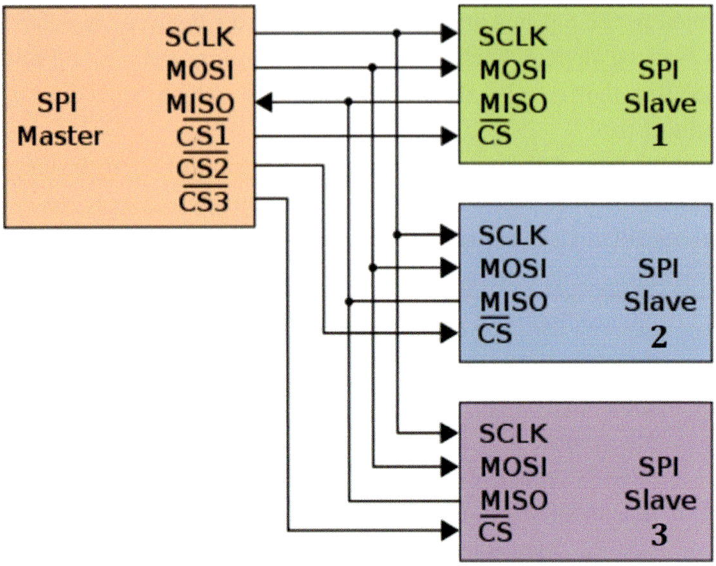

Figure 5.4 SPI link between the master and multiple slaves.

The peripheral is disconnected from the SPI bus when the chip select line is held high. This kind of logic is called "active low," and it is frequently utilized for resetting and enabling lines. The peripheral is activated when the CS line is brought low, right before data is transferred to it. The line is raised once more when we have finished communicating with the peripheral. This is equivalent to the "latch" input in a shift register, which sends the data that is received to the output lines.

Data transfer: The master must provide the clock signal and enable the CS signal to pick the slave node in order to initiate SPI communication. In order to pick the slave, the master must provide a logic 0 on the chip select signal, which is often an active low signal. Since SPI is a full-duplex interface, data can be sent simultaneously via the MOSI and MISO lines by the master and slave, respectively. Data is sent concurrently (shifted out serially into the MOSI/SDO bus) and received (the data on the bus MISO/SDI is sampled or read in) during SPI communication. The data shifting and sampling are synchronized by the serial clock edge. The rising or falling edge of the clock can be chosen by the user to sample and/or shift the data thanks to the SPI interface. To find out how many data bits are sent via the SPI interface, please consult the device data sheet.

SPI's benefits:

(a) *High-Speed Data Transfer*: SPI is appropriate for applications requiring quick data transfers since it offers high-speed communication. Peripheral devices and certain microcontrollers are required for this data transfer. When fast communication is required, such as for real-time data acquisition or high-performance sensor interfacing, high-speed data transfer can be employed.

(b) *Basic Hardware Requirements*: A few hardware components require SPI. It is used for data transfer using simple protocols and consists of four signal lines: SCLK, MOSI, MISO, and CS. Compared to other communication protocols, these requirements can be implemented more easily and with lower hardware costs.

(c) *Full-Duplex Communication*: SPI facilitates bidirectional communication by allowing data to be transferred and received in both directions. It is possible to enable this bidirectional communication to move data enough and make it appropriate for applications that need high-speed communication.

(d) *Several Slave Devices*: With a single master device, SPI can support several slave devices. Every slave device is linked to a slave select (CS) line, which enables the master to choose and interact with individual slave devices as needed. This feature allows the SPI to interact with the many peripheral devices in the system without requiring any additional hardware.

(e) *Broad Adaptation*: SPI is a widely used communication protocol in the electronics embedded systems industry, and it is most readily available for support of microcontrollers and peripheral devices. By making this adjustment, the interoperability of the various hardware parts may be guaranteed. Because of this adaptation, SPI is now a depend-able option for communication in a wide range of applications.

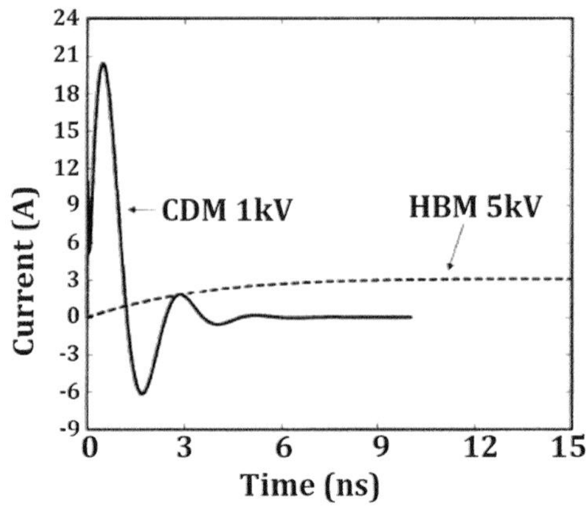

Figure 5.8 Event response comparison between HBM and CDM.

5.1.2.1.3 *The machine model*

A charged capacitor is connected in series with a 10 Ω resistor, a 700 nH inductor, and the DUT in the machine model that is seen in Fig. 5.9. With five stress levels ranging from 25 V to 400 V, the machine model has the lowest voltage levels among the three types of ESD [5].

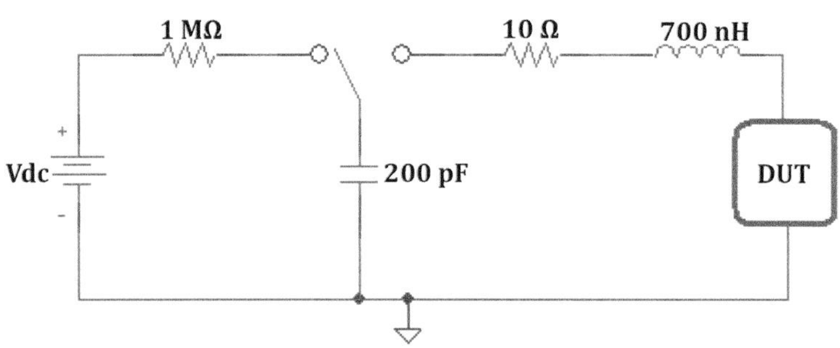

Figure 5.9 MM ESD test equivalent schematic.

Similar to a CDM event, the DUT response to an MM event is followed by an LCR response, as seen in Fig. 5.10. This circuit, on the other hand, has substantially greater capacitances, which cause a delayed rise time of about 20 ns.

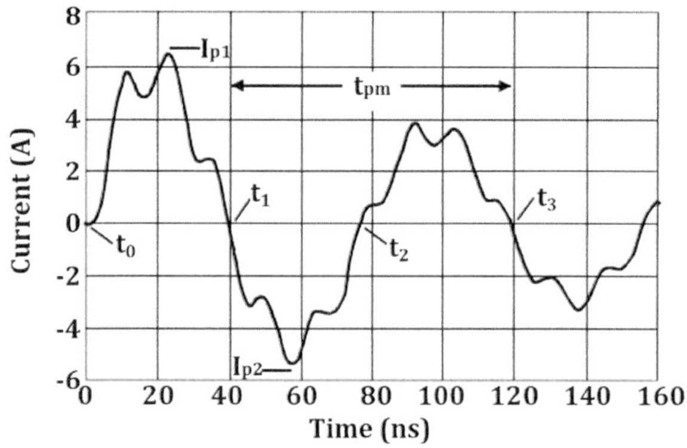

Figure 5.10 Event response of MM.

5.1.2.2 Basic ESD bond pads

Connectors between integrated circuits and the ESD-filled environment outside the chip are provided via bond pads. Bond pads are immediately in the discharge path of an ESD event since they are the gateway to and from an IC. Therefore, the first line of defense against an impending electrical shock is to install safety circuitry right next to an IC's bond pads. Before it can destroy vital circuitry, the hazardous discharge current that is flooding the system needs to be diverted here, at the gateway.

Two diodes in a CMOS process offer what is arguably the most fundamental ESD protection. When the I/O pad climbs above the supply voltage, as seen in Fig. 5.11, a diode positioned between the I/O pad and the Vdd pad becomes active. Thus, a positively charged ESD event that strikes an I/O pad causes the chip's positive supply line to discharge back out. As the I/O voltage falls below ground, a diode between an I/O pad and ground turns on to protect the pad against negatively charged ESD occurrences.

The two safety diodes are useful in preventing overvoltages on I/O pads, but they are ineffective against a supply line that reaches hazardous levels. For instance, Vdd may pulse to a dangerously high level if the supply line circuitry is unable to absorb the event. A typical ESD pad inserts a voltage clamp between the source voltage line and ground to address this problem.

To solve this problem, Fig. 5.14 depicts a simple ESD-transient detection circuit coupled to an NMOS clamp's gate [7]. The ESD-transient triggered clamp only turns on when there are abrupt fluctuations in the supply voltage, lowering the clamp's turn-on voltage when it is off in DC conditions. Capacitor C is an open circuit when working in DC mode; it charges to VDD, maintaining the inverter's low output and turning off the clamp. However, the capacitor becomes low impedance in relation to the resistor with a quick VDD spike. As voltage builds up across the resistor, *Vx* becomes low in comparison to the spiked VDD. By bringing the inverter output high and turning on the clamp, the voltage differential activates *Mp*. The ESD-transient triggered clamp is able to provide improved protection for minimum-sized FETs by identifying abrupt changes in VDD [8].

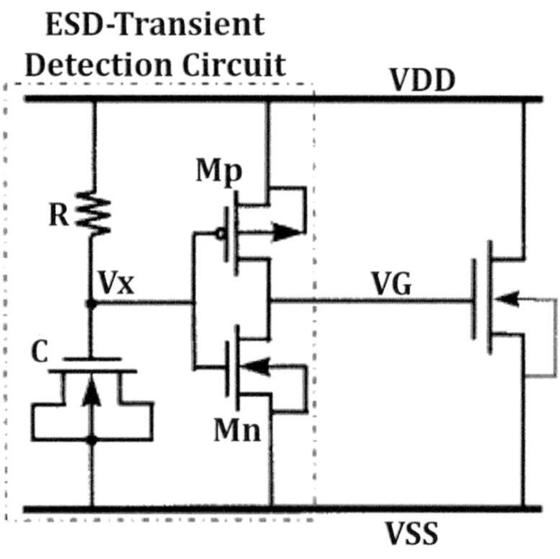

Figure 5.14 Clamp circuit with ESD-transient triggered event.

This method has additional benefits. NMOS clamp FETs do not require breakdown to operate, in contrast to gate-grounded clamp FETs. This not only increases the clamp's durability but also simplifies its construction. However, the parameters of a process determine how a gate-grounded clamp is designed, therefore altering to a new process necessitates a total redesign of the clamp.

The disadvantage of this method is that it offers no protection against unintentional DC overvoltage as the triggering circuitry responds quickly to voltage fluctuations. To guard against DC overvoltages, more gate-grounded clamps would be needed if the ESD-transient triggered clamps were to be used. On the other hand, a well-designed gate-grounded clamp can quickly activate to safeguard the internal circuitry of the integrated circuit while also providing protection against DC overvoltages. A further disadvantage is the additional circuitry required. Enough size is already required for the NMOS clamp to withstand elevated current levels. Even more chip space is occupied by the detection circuitry, particularly if it has an exponential horn driver circuit for the NMOS clamp to increase turn-on speed.

5.1.2.5 Resonant ESD pad

The addition of parasitic capacitance by any ESD protection circuitry coupled to a pad can reduce the frequency responsiveness of the circuit. It is not possible for circuits intended for gigahertz operation to lose frequency responsiveness. Pads that are attached to high-frequency circuits—like the input of an LNA—must thus be able to both minimally impact the performance of high-frequency circuits and provide adequate protection against ESD incidents. To account for the additional parasitic capacitance, a resonant ESD pad is implemented at the input of an LNA according to a design described in [6].

The design is depicted in Fig. 5.15. A varactor is used to fine-tune the resonant circuit, while an inductor is used to resonate out the capacitance. A gate-grounded clamp is used in this specific design to connect the pad to the ground. It might have used the diode protection shown in Fig. 5.11 just as easily.

While the design shown in Fig. 5.16 is frequently utilized in low-frequency applications, the designs shown in Figs. 5.12 and 5.15 were developed especially for high-frequency applications. The use of a series resistor between the pad and the inner circuitry shown in Fig. 5.16 is a minor but significant difference between Figs. 5.12 and 5.16. Since a series resistor causes a voltage drop between the pad and the internal circuitry, high-frequency ESD pads do not employ them. Removal of the resistor avoids needless signal loss and related additional noise in high-frequency circuits, such an LNA circuit, where input sensitivity is a crucial parameter.

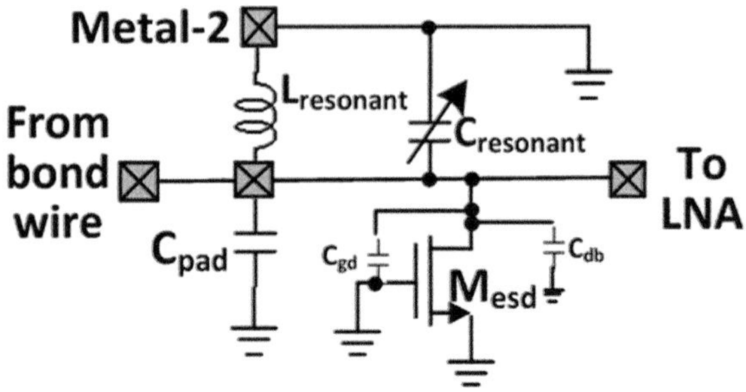

Figure 5.15 ESD pad with resonant circuit.

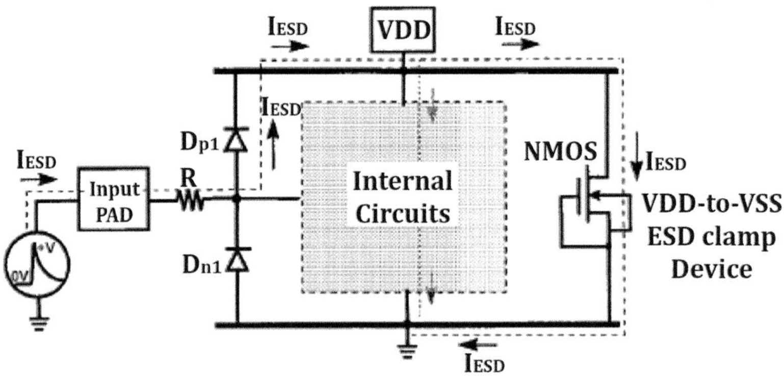

Figure 5.16 ESD pad suitable for low-frequency designs.

The resonant ESD pad protection circuit has no obvious disadvantages, although it is not the best idea to add another inductor to an LNA circuit. To minimize the number of components required, the resonant tank on the ESD pad and the LNA matching network should be merged in practice.

5.2 Prototyping

An IC is designed, fabricated, tested, and packaged as part of the integrated circuit design and manufacturing process. The process of designing and producing integrated circuits involves multiple processes, ranging from specification and design to system integration. Different facets of electronic design and production are represented by the design and manufacturing of ICs and printed circuit boards (PCBs) [9–12]. A sample packaged IC prototype as PCB is shown in Fig. 5.17.

Figure 5.17 Sample IC prototype as PCB.

The process of designing and producing ICs entails building intricate electronic circuits on tiny silicon chips. It covers the IC's design, manufacture, testing, and packaging. In order to increase performance, yield, and reliability, designers and manufacturers continuously refine and optimize each step of the highly iterative process. The types and sizes of ICs vary widely. The industry has started to produce high-spec artificial intelligence (AI) chips recently; these chips are significantly more capable than standard chips and can carry out the intricate calculations and data processing needed for AI tasks. All IC chips, regardless of size or kind, go through a set of clearly defined phases during the design and manufacturing process.

Wafer Fabrication: Various layers are deposited and etched, doping is applied, and transistor structures are created on the silicon wafer using the masks to form the appropriate patterns. One wafer is used to build many integrated circuits.

Packaging: Following the completion of the wafer fabrication process, the individual integrated circuits are separated and placed into appropriate housing to offer protection and electrical connections. Leads or balls are also attached for external connections during packaging.

Assembly and PCB Design: At this point, PCBs, which offer mechanical support and electrical connectivity, are used to install the ICs. Component placement, electrical connection routing, signal integrity, power distribution, and thermal management are all part of the PCB design process.

System Integration: To build the finished electronic product, the entire system is integrated with additional parts and sub-systems once the ICs are placed on the PCB.

We will try to understand the post-IC fabrication steps starting with dicing, wire bonding and chip-on-board (COB) integration, bumping and flip-chip integration, IC packaging and PCB assembly in the subsequent sub-sections.

5.2.1 Dicing

To reduce the mass production cost, the IC designs are fabricated as dice of even shapes typically square or rectangle arranged as rows and columns on a wafer. Optimum spacing between the dice is determined such that we get maximum dice per wafer and also spacing large enough to ensure the dice are not damaged during the dice sawing process. To mark the spacing required for dicing or saw cutting the dice, usually the semiconductor foundries will provide a seal ring and scribe seal which the system integrator will include as shown below to guide the wafer fabrication process. The critical spacing and dimensions with reference to the die corner are shown in Fig. 5.18.

5.2.1.1 Diced silicon wafers

The process of sawing or laser cutting silicon wafers into smaller pieces, known as dies, is known as silicon wafer dicing. Depending on the use, silicon wafers are usually cut into tiny pieces that range in size from a few millimeters to a few hundred micrometers.

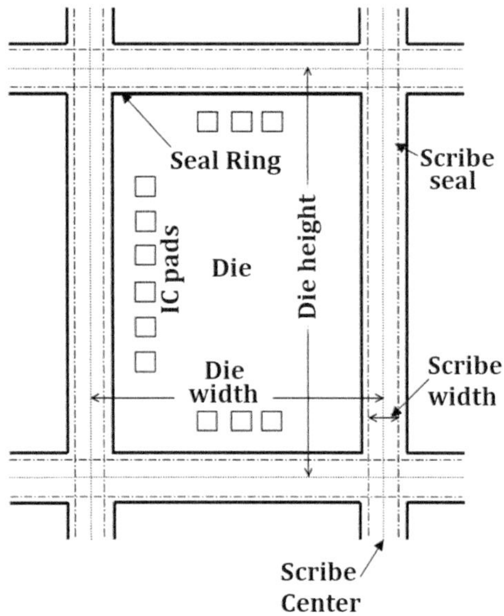

Figure 5.18 Dice arrangement in wafer with scribe seal and seal ring.

The ability to divide a wafer into separate devices that can be installed onto a PCB or other substrate makes the dicing process a crucial step in the fabrication of microelectronic devices, including memory chips, microprocessors, and other integrated circuits.

The semiconductor packaging industry mostly uses the dicing technique to divide silicon wafers into individual chips. The production of integrated circuits involves a series of steps that transform raw materials into packaged devices that may be assembled into electronic circuits. Put another way, it's a procedure known as the "dicing process" that separates the die from the silicon wafer. Wafers are fixed on a metal frame during the dicing process using PVC or Mylar tape, which has an adhesive backing. The procedure takes place in a clean environment with regular air circulation since even the smallest dust particles have the potential to harm the intricate circuits imprinted on a wafer.

Depending on what it is used for, this tape has various qualities. UV curable tapes are utilized for lower die sizes and non-UV dicing tape for larger die sizes. When a wafer is diced, the die stays on the Mylar tape until a die bonder or other piece of equipment retrieves it. Typically, the dies are square or rectangular in shape. They can also take on different shapes based on the dicing technique employed.

This procedure includes breaking, scribing, laser cutting, and mechanical sawing. The two main restrictions on the dicing process are the development of chipping and blade yielding. It is advised to use a specific monitoring technique to figure out the maximum feed rate while staying within these bounds. For the purpose of quality, accuracy, and precision, all techniques are typically automated. To satisfy client requests, the company has embraced ongoing development and progress. Precise and reliable procedures are employed to guarantee that the actual units produced are of the highest caliber and dependable to stay competitive in the face of change. To improve die strength during package building, pre-assembly processes must optimize chippings, which directly affects the dependability and quality of the packaged goods.

The analyst presents a comprehensive picture of the market by researching, synthesizing, and summarizing data from numerous sources in addition to analyzing important criteria.

By offering a dependable and consistent set of parameters, issues linked to chipping may be minimized and their effect on overall package dependability might be diminished. Moreover, there may be no chance that related flaws will be transferred to different wafer technologies and later packages. Wafers are categorized based on the technological family to which they are assigned. Different indirect materials and characteristics are used for each wafer technology categorization. Therefore, based on prior experience and data-driven research, established procedures within comparable wafer technology must be adhered to in order to guarantee a favorable reaction.

The increasing integration of microelectronics in consumer electronics is driving the need for dicing equipment. The need for technologies like micro-mechanical element switch (MEMS) devices and power devices is driving the demand for thin wafers. Moreover, this requirement bolsters the aim for enhanced production techniques, an essential phase in the production of ultra-thin wafers. MEMS devices, power devices, and radio frequency identification are thought to be the main markets for very thin wafers. Due to this situation, there is a growing need for improved manufacturing procedures, particularly for the processing and dicing steps that are essential to producing ultra-thin wafers.

Silicon wafers can be diced using a variety of techniques, such as waterjet, laser, and mechanical sawing. Whereas laser dicing employs a concentrated laser beam to cut through the wafer, mechanical sawing uses a revolving blade or diamond saw. In waterjet dicing, the wafer is cut through using a high-pressure water stream.

The exact specifications of the application, such as the thickness and substance of the wafer, the size and form of the diced, and the level of precision and accuracy needed, determine which dicing process is best. Wafers are normally designed and diced after which the dies are inspected to make sure they satisfy the necessary requirements and are separated.

The markets for semiconductors and IC packaging are being driven forward by the following considerations. This article describes the pre-assembly knowledge acquired to enhance silicon wafer backside and topside chip-out during the qualification run and evaluation.

1. Consumer expectations for the features of new electronic devices have expanded, driving the electronic packaging industry. This is due to the rising demand for electronic devices.
2. The electronics market has been entirely dominated by the rise of IoT and connected gadgets, together with smart devices. Wearables with connectivity are expected to become more and more common, in addition to smartphones.

5.2.2 Wire Bonding and Chip-on-Board Integration

During the manufacture of semiconductor devices, wire bonding is a technique used to attach an IC or other semiconductor device to its packaging. Although these uses are less prevalent, wire bonding can also be used to link an IC to other devices or to connect one PCB to another. The great majority of semiconductor packages are assembled via wire bonding, which is usually regarded as the most versatile and affordable connectivity approach. Frequencies higher than 100 GHz can be utilized with wire bonding.

Before wire-bonding the die is attached to the PCB using either conductive or non-conducting epoxy resin. The wire bonding of a sample IC on the PCB is shown in Fig. 5.19.

Figure 5.19 Dice attach and wire bonding.

Materials for bondwire are often chosen from aluminum, copper, silver, and gold based on the intended use and specifications of the final product. For high-powered applications, wire diameters can reach several hundred micrometers, although they can start at less than 10 μm. In the wire bonding business, copper is replacing gold. The rising price of gold and the relatively stable (and much lower) price of copper have caused this shift. Due to its hardness and corrosion susceptibility, copper was often thought to be less dependable even if the metal possessed higher thermal and electrical conductivity than gold. Subsequently, it is anticipated that over one third of all wire bonding equipment will be configured for copper.

For wire bonding interconnects in many semiconductor and microelectronic applications, copper wire has emerged as a material of choice. For fine wire ball bonding, copper is utilized in sizes ranging from 10 μm (0.00039 in) to 75 μm (0.003 in). When utilized at lesser diameters, copper wire can function similarly to gold without having to pay as much for the material. Higher electrical conductivity in copper makes smaller dimensions possible. When compared to gold wire bonds, copper wire bonds are just as dependable, if not more so.

Wedge bonding of copper wire up to 500 μm (0.02 in) is feasible. When there are issues with complex geometry or a necessity for a high current-carrying capability, large-diameter copper wire can and will replace aluminum wire. Manufacturers' use of annealing and other manufacturing stages improves the ability to wedge bind silicon using large-diameter copper wire without causing die damage.

Because copper wire is tougher than both gold and aluminum, bonding factors need to be closely monitored. This presents some complications. Because copper has a higher fusing current and a higher current-carrying capacity, ultrasonic bonding requires a higher power input. Since this substance naturally forms oxides, shelf life and storage are important considerations. To extend the shelf life and safeguard copper wire, special packaging is needed. A popular substitute that has demonstrated notable resistance to corrosion is palladium-coated copper wire, despite having a harder texture than pure copper and costing more than gold.

In order to prevent corrosion, copper wire, including its plated variants, needs to be worked in the presence of forming gas, which is a mixture of 95% nitrogen and 5% hydrogen, or a comparable anoxic gas, when fabricating wire bonds. Using high-purity types of copper is one way to deal with its relative hardness. When employed in automotive applications, long-term corrosion effects and other stability issues led to higher quality standards [11].

Surface-mount LED package in red, green, and blue with gold wire bonding details.

Ball bonding typically uses pure gold wire doped with small quantities of beryllium and other elements. Thermosonic bonding is the technique of joining two materials together by means of heat, pressure, and ultrasonic energy. Ball bonding to the chip and subsequently stitch-bonding to the substrate is the most used method for thermosonic bonding. Sagging is eliminated and looping features are enhanced by extremely strict controls during processing.

The ideal wire size for a given wire bonding application is usually determined by the junction size, bond strength, and conductivity requirements. Gold wire is typically produced by manufacturers in sizes starting at 8 μm (0.00031 in) and up. The diameter of gold wire has a production tolerance of +/–3%.

Because alloyed aluminum wires are easier to draw to fine diameters and have better pull-test strengths in finished devices, they are typically favored over pure aluminum wire, except for high-current devices. Sizes greater than 100 μm (0.0039 in) are most frequently used for pure aluminum and 0.5% magnesium-aluminum.

In semiconductor production, the "purple plague" (a brittle gold-aluminum intermetallic compound) that is occasionally connected to pure gold bonding wire is eliminated by using all-aluminum systems. Particularly well suited for thermosonic bonding is aluminum.

Special controls are employed in the 1% silicon-aluminum wire manufacturing process to ensure that good quality bonding can be achieved at high production speeds. The uniformity of the alloy system is one of the most crucial features of this kind of high-grade bonding wire. Special consideration is given to homogeneity during the manufacturing process. Regular microscopical inspections are carried out to verify the alloy structure of final quantities of 1% silicon-aluminum wire. Additionally, processing is done in a way that produces the smoothest possible surface finish and maximum cleanliness, allowing for completely snag-free de-reeling.

The main classes of wire bonding can be classified as ball bonding, wedge bonding and compliant bonding. Ball bonding often requires heat and is limited to copper and gold wire. Heat is only needed for gold wire in wedge bonding. For power electronics applications, wedge bonding can make use of wire ribbons or large-diameter wires. Only wires with a small diameter are appropriate for ball bonding in interconnect applications.

In both types of wire bonding, the wire is joined at both ends by applying heat, ultrasonic radiation, and sometimes downward pressure to create a weld. The metal is softened by heat. The most dependable and robust wire bond can be achieved by using the ideal temperature and ultrasonic energy combination. Thermosonic bonding is the term for the technique that uses both heat and ultrasonic energy.

The wire in wedge bonding needs to be drawn in a straight line in accordance with the first bond. Because of the time required for tool alignment, this slows down the process. In contrast, ball bonding, which has no preference for orientation, forms its initial bond in the shape of a ball with the wire protruding at the top. It is therefore faster because the wire can be drawn in any direction.

Compliant bonding [12] is a technique used to attach silicon integrated circuits (sometimes called beam-leaded integrated circuits) to gold wires and electroformed beam leads. It works by transmitting heat and pressure through an indentable or compliant aluminum tape.

5.2.2.1 Chip-on-board

This approach of making circuit boards involves directly bonding or wiring integrated circuits, including microprocessors, to a printed circuit board, then covering it with a blob of epoxy [9]. By doing away with the need to package separate semiconductor components, chip-on-board products can be lighter, more afford-able, and more compact. When integrated circuit leads' inductance and capacitance are decreased, radio frequency systems can function better thanks to COB construction. COB—also known as "level 1.5"—effectively combines two layers of electronic packaging: level 1 (components) and level 2 (wiring boards) [12].

A cross-sectional view of the PCB with the wire bonding as COB is shown in Fig. 5.20.

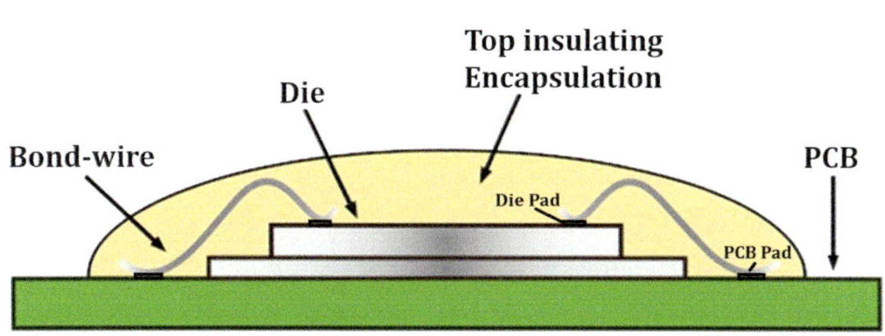

Figure 5.20 COB cross-sectional view.

Semiconductor wafers are completed and sliced into dies. After that, each die is attached to the PCB physically. The integrated circuit's (or other semiconductor device's) terminal pads are connected to the printed circuit board's conductive traces using three distinct techniques.

Electronic components are assembled on a PCB using a packaging technique called chip-on-board (COB) as shown in Fig. 5.21. This method connects bare integrated circuits on the board's surface instead of configuring individual components on the board. Because less traditional packing methods like ceramic or plastic packaging are used, electrical equipment and projects can be made smaller and lighter.

Figure 5.21 Wire bonding and chip-on-board assembly PCB.

Solder bumps or adhesives are employed to mount the chips, resulting in a small and efficient assembly. Direct bonding creates fewer electrical lines between the chip and the board, which results in better electrical performance and less signal losses.

Since chips in this technology are directly connected to heat sinks or thermal pads on the board, it also offers effective thermal control.

Wearable technology, mobile phones, LED lights, and power electronics are examples of projects where this technology works best due to its good electrical performance, compact size, and good thermal properties. Additionally, the integration and miniaturization of electronic systems are advanced by this technique.

The making procedure of chip-on-boards

1. *Substrate Preparation*: The PCB board is cleaned on the surface before an adhesive layer of conductive material is put, which is where the chips are bonded.
2. *Die Attach*: The bare chips are placed on the board's adhesive-coated sections. This procedure is carried out using specialist equipment or pick-and-place devices.
3. *Bonding*: Configured chips are joined to a board by means of conductive solder bumps. A strong connection between the chip contact pads and the board's conductive traces is ensured by this bonding procedure.
4. *Wire Bonding*: In certain circumstances, bonding pads and board traces are connected via wire bonding, which uses thin wires. This procedure facilitates the transmission of electrical signals between the board and the chip.
5. *Encapsulation*: The entire assembly can be coated with an encapsulant substance to shield chips and wire bonds from external components. Additionally, this substance has a clear epoxy coating.
6. *Testing*: Various testing techniques are applied during COB construction to guarantee appropriate performance and dependability. To confirm the COB's functionality, tests like temperature cycling, electrical testing, and visual inspection are carried out.
7. *Final Assembly*: The chip-on-board assembly is prepared for integration into the finished electronic products, such as LED lights, phones, or other projects, after it passes all testing.

Additionally, PCB also can employ flip-chip technology, which involves using our internal bumping process or pre-applied solder balls at the wafer level, to attach the chip face down to the PCB. The details are discussed in the next sub-section. The chip can then be protected using the glob top or Dam/Fill packaging method. One can also combine the quick turnaround service with many chip-on-board assemblies.

Adhesives are used in PCB chip technology to directly attach semiconductor chips to the PCB substrate. It is then packaged by wire connecting it to an existing circuit layout on the board. Surface-mount technology (SMT) is advanced by chip-on-board. The following is the primary distinction between SMT and COB: COB, gadgets typically have a high lead count, are active, and do not need to be packaged externally in molded plastic or ceramic. The on-board chip is a plain chip that is put straight onto the PCB. Once the wires are connected, cover the chip to complete the connection with a ball of plastic or epoxy. After the bare PCB chip is wire-bonded and attached to the board, epoxy resin is poured over it to cover and insulate it.

Mounted on a laminate substrate with signal conditioning or support circuits is an unpackaged IC. An electrical connection is created when the IC is gold wire bonded to the matching substrate connector. The chip and wire bonds can then be shielded by applying a junction coating material on top of the chip. Therefore, PCB chips are a great option for circuits that are smaller. A COB solution emerges when the design parameters are not met by conventional assembly techniques. One way that a chip on a board is advantageous is that it makes the circuit lighter and of higher quality. Chip-on-board technology is the best option for circuit miniaturization when this is a major problem.

COB technology also gives system designers new possibilities for assembly. In this technology, electrical connections are made by directly adhering the silicon chip to the PCB; to shield the chip from light and damage, an opaque epoxy resin covering is used.

Benefits and characteristics of Chip-on-board

1. The design of high and low pressure
2. Personalized coating
3. Double-sided, multi-layer
4. Test of the functional board
5. The volume: high or low
6. Broad range of temperatures
7. An affordable cost option
8. Application that is turnkey

COB technology has transformed the electronic industry and offers unique benefits over traditional packaging methods. Because bare chips are directly bonded to the board, the resulting electrical devices are smaller, lighter, and more efficient. Good heat management, reduced signal losses, and improved electrical performance are achieved by eliminating bulk packaging and shortening electrical paths. Additionally significant to the shrinking of electrical components, this technique cleared the path for innovative, small-sized technology across various industries. COB offers a variety of applications, ranging from lighting systems and automotive applications to small mobile phone devices. Embracing COB's potential empowers engineers, designers, and consumers alike as we advance toward a future where electronics are seamlessly integrated into our lives and offer enhanced functionality and performance.

5.2.3 Bumping and Flip-Chip Integration

The markets for consumer electronics, and the mobile phone sector in particular, are quite competitive. Their motivation stems from the aim of cramming an increasing number of features and improved value into a mobile device of the same size, frequently at a reduced cost. The desire for consumer electronics to be more affordable, smaller, and thinner has accelerated the development of highly integrated circuits and encouraged the creation of smaller, thinner containers to go along with this cutting-edge technology. Leading IC post-processing firms and semiconductor foundries have developed a variety of wafer-bumping packaging solutions as a result. Die-in-package assembly has evolved from wire bonding the chip into a sizable package to COB or even die-embedding into the PCB assembly, which is currently heading toward eliminating the packaging entirely.

Removing the wire bonds and replacing them with bumps was a crucial stage in this process. Before the wafer is divided into individual chips, a technique known as "wafer bumping" attaches solder spheres to the chip's I/O pads. After that, the bumped dies can be inserted into packages or soldered straight to the PCB, or the previously mentioned COB.

Usually, the bumps are positioned on a substance called under bump metallization (UBM) that is plated onto the die pads.

Depending on the purpose and application, the bumps can be made of a wide variety of materials. Lead-free bumps are typically used, while copper pillar bumps have a higher density.

With regard to packaged bumped dies, there are several package types available in the industry:

(a) *Flip-chip ball grid array* (FCBGA): The most popular kind of package used with bumped dies is still the ball grid array (BGA) flip-chip package, in which the bumped die is flipped onto a substrate that directs signals to the package balls. Benefits include scalability for physically big and complicated dies and high thermal performance. The substrate used in low-cost FCBGAs is laminate, similar to PCBs; however, build-up substrates can be purchased, which provide finer pitch routing, improved signal and thermal performance, and a reduced profile. When it comes to high-power designs and designs with a lot of balls (more than 100, for instance), an FCBGA is the recommended flip-chip solution and is shown in Fig. 5.22(a).

(b) Wafer-level chip-scale package (WLCSP): The WLCSP is essentially a die-sized container with balls-like bumps that can be glued straight to a PCB, is actually a chip-scale package as shown in Fig. 5.22(b). Two primary variations exist: the first involves directly attaching balls to the die pad openings, which optimizes signal performance, minimizes inductance, and maximizes speed by taking the shortest path from the die circuit to the PCB. The other version, known as fan-in, uses a redistribution layer to route chip pad openings, such as those from the die's edges to balls that can be arranged in a regular pattern across the die. The WLCSP package is the same size as the die in its final configuration. Copper (Cu), aluminum (Al), or a mixture of copper and aluminum (AlCu) could be the redistribution layer. The die's reverse might be left uncovered, coated in metal, or covered in a layer of protection. For low-power, low-ball-count devices, when the tiny form factor is advantageous, this is the recommended approach.

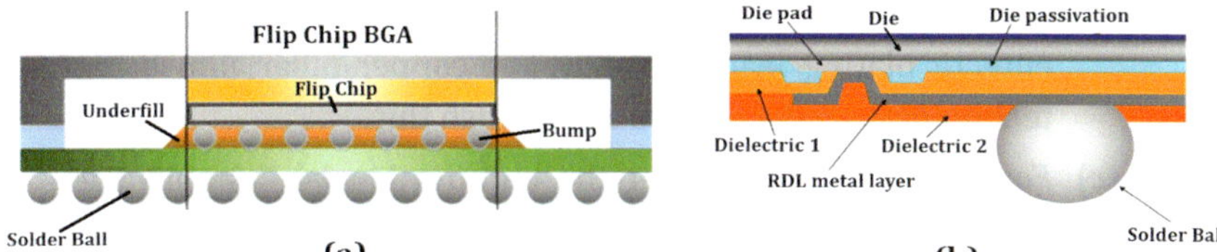

Figure 5.22 Packaged bumped dies types: (a) Flip-chip BGA; (b) wafer-level chip-scale package.

5.2.3.1 The redistribution layer (RDL)

In wafer-level packaging techniques, the redistribution layer (RDL) technology is essential for optimizing signal wiring and transmission pathways between bare dies and solder balls. RDL makes the product more flexible overall and increases the signal connectivity density by adding one or more layers of lateral connections to the original wafer. In addition, dielectric materials are used in RDL technology to preserve the RDL layer by serving as a stress buffer and to guarantee that the UBM technology, which supports the contact pad, solder ball, RDL, and dielectric, operates as intended. Additionally, fan-in WLCSP, fan-out wafer-level ball grid array (eWLB), integrated passive devices (IPD), and system-in-package (SiP) solutions are just a few of the advanced fan-out wafer-level technologies that RDL is a critical enabler for.

5.2.3.2 Flow of wafer-bumping process

The metal bumps on the wafer that are used to link the IC signal connections are created through a series of processes in the wafer-bumping process. Wafer bumping involves the following steps:

(a) *Preparing Wafers*: Wafer preparation involves cleaning and examining the wafer for flaws before beginning the wafer-bumping process. This guarantees that there are no impurities on the wafer that can interfere with the bumping process.

(b) *Deposition of Bump Material*: The next stage is to cover the wafer surface with the bump substance, which is usually

approach. The formal planning process and plan implementation will ensure that the package and product operate as intended for the duration of its expected lifetime in the end user's system, under the suggested application conditions.

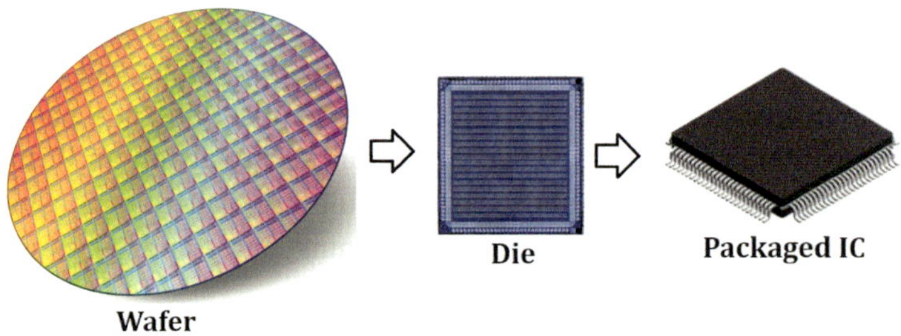

Figure 5.23 Wafer photo to IC package.

The first problem is most likely to be fully and thoroughly understanding the application and the system into which the component or module will be incorporated. Package design may begin after their comprehension. Just one of the elements that must be considered is the circuit's proper electrical performance within the recommended package. The mechanical aspects of the design need to be closely scrutinized to make sure the package will not crack under the specified living conditions.

Second, none of the substrates, components, or dies inside the package may break or lose their connection. Third, all solder, epoxy, and wire connections need to be strong enough to endure the application's anticipated temperature swings and mechanical stresses. After these elements have been thoroughly explored, the thermal properties of the package must be evaluated and simulated to effectively accommodate heat transmission to the system. Thermal management is probably one of the most crucial aspects of package design since it can cause catastrophic circuit overload and failure in out-of-control circumstances, fatigue failures over time, and a shorter product life.

Thermal interactions between the various materials used for and within the package may increase the mechanical stress of the entire package system, increasing the risk of failure.

After comprehensive mechanical and thermal modelling and analysis, the real package design may be ascertained. At that point, material and electrical dimensions and quality become the primary factors to consider. Understanding the package's circuit isolation and electromagnetic propagation channels in full is required. Impedance levels for input and output to and from the package must also be established and designed.

Developments in package design systems have allowed computer integrated manufacturing systems to generate package prototypes rapidly by tying the package's design to the machining machines that will create it. These technologies can produce prototypes in plastic for further investigation, or they can produce prototypes in metal to provide prototype samples.

5.2.4.1 Electrical

In contrast to on-chip signals, the current-carrying traces that emerge from the die, travel through the package, and enter the PCB have completely distinct electrical characteristics. They demand a lot more electrical power and unique design methods than signals that are limited to the chip itself. Low resistance, low capacitance, and low inductance are therefore crucial properties for materials used as electrical contacts. Prioritizing signal transmission qualities while reducing any parasitic elements that can adversely affect the signal is a need for both the construction and the materials.

As other aspects of technology accelerate, managing these traits becomes more and more crucial. A high-performance computer's delay can be almost entirely accounted for by packaging delays, and this speed barrier is predicted to worsen.

5.2.4.2 Mechanical and thermal

The integrated circuit package needs to be able to withstand physical damage, keep out moisture, and effectively dissipate heat from the chip. Furthermore, the package is frequently needed for RF applications to block electromagnetic interference, which could negatively impact adjacent circuits or impair the operation of the circuit. Lastly, the package needs to make it possible to connect the chip to a PCB. The packaging is made of ceramic, metal (usually Kovar), or plastic (thermoset or thermoplastic). Useful mechanical strength, moisture and heat resistance are provided by all three material kinds. However, because of their greater robustness (which also permits higher pin-count designs), heat dissipation, hermetic performance, or other factors, metallic and ceramic packaging are frequently chosen for higher-end electronics. Ceramic packaging typically costs more than comparable plastic packaging.

To improve heat transfer, some packages include metallic fins; however, these take up room. More interconnecting pins can also be accommodated in larger packages [9].

5.2.4.3 Economic

When choosing integrated circuit packaging, cost is a concern. Although a comparable ceramic package can dissipate up to 50 W in the same scenario, an affordable plastic package can typically dissipate heat up to 2 W, which is suitable for many simple applications [9]. The chips inside the container tend to get hotter as well as smaller and faster. Packing gets more expensive as the requirement for more efficient heat dissipation flows from it. In general, the cost of manufacturing a package increases with its required size and complexity. To cut expenses, wire bonding can be utilized in place of methods like flip-chip.

5.2.4.4 Verification

Lastly, design verification is required. The verification phase often includes numerous long-term reliability tests that provide the designer and the end user with confidence that the package and its contents will resist the predicted lifetime and application conditions. Exact design tolerance testing, material characteristics testing, and parameter tolerance testing are examples of more specialist testing.

A sizable amount of the final testing may also consist of system-level integration tests. Typically, packages are tested to specific power levels under specific environmental conditions after being fully integrated into the system and subjected to set power levels.

These are a few of the key elements that go into creating a package.

5.2.4.5 Common IC packaging types

Packages come in a wide variety, with variations in their sizes, mounting styles, and pin counts. Each pin has a distinct position and purpose, and all ICs are orientally polarized. This means a mechanism of identifying which pin is which must be included in the box. To identify which pin is the first pin, the majority of ICs will employ a dot or a notch. Moving counterclockwise around the chip, the pin numbers grow progressively after you know where the first pin is.

The method by which a package type mounts on a circuit board is one of its primary differentiators. There are two forms of mounting for packages: surface-mount (SMD or SMT) and through-hole (PTH). Typically, through-hole containers are larger and considerably simpler to handle. They are intended to be soldered to the other side of a board after being inserted through one side.

Surface-mount packaging comes in tiny-to-tiny sizes. All of them are intended to be soldered to the surface and positioned on one side of a circuit board. An SMD package's pins can occasionally be arranged in a matrix on the chip's bottom, or they can protrude from the side or perpendicular to the chip.

This form factor of ICs is not very "hand-assembly-friendly." Typically, they need specialized equipment to help with the process.

Several conventional package types have been important to the electronics industry in the realm of integrated circuit packaging. These consist of ball grid array, quad-flat package, small-outline package, and dual inline package. Every form of package has unique properties, benefits, and drawbacks and is used in a variety of electrical gadgets.

(a) Dual Inline Package (DIP)

The most typical through-hole integrated circuit package that you will come across is the DIP, or dual inline package. These tiny chips have a rectangular, black plastic shell with two parallel rows of pins that extend perpendicularly out of it. A breadboard's center can be crossed by DIP ICs thanks to the precise spacing between each row of pins. This ensures that the pins don't short to one another and gives each pin its own row on the board. DIP ICs can be soldered into PCBs in addition to being utilized in breadboards. On one side of the board, they are placed, and on the other, they are soldered into position. Placing the chip in a socket is sometimes a better idea than soldering directly to the integrated circuit. Sockets make it possible to easily remove and replace a DIP IC compared with permanently soldering.

One of the first and most popular types of integrated circuit packages is DIP as shown in Fig. 5.24. Its package body has two rows of parallel pins that expand, making it simple to install and solder into a PCB. The ease of use, affordability, and compatibility with through-hole mounting technology of DIP packages are well recognized. They are frequently utilized in automotive, industrial, and consumer electronics applications.

Figure 5.24 DIP ICs.

These days, there is an enormous range of surface-mount package types known as surface-mount (SMD/SMT) packages. Working with surface-mount packaged ICs typically requires a special PCB with a copper design that matches the ICs' soldering location. The hand-solderability of these more popular SMD package types ranges from "doable" to "doable, but only with special tools" to "doable only with very special, usually automated tools."

(b) Small-outline Packaging (SOP)

The surface-mount alternative to the DIP is the SOP, which is a small-outline integrated circuit (SOIC) and is shown in Fig. 5.25. It is the result of shrinking a DIP to the appropriate size and bending every pin outward. These packages are among the simplest SMD parts to hand solder provided you use a steady hand and a close eye. Each pin on an SOIC package is often separated from the others by approximately 0.05" (1.27 mm).

Figure 5.25 SOICs.

(c) Shrink Small-Outline Package (SSOP)

An even more compact form of SOIC containers is the SSOP shown in Fig. 5.26(a). The thin small-outline package (TSOP) and the thin-shrink small-outline package (TSSOP) shown in Fig. 5.26(b) are two further, comparable IC packages. Compared to DIP packages, SOP packages are more compact and smaller.

(a) (b)

Figure 5.26 (a) SSOP and (b) TSSOP IC packages.

They are compatible with surface-mount technology (SMT) and provide increased space efficiency. Direct surface attachment of SOP packages is common on the PCB. They are frequently found in computer peripherals, communication devices, and portable electronics.

(d) Quad-Flat Package (QFP) and Quad-Flat No-leads (QFN)

If you arrange IC pins in every direction, you will have what appears to be a QFP shown in Fig. 5.27(b). QFP ICs can have as few as eight pins on each side (for a total of 32) to as many as 70 pins (for a total of 300+). On a QFP IC, the pin spacing typically ranges from 0.4 mm to 1 mm. Low-profile (LQFP), very thin (VQFP), and thin (TQFP) packages are smaller versions of the conventional QFP package. An appearance akin to a QFN package shown in Fig. 5.27(a) could result from sanding the legs off a QFP IC. On QFN packages, the connections are small, visible pads on the IC's bottom corner edges. In certain packages, the pad is only visible on the bottom of the chip; in other cases, they wrap around and are visible on both the side and bottom.

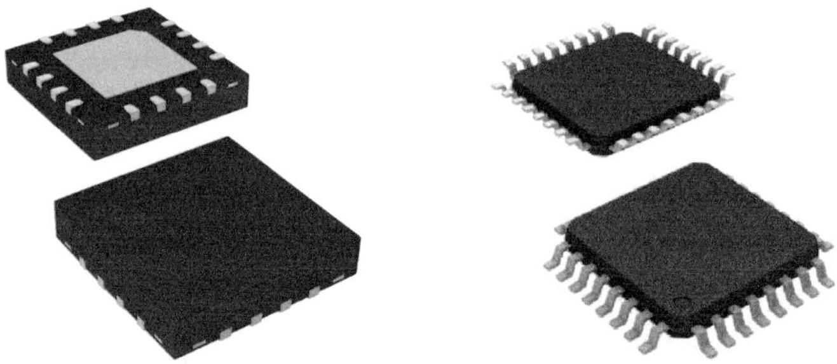

(a) (b)

Figure 5.27 (a) QFN and (b) QFP IC packages.

QFP packages have more pins than DIP and SOP packages because they have a square or rectangle body with leads coming out on all four sides. QFPs have a high integration density, superior heat dissipation, and outstanding electrical performance. High-performance electronics such as digital signal processors and microcontrollers find use for them.

The conventional QFN package has smaller variants called thin (TQFN), very thin (VQFN), and micro-lead (MLF) packages. Packages with pins on just two sides are called thin-dual no-lead (TDFN) and dual no-lead (DFN). Today's ICs, including many microprocessors and sensors, are packaged in QFP or QFN formats.

(e) Ball Grid Array (BGA) Packaging

Lastly, BGA packaging shown in Fig. 5.28 is for extremely sophisticated integrated circuits. On the bottom of the integrated circuit, tiny solder balls are stacked in a two-dimensional grid to create these incredibly complex small packages. There are instances when the solder balls are affixed straight to the die. Rather than using conventional pins or leads, BGA packages employ a range of tiny solder balls on the bottom surface. As a result, there are more pins, better electrical performance, and better thermal management. BGAs are frequently employed in devices like microprocessors, graphics cards, and networking that require a high pin density. Consider yourself a master solderer if you can solder an IC that is packaged in a BGA by hand. Usually, an automated process utilizing pick-and-place equipment and reflow furnaces is needed to attach these packages to a PCB.

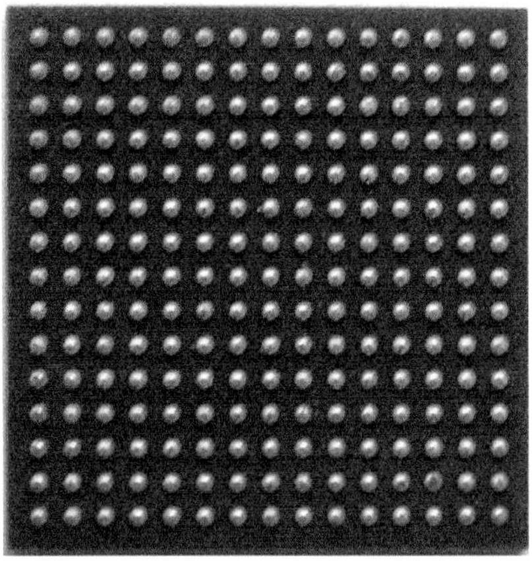

Figure 5.28 Bottom view of the BGA.

(f) Common ICs

It's difficult to discuss all the ways in which integrated circuits are used in electronics because they are so common. These are some of the more typical ICs that are used in educational electronics.

The building blocks of many more integrated circuits, such as shift registers, timers, and digital logic gates, can be assembled into separate integrated circuits. Like this quad-input AND gate, some logic gate ICs may have several gates packaged together. An IC can be used to connect logic gates to produce simple logic circuits such as shift registers, counters, latches, and timers. Along with SOIC and SSOP packaging, DIP packages contain many of these straightforward circuits.

Integrated circuits include microcontrollers, micro-processors, and FPGAs, which may fit hundreds, millions, or even billions of transistors onto a single tiny chip. These parts come in a variety of sizes, functions, and levels of complexity; an 8-bit microcontroller can be as complicated as a 64-bit multi-core microprocessor that manages computer activities.

These are often the biggest ICs in a circuit. Basic micro-controllers come with anything from eight to one hundred pins and are packaged in a variety of formats, from DIP to QFN/QFP. The package becomes similarly complex as the intricacy of these components increases. Complex microprocessors and FPGAs are only accessible in sophisticated packages such as QFN, land grid array (LGA), or BGA, and can have up to a thousand pins.

5.2.4.6 Advanced and emerging IC packaging types

The electronics industry has seen the emergence of innovative packaging technologies in addition to conventional IC package types to meet changing needs for increased integration density, enhanced performance, and downsizing. Now, let us examine a few of these cutting-edge integrated circuit package types: the

3D packaging, the chip-scale packages, the system-in-package, and the multi-chip modules.

(a) Chip-scale packages (CSP) shown in Fig. 5.29(a) are a type of miniaturized package in which the tiny form factor is achieved by closely matching the package size to the semiconductor die. Benefits of CSP include shorter communication pathways, better electrical performance, and a smaller environmental imprint. It finds use in wearables, mobile devices, and other electrical goods with limited space.

(b) A system-in-package (SiP) is a semiconductor device package that combines passive components and several ICs into one package as shown in Fig. 5.29(b). Because SiP allows different components to be close to one another, it provides increased functionality, lower power consumption, and better performance. SiP is frequently utilized in wireless communication systems, IoT devices, and smartphones.

(c) In order to achieve high levels of integration and performance, multi-chip modules (MCM) integrate numerous semiconductor chips or dies into a single package. MCM provides optimal power distribution, shorter connector lengths, and better system performance. It finds use in high-speed data processing, telecommunications, and sophisticated computing.

Using 3D packaging techniques, a structure with three dimensions is created by stacking many semiconductor dies vertically. Significant benefits can be obtained from this method in terms of lower form factor, better electrical performance, and higher integration density. In applications like high-performance computing and advanced memory systems, it makes it possible to integrate diverse components and integrate systems efficiently.

Fan-Out Wafer-Level Packaging (FOWLP), WLCSP, and chiplet-based packaging are some of the IC packaging trends that are just getting started.

Packages can be made smaller thanks to FOWLP, which distributes the input/output connections across the package's surface. WLCSP reduces package size and facilitates high-volume manufacturing by directly packing individual semiconductor dies on the wafer. Conversely, chiplet-based packaging combines several tiny semiconductor chips, or chiplets, into a single package to facilitate effective system assembly and design for highly tailored and optimized electronic systems. With their increased cost-effectiveness, increased performance, and superior integration, these trends have the potential to completely transform IC packaging.

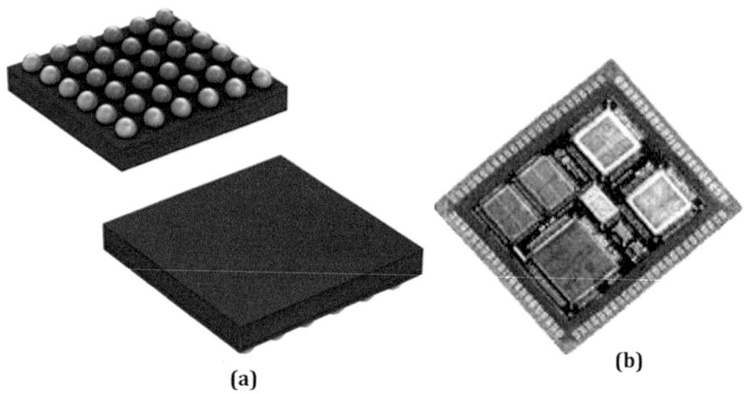

Figure 5.29 (a) CSP and (b) SiP IC packages.

Summary of IC packages

(a) Semiconductor components are shielded and made possible to be integrated into electrical devices using IC packaging. It protects the parts from the elements and makes it easier for them to connect and work together inside the gadget.

(b) IC packages in the common formats are DIP, SOP, QFP, and BGA. Every variety has unique qualities, benefits, constraints, and uses. QFP offers a large pin count, BGA offers better electrical performance and thermal control, SOP delivers space efficiency, and DIP is straightforward and affordable.

(c) SiP, MCM, CSP, and 3D packaging are examples of advanced packaging technologies. Miniaturization is possible with CSP, multi-component integration is possible with SiP, high integration and performance are made possible with MCM, and enhanced performance and density are achieved with 3D packaging.

(d) In the electronics sector, integrated circuit packaging plays a vital role as the protective and connecting link between fragile semiconductor components and the outside environment. It includes the steps involved in enclosing integrated circuits in a protective shell and offering heat control, electrical connections, and mechanical support. Comprehending the function and significance of integrated circuit packaging is imperative in order to appreciate its influence on electronic gadgets.

5.2.5 PCB Assembly

PCB design is becoming a separate and specialized area within the electronics industry. PCBs are essential because they provide electrical connections between electronic components, a compact package that can be included into a completed device, and rigid support for holding components. They are the main component of an electronic device that defines its design and operation and allow very high data rates of communication amongst current semiconductors. Copper cables are used in an electrical arrangement known as a PCB to connect components. PCBs also provide mechanical support for electronic components so that a device can be placed inside an enclosure.

A solid structure known as a printed circuit board houses electrical circuitry composed of bigger metal sections known as planes and embedded metal surfaces known as traces. The board's circuitry is connected to the metal pads that components are soldered on. This makes it possible to link components. Circuitry can be arranged in one, two, or more layers on a board. To ensure pure circuitry transmission, circuit boards are constructed with a dielectric core material that has low electrical conducting capabilities. Additional layers of metal and dielectric are inserted between the core material as needed. Copper is typically utilized for the metal lines and planes of the circuitry, whereas FR-4, a flame-resistant mixture of woven fiberglass fabric and epoxy resin, is the common dielectric material used for circuit boards.

Each PCB is made up of copper layers that are conductive and layers of electrically insulating material. Conductive structures on printed circuit boards include copper traces, pads, and conductive planes. The mechanical framework is made up of the insulating material laid in between the conductor layers. After the structure has been plated, a non-conductive solder mask is applied over it, and an electrical component legend is created by silk screen printing over the solder mask. The printed circuit board assembly (PCBA) receives the bare board after these fabrication steps, which is used to solder components to the board and test the PCBA.

In order to leave the appropriate copper traces for connecting the circuit board's components, the inner copper layers are removed during the manufacturing process. Several etched layers are laminated one after the other until the printed circuit board stack-up is complete. This is how PCBs are generally made: the bare board is made, and it is then assembled.

Before delving into PCB design, it is critical to understand PCB history.

In the past, electronics were designed and assembled using discrete parts and small ICs. The parts were connected together after being fastened to a rigid substrate. Originally, the top ply of a plywood sheet was replaced with this original substrate, which is a material called bakelite. The sheer number of wires could cause them to become tangled or take up a lot of space within a design. Dependability declined, and troubleshooting became difficult. It took a protracted production procedure to physically solder many components and their electrical connections.

Older PCBs typically had the configuration seen below , which includes exposed copper traces and many through-holes for large, bulky components.

These days, typical designs can include a very high pin count on sophisticated chips, tiny integrated circuits, and very small passive components, among many other tiny components. Since all these components cannot be manually joined using soldered wires, the manufacturing process described above is followed to deposit the copper connections directly onto insulating surfaces. Many of the modern technologies are powered by sophisticated high-density interconnect (HDI) designs, which underpin everything from cellphones to heart rate monitors to rockets. Several electrical connectors and thousands of connections are found in HDI designs.

5.2.5.1 PCB design procedure

Printed circuit boards go through several stages when a new design is started. With computer aided design (CAD) software, which has numerous tools specifically for circuit board design and layout, production-grade circuit boards are designed. CAD software is designed to guide designers through a specific circuit board design process, starting with the creation of basic electrical drawings and concluding with the preparation of manufacturing files. The basic steps in circuit board design are as follows:

(a) *Front-end development*: To develop the board's functioning, the main components are chosen at this point, and some basic circuit diagrams are usually created.

(b) Schematic capture is the process by which basic circuit diagrams are converted into electronic drawings that specify the electrical connections between components using CAD software. Design elements are represented by schematic symbols.

(c) *Material selection and PCB stack-up design*: In this step, the required plane layers, signal layers, dedicated routing channels, and particular material qualities are considered while choosing laminate materials for the stack-up.

(d) *Component placement*: Following the establishment of the board shape and the importation of components into a fresh PCB layout, the components are positioned within the layout in accordance with the mechanical specifications of the design.

(e) *Routing*: The next step is to route traces between components when the component placement has been approved. CAD software's routing functions are used to ensure impedance control, trace geometry may be chosen at this step (for high-speed signals).

(f) *Design review and verification*: It's usually a good idea to examine and assess the design after routing is finished to make sure there are no errors or unsolved issues. Both manual inspection and post-layout simulation tools can be used for this.

(g) *Getting Ready for Production*: After the design is finished, standard manufacturing files need to be created to get ready for production. Automated fabrication and assembly equipment uses these files.

5.2.5.2 PCB design structure and applications

The stack-up, or layer arrangement, of a printed circuit board defines many of its key performance attributes. Alternating layers of prepreg (two types of dielectrics utilized in the layer stack-up) and conducting and insulating material are used to build the layer stack-up. When designing for high-dependability applications, care should be taken in the selection of the prepreg and core components because their mechanical and dielectric characteristics will impact the design's reliability and signal/power integrity. For instance, highly dependable designs that may be used in challenging environments are needed for military and medical applications, while a telecom system PCB may need a small-sized, low-loss PTFE laminate.

Below is a description of a PCB stack-up. A four-layer structure with two internal plane layers (L2_GND for ground and L3_PWR for power) is implemented by the stack-up in this example. IoT devices, lightweight embedded systems, and numerous more designs utilizing high-speed protocols are suitable for this kind of stack-up. The internal plane configuration offers some protection against external EMI and aids in maintaining power integrity. Additionally, the internal plane layers offer a reliable reference for signals with regulated impedance. This kind of stack-up is common in many designs and is frequently where contemporary printed circuit boards begin.

Modern PCB design devices use HDI design techniques, which allow for incredibly small feature sizes on a PCB. These circuit boards are used in more complex systems that need a lot of connections and components, although they generally follow the same engineering procedure as any other circuit board.

Ball grid array packaging, along with related packaging types such as LGA, is the primary component packaging type that can drive the adoption of HDI design and production in a PCB build. BGA packaging is commonly used for popular microcontrollers, MPUs, and FPGAs because it can accommodate many I/O pins into a little component package size. BGA packages are the primary force behind device downsizing, with processors found as BGA packages in a wide range of consumer goods. For instance, the bottom side of an earlier CPU is depicted below, but the supporting parts on the IC substrate are visible from the bottom side.

The component packaging that holds the semiconductor chips in ICs represents a convergence point for the design and production of PCBs in the next generation. Advanced processor and digital/RF module chips are positioned on an IC substrate, which facilitates the connections between the semiconductor die's electrical contacts and the printed circuit board. The most sophisticated ICs are using feature sizes that are comparable to those of IC substrates, and UHDI PCBs and IC substrates are made using the same manufacturing techniques.

Many manufacturers are finding it difficult to produce UHDI PCBs because of the small feature sizes, and these capabilities are only currently available at scale. The manufacturing skills needed to make these boards are entirely additive; instead of using the subtractive etching techniques used to make traditional PCBs, metal deposition technologies are employed to construct the traces and vias. Additionally, these PCBs will power the most sophisticated systems running at the fastest data rates. The procedure for designing UHDI PCBs is quite similar to that of conventional PCBs, despite the significant differences in the production processes.

5.2.5.3 Types of PCBs

(a) *Single-sided*: Components on this board are mounted on a single surface, making it single-sided. Usually, the back surface is covered with a solder mask and is completely copper (ground).

(b) *Double-sided*: Components are installed on both surfaces of this kind of circuit board. Since each surface in the PCB stack-up is a signal layer, the surfaces will all have traces on them that transmit signals from one component to another.

(c) *Multi-layer PCBs*: These boards may have conductive plane layers on the interior layers, or they may have conductors on the internal layers that transfer electrical signals between components. Single-sided or double-sided multi-layer PCBs are both possible.

(d) *Rigid-flex PCBs*: In a PCB assembly, a flexible polyimide ribbon is used to join two or more rigid parts. If there is a design requirement for a movable feature, like a folding or bending enclosure, a rigid-flex board may be utilized.

(e) *Flex PCBs*: Made entirely of flexible polyimide ribbons, fully flexible PCBs are constructed without the use of any rigid materials. Similar to rigid and rigid-flex printed circuit boards, these boards can have components mounted and soldered on them.

(f) *Printed flex PCBs*: These PCBs have a flexible base material on which copper conductors are produced using an inkjet printer or a comparable additive method. The final boards have a striking resemblance to flex PCBs.

(g) Metal-core PCBs, also known as insulated metal substrate (IMS) PCBs, incorporate a metal slab—typically aluminum—in the core layer to give far more rigidity and heat dissipation than are found in conventional rigid printed circuit boards. A few design considerations are necessary to achieve solvability in the metal-core PCB design manufacturing process, which differs significantly from the conventional rigid PCB design manufacturing method. High-power lighting and certain industrial applications frequently use these boards.

(h) *Ceramic PCBs*: Less frequently found ceramic PCBs are utilized in situations where a high level of thermal conductivity is necessary for the board to be able to disperse a significant amount of heat away from components.

(i) *HDI PCBs*: To support an extremely high density of copper connections, these PCBs feature extremely high-pin-count components that call for a specific manufacturing procedure and materials.

(j) *UHDI and substrate-like PCBs*: These PCBs are so dense and compact that they can't be built using subtractive etching; instead, they need to be manufactured using a specific type of additive manufacturing technique that creates IC packages.

These above-mentioned PCBs have different manufacturing and assembly procedures, but as long as the appropriate PCB design guidelines are followed by the software program, designers can produce any of these boards with the use of modern CAD software.

The circuit board shown above is an older model that primarily makes use of through-hole components to achieve the necessary functionality. Due to their greater utility in high-density, double-layer designs, surface-mount (SMD) components have become the predominant choice for modern printed circuit boards. Nowadays, SMD components are the industry standard, utilized in the majority of applications that need low power, low cost, and compact form factor. Through-hole components are still used in some applications, though, because they are more dependable and simpler to assemble—even by hand. An illustration of a contemporary PCB with high-density SMD components may be seen in Fig. 5.30.

Figure 5.30 High-density SMD components on a portion of PCB.

5.3 Antennas

An overview of modern antenna types used in radar, satellite communication, cellular, communication lines, and other microwave and millimeter-wave systems is provided in this chapter. The microwave is assumed to encompass the frequency range of 800 MHz to 94 GHz. The mathematical description of non-ionizing electrodynamic wave propagation and radiation using Maxwell's equations of electromagnetic theory and the wave equation is discussed.

5.3.1 Fundamental Antenna Parameter Definitions [13–15]

In essence, an antenna is a device that converts directed wave energy into energy that is released into space. Antenna physics displays both spatial and frequency (time) dependencies concurrently, in contrast to RF/microwave transmission lines. In his outstanding phenomenological description of an antenna device, Schantz characterizes the functionality of antenna from several angles as follows:

(a) Electromagnetic (EM) radiation

The release of electromagnetic waves as energy from a device is called as EM radiation. The antenna functions as dual conversion transducer such that it can convert the EM radiation to electrical signals in the receiver and from electrical to EM radiation in the transmitter.

(b) Radiation pattern

An explanation, either mathematical or pictorial, of how an antenna's radiation characteristics change with space coordinates is called as radiation pattern. Usually, one uses the usual (r, θ, φ) spherical coordinate system, as Fig. 5.31 illustrates. The next paragraphs will cover each of the radiation properties: gain, directivity, radiated phase, polarization parameters, radiation density (Watt/m^2), radiation intensity (Watt/solid angle), and so on. Since radiated power is easily measured with modern antenna measuring equipment, it is frequently utilized at microwave frequencies instead of electric field. An expression of fluctuation in transmitted or received power density quantified at a constant radius around the antenna is, for instance, a power radiation pattern. A common graphic representation of power patterns is normalized to the primary beam's peak.

The two-dimensional cross sections of the three-dimensional power radiation pattern are called radiation pattern cuts. As illustrated in Fig. 5.32, a pattern cut in the direction of the phi unit vector is called a phi cut, whereas a cut in the theta unit vector's plane is called a theta cut. As shown in Fig. 5.33, the major plane cuts are two-dimensional, orthogonal antenna cuts made in the power radiation patterns' E and H field planes.

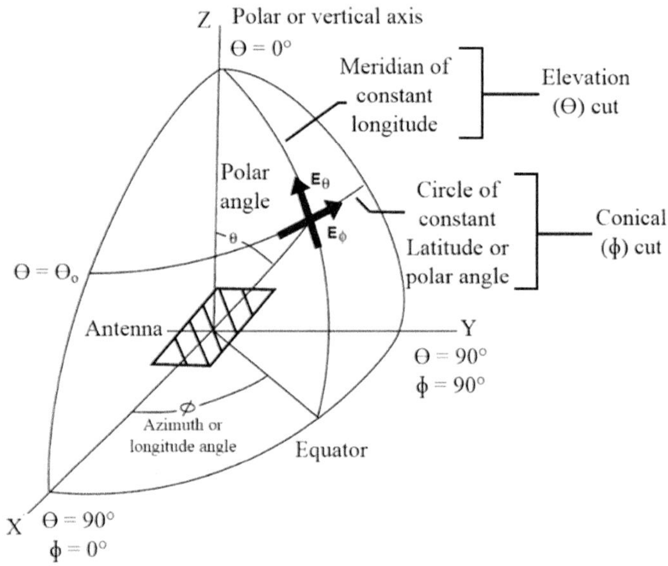

Figure 5.31 Radiation pattern in the coordinate system.

(c) Isotropic Radiator

An imaginary lossless antenna with uniform radiation in every direction. An isotropic radiator radiates in three dimensions as a sphere.

A circular aperture intended for a −25 dB side lobe level is assumed to have a 22.5° phase error in the far-field formulation, resulting in an inaccuracy of ± 1.0 dB at the 25 dB side lobe level [13]. Lower side lobe levels are subject to a stricter far-field criterion [15]. For radar applications and standard communication systems, antennas function in the far field.

(f) Directivity

An antenna's directivity can be defined as the ratio of the radiation intensity in a particular direction to the antenna's average radiation intensity across all directions. Usually, this ratio is stated in dB. Most of the time, the highest directivity is implied when the term directivity is used. An antenna's directivity is a function of its spatial coordinates and describes its capacity to focus power density (transmitting) or to accept an incoming wave's power density preferentially. Energy conservation prevents a passive, lossless antenna from amplifying its input signal; instead, it redistributes input energy according to spatial coordinates.

Think about a highly directed antenna, like a parabolic reflector, and an isotropic radiator. Assume that all the antennas have the same power at their input terminal, are lossless, and have matched impedance and polarization. The power density as a function of spatial coordinates is significantly different for the matched isotropic radiator, even though it couples the same total input power into free space as the parabolic reflector. The parabola is proportionally more "sensitive" in this area because it concentrates much of the input power into a very small spatial sector. On the other hand, the isotropic radiator does not have directional sensitivity and evenly distributes the input power in all directions. Figure 5.34 provides illustrations of these ideas.

In an antenna, resistive loss mechanisms, polarization mismatch, and impedance mismatch variables are not taken into consideration by directivity. As was previously said, a high directivity rating indicates that a significant portion of the antenna's input power is concentrated in a narrow angular area. For instance, directivities of parabolic reflector antennas used in radio astronomy might be higher than 80 dB. In comparison, an isotropic radiator radiates equally in all directions, hence its directivity is 0 dB (1.0 numerical). The theoretical directivity of a half-wave resonant dipole radiating in free space is 2.14 dB.

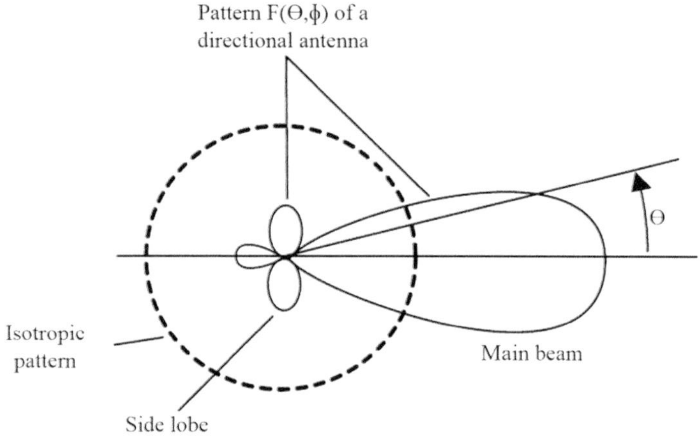

Figure 5.34 Evaluation of directional pattern cut view against the 2-D isotropic view.

(g) Half-power beam width (HPBW)

As shown in Fig. 5.32, half-power beam width is the angle formed by two directions in the greatest lobe of the power radiation pattern where the directivity is half that of the peak directivity, or 3 dB down. An antenna's half-power beamwidth is a gauge of its spatial selectivity. Over a limited angular range, low beam width antennas exhibit high sensitivity. Directivity and half-power beamwidth are negatively correlated for antennas where much of the radiation is contained inside the primary beam.

(h) Power gain

The relationship between the total input power accepted by the antenna input port and the radiation intensity in a particular direction is known as the antenna's gain. Decibels above an isotropic radiator are typically used to indicate this ratio (dBi for linear polarization and dBic for circular polarization).

Through the radiation efficiency parameter, gain and directivity are connected. The internal insertion dissipative power losses of the antenna are measured by radiation efficiency:

$$G = \eta_{\mathrm{cd}} \times D, \tag{5.3}$$

where D is the directivity of the antenna, G is the power gain, and η_{cd} is the antenna efficiency owing to conductor and dielectric dissipative losses.

Peak power gain is typically implied when discussing power gain. Polarization and impedance mismatch effects are not included in the gain parameter, but dissipative losses are.

(i) Antenna Polarization

The electric field polarization characteristics of the propagating wave that the antenna receives are referred to as the antenna's polarization. The contour that the radiated electric field vector follows as the wave moves across space is described by wave polarization. Elliptical wave polarization is the most prevalent type. Elliptical polarization has particular situations, such as circular and linear polarizations.

Figure 5.35 shows examples of linear polarized, circular, and elliptical propagating electric field vectors. In addition, the terms "vertical" and "horizontal" polarizations are occasionally employed to allude to the direction of the linearly polarized electric field vector at the antenna plane.

The polarization ellipse, or planar projection of the contour that the wave's electric field vector sweeps out as it travels through space, characterizes an elliptically polarized wave. The tilt angle indicates the polarization ellipse's direction, and the axial ratio is the ellipse's main diameter divided by its minor diameter. As the wave propagates from a reference point, the direction of rotation of the radiated field can be described as either left-handed or right-handed. In modern microwave systems, left-hand circular polarization (LHCP) and right-hand circular polarization (RHCP) are frequently employed. For most applications, it is preferable if the polarization characteristics of the transmitting and receiving antennas are same.

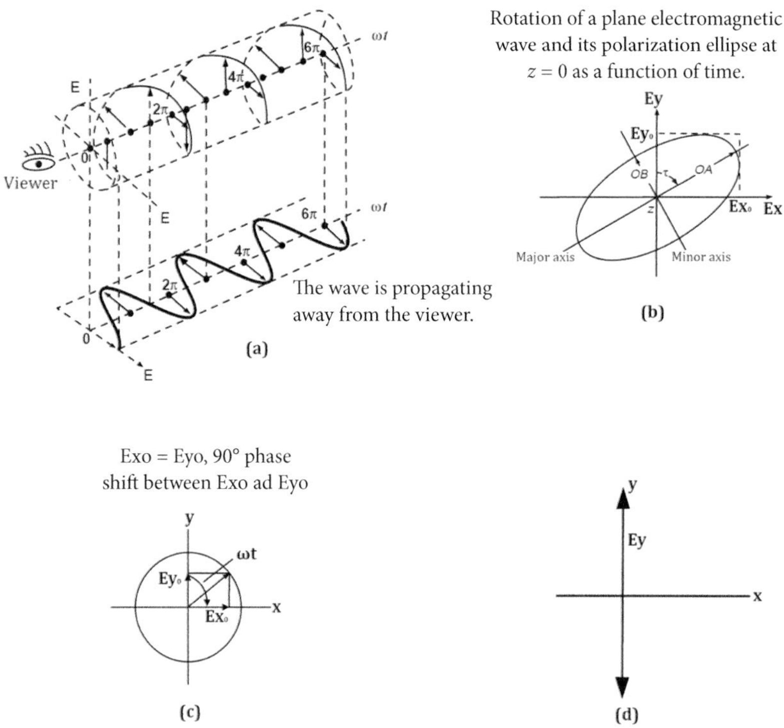

Figure 5.35 Polarization of the antenna (a) elliptically polarized wave, (b) polarization ellipse, (c) circular polarized wave, and (d) linear polarization.

An antenna's power reception will be less than maximum if the wave it is trying to receive is polarized differently than the antenna. This effect is measured by a polarization mismatch loss, which is described by

$$\mathrm{PML} = \left|\rho_r \cdot \rho_t\right|^2 = \left|\cos\psi_p\right|^2, \tag{5.4}$$

The receive polarization unit vector is denoted by ρ_r, the transmit polarization unit vector by ρ_t, and the angle between the two-unit vectors is represented by Ψp.

This formulation works with circular, linear, and elliptical polarizations. Figure 5.36 shows a few polarization mismatch losses that are frequently observed. It shows that there is no interaction between an LHCP wave and a pure RHCP wave. Systems that use polarization diversity, like cellular radio and satellite communications, take advantage of this feature. The ellipticity of a polarized wave is measured by the axial ratio, which is commonly represented in dB. Pure circular polarization is defined by an axial ratio of zero decibels. As a rule of thumb, the greatest axial ratio at which an elliptical wave is deemed circular is six decibels. The wave's sensation is not described by the axial ratio. An antenna's axial ratio is a function of spatial coordinates in three dimensions.

Wave polarization (rows) / Wave polarization (columns) — Polarization Loss	Vertical	Horizontal	Right hand circular	Left hand circular	45° Right linear	45° Left linear
Vertical	0 dB	o/c	3 dB	3 dB	3 dB	3 dB
Horizontal	o/c	0 dB	3 dB	3 dB	3 dB	3 dB
Right hand circular	3 dB	3 dB	0 dB	o/c	3 dB	3 dB
Left hand circular	3 dB	3 dB	o/c	0 dB	3 dB	3 dB
45° Right linear	3 dB	3 dB	3 dB	3 dB	0 dB	o/c
45° Left linear	3 dB	3 dB	3 dB	3 dB	o/c	0 dB

Figure 5.36 Summary of wave polarization between receive and transmit antenna and the polarization loss (with direction of polarization is into the page).

One may theoretically separate an arbitrary elliptically polarized wave into a RHCP component and an LHCP component. Imagine you have two circular antennas, one in the transmit mode and the other in the receiving mode, with opposite polarization senses. The orthogonal (unwanted) polarization components of each antenna grow more significant as they diverge from pure circular polarization toward elliptical polarization, which ultimately lessens the isolation between the antennas. Similar to this, an increase in loss results from a departure from pure circular polarizations between transmitting and receiving antennas of the same polarization sense. Given the axial ratio and polarization sense of each antenna, Fig. 5.36 quantifies the lowest polarization isolation [13].

Losses within the structure and at the antenna's input terminal are taken into account by radiation efficiency. Efficiency resulting from dielectric and conductor losses is described as

$$\eta_{\mathrm{cd}} = \eta_c \times \eta_d, \tag{5.5}$$

where η_d represents the efficiency due to dielectric loss and η_c represents the efficiency due to conduction loss.

The conductivity of the metal being used determines the amount of conduction loss. Anisotropic dielectric loss can occur, and it is commonly expressed in terms of the loss tangent parameter. The literature has tabulations for both parameters [15]. It is also possible to determine conduction and dielectric loss efficiency through experimentation.

(j) Input Impedance

The impedance at the input terminal reference plane is the antenna's input impedance. The ratio of voltage to current at the terminal plane, which is correlated with the ratio of the magnetic and electric fields there, is known as impedance. You can think of an antenna as an electromagnetic wave mode translator. Unbounded (radiation) electromagnetic waves are created when guided waves move from the input generator to the antenna input terminal via the input transmission line. As a result, the input port of an antenna that radiates into space can be compared to an impedance that has real and imaginary components that change with frequency.

In order to start an average power flow out of the antenna, real power from the input generator must be transferred into the radiated electrical and magnetic fields through the real or resistive component of the input impedance. The radiation resistance is the name given to this resistive element. Reactance at the antenna input terminal is a manifestation of the reactive near-field region, which is directly adjacent to the antenna structure. To convert input generator power into radiated waves as efficiently as possible, matching impedance conditions over the desired frequency range are necessary. The input impedance's real and imaginary components can both change with frequency, which makes the impedance-matching issue more challenging. To guarantee the best possible energy transfer between the input transmission line and the antenna's input terminal, standard transmission line and broadband circuit matching techniques can be used [13].

(k) Impedance mismatch loss

The quantity of power from the input generator that flows through the antenna's input terminals is measured as impedance mismatch loss. The quantity of power that is reflected back into the generator from the antenna input is measured by the input reflection coefficient. Mismatch loss results from guided electromagnetic wave reflections back to the input generator rather than from dissipative or resistive losses.

The definition of impedance mismatch loss is given by

$$\mathrm{MML} = (1 - |\Gamma|^2), \qquad (5.6)$$

where Γ = the input terminal voltage reflection coefficient of the antenna.

Typically, the voltage reflection coefficient is expressed and quantified as either the voltage standing wave ratio (VSWR) or the return loss:

$$\text{Return loss} = 10\log(|\Gamma|^2) \tag{5.7}$$

$$\text{VSWR} = \frac{1+|\Gamma|}{1-|\Gamma|} \tag{5.8}$$

(l) Overall antenna efficiency

The polarization loss, impedance mismatch factors, and conductor/dielectric losses that were previously discussed metrics have little bearing on an antenna's overall efficiency. Other efficiency metrics that are unique to particular classes of antennas also exist. The following terms are included:

- Spill over efficiencies (reflector and lens antennas)
- Illumination efficiencies
- Efficiency of blocking (reflector antennas)
- Aperture taper efficiencies (arrays, lenses, apertures, and reflector antennas) as well as random phase and amplitude errors

(m) Effective Area

It is the product of the power density of a plane wave incident on the antenna from a given direction and the available power at the antenna input terminals because of the plane wave's polarization matching. The ability of an antenna to capture power under plane wave illumination is measured by its effective area. An aperture antenna's aperture efficiency determines how closely its effective area and physical area are related, as demonstrated below:

$$\eta_{\text{ap}} = \frac{A_{\text{em}}}{A_p} = \frac{\text{Maximum effective area}}{\text{Physical area}} \tag{5.9}$$

$$A_{\text{em}} = \frac{\lambda^2}{4\pi} \times D_0 \eta_{\text{cd}} (1 - |\Gamma^2|) \times |\rho_r \cdot \rho_t|^2 \tag{5.10}$$

(a) Wire Antennas

Arguably the most popular kind of antenna, wire antennas are widely employed in modern microwave systems. The traditional microwave frequency bands see widespread use of the dipole, monopole, and whip antennas. The electrically thick monopole and dipole, biconical dipole, bow tie, coaxial dipole and monopole, folded dipole, discone, and conical skirt monopole are examples of broadband versions of the traditional wire antenna. Foreshortening the antenna's electrical length and compensating with lumped impedance loading, top-hat capacitive loading, dielectric material loading, and helical winding of the monopole element to retain electrical length but with foreshortened effective height are some of the methods used to reduce the size of this type of antenna for a given resonant frequency. In [15] a great discussion on electrically tiny antennas is included.

(b) Loop Antennas

Loop antennas are used as field probes at the L band and higher and are widely utilized in the HF through the UHF band. The most popular shape is the circular loop; however, other geometric shapes are sometimes utilized, including the rectangular one. Loops are categorized as electrically large when the loop circumference is roughly one wavelength and electrically tiny when the overall wire length (circumference multiplied by the number of turns) is less than 0.1 wavelength. Despite having a low radiation efficiency, electrically tiny antennas are nevertheless utilized as field probes for measuring the strength of electromagnetic fields and in pagers and portable radio receivers. A multi-turn loop antenna filled with ferrite material is the common AM portable radio antenna. Large electrical loops are frequently employed as radiating elements in array antennas. In this instance, the radiation pattern is end-fire in the direction of the loop's axis.

(c) Slot Antennas

Slot antennas are widely employed in missiles, aircraft, and other applications where robustness and a low profile are necessary. Slot antennas are fed by passing an excitation electric field across their gap; they are typically half-wave resonant. There are several applications for cavity-backed slots in the UHF frequencies. The radiation pattern of the extremely low-profile annular slot antenna is similar to that of a monopole wire. It is feasible to achieve circular polarization when fed in phase quadrature with crossed slots. Phased array applications in the microwave frequencies often use stripline-fed slots.

Microwave and millimeter-wave bands are two areas where waveguide-fed slotted array antennas are widely used. Two-dimensional planar-phased arrays are created by joining several waveguide linear arrays ("sticks"), which are made of resonant slots machined along the length of conventional waveguide transmission lines. Now, the edge slot and the longitudinal broad wall slot are the most often used waveguide slots. Building high-efficiency, regulated side lobe level arrays at a reasonable cost is possible with this sort of array. By adding phase shifters to the feed manifold that activated each linear waveguide array stuck inside the two-dimensional aperture, one-dimensional electrically scanned phased arrays can be achieved.

(d) Helical Antennas

In the UHF to microwave bands, the helical antenna has been widely used for both single radiating elements and phased array antenna elements. Balanis [15] has a great discussion on helical antennas. The helical antenna may function in both the conventional mode and the axial (end-fire) mode. A directional, circularly polarized pattern with a 2:1 frequency bandwidth can be produced by the axial mode.

With very good axial ratio performance, the normal mode helix has found significant application in broad-beam, cardioid-shaped radiation patterns. Kilgus created quadrafilar and bifilar helical antennas for use in receiving satellite signals [13]. Energizing every quadrafilar element in the helix in phase quadrature produces circular polarization.

In addition, helical antennas are utilized as feeds for parabolic reflector antennas and as radiating elements for phased arrays.

(e) Yagi–Uda Array

A popular directional antenna for the VHF and UHF bands is the Yagi–Uda array. A complicated mutual coupling (surface wave) mechanism involving the driven radiation element, a reflector element, and one or more directing elements allows for the realization of end-fire radiation. Although printed dipole elements and loops are also utilized, dipole elements are used more frequently. A thorough, step-by-step design process for this kind of antenna is provided by Balanis [15]. Computer simulations of the method-of-moment are frequently used to validate and improve the design.

(f) Frequency-Independent Antennas

When the physical dimensions of traditional antenna radiator types, such the wire dipole antenna, are a specific fraction of the operational wavelength, they radiate efficiently. For instance, the half-wave dipole can achieve respectable performance for finite bandwidths and is strictly resonant at only one frequency. The designs that form the basis of the frequency-independent antenna are expressed in terms of geometric angles.

Rumsey has demonstrated that an antenna might hypothetically function over an unlimited bandwidth if its shape could be entirely determined by angles [14]. In actuality, the physical truncation of the antenna construction limits the lower operational frequency, while the antenna feed structure determines the higher operating frequency limit. This idea is utilized by the cavity-back planar spiral, conical spiral, planar spiral, and planar slot spiral antennas.

The log-periodic antenna is a closely related and highly practical antenna. The electrical periodicity of this kind of antenna is a logarithm of frequency. Although the antenna's form cannot be entirely determined by angles, it is nevertheless incredibly broadband even though it is not fully frequency independent. The log-periodic dipole array is the most widely used architecture. For building this antenna, Balanis has recorded a thorough design process [13].

The electrically active portion of the antenna is a portion of the whole antenna structure, and it moves to different parts of the antenna depending on frequency [15]. This is an important characteristic of frequency independent antennas.

(g) Aperture Antennas

In the millimeter-wave and microwave frequency ranges, horn antennas are the most widely utilized aperture antennas. The literature contains a wealth of information about horn antenna design. The fan beam radiation produced by the E plane and H plane sectoral horns has a low beamwidth in one primary plane and a broad beamwidth in the principal plane that is orthogonal to it. A narrow beam is provided in both primary planes by the pyramidal horn. Sector and pyramidal horns are usually linearly polarized and fed via rectangular waveguides. A broadband variant of the pyramidal horn is the ridged horn. Broadband, circularly polarized radiation patterns can be produced by combining an orthomode transducer (OMT) with a dual ridge square aperture horn.

Circular waveguides are usually used to feed conical horns. The corrugated horn is a significant variant on the basic conical horn and is commonly used as a feed for electrically large reflector antenna systems with high efficiency. In order to prevent spurious radiation from entering the back and side lobe regions, corrugations are utilized to suppress field diffractions of the aperture plane's edge. By blending the aperture edge to create a gentle transition into free space, the aperture-matched horn also reduced diffraction from the aperture edge. Dual-mode conical horns reduce cross-polarization, decrease beamwidth, and control orthogonal beamwidths by superposing waveguide modes in the throat region. When a mode-generating/combining feed structure is utilized to generate a sum and difference beam, multi-mode horns can be employed in monopulse radar applications.

(h) Reflector Antennas

When very high gain, great radiation efficiency, and narrow beamwidths are needed, reflector antennas are widely utilized. Radio astronomy, satellite communication systems, deep space communication systems, terrestrial communications systems, and navigation systems are a few examples of applications. Applications in radio astronomy have shown gains of 80 dBi. In prime focus reflector systems, the feed antenna, usually a paraboloid, is located at the reflector's focal point. The obstruction of the feed and its mounting struts produce is a drawback of this method. In some cases, the performance of multiple reflector systems is better than that of prime focus designs.

A multiple reflector system offers a number of benefits, such as: (a) the ability to mount the feed behind the main reflector, eliminating the need for an interconnecting transmission line between the feed and the receiver front end; (b) a reduction in side lobe and spill over radiation; (c) the ability to achieve a large focal length with a shorter physical length; and (d) the ability to achieve beam shaping by sculpting the surfaces of the main reflector and sub-reflector. Offset and Cassegrain sub-reflector systems are widely used. The offset reflector offers the benefits of avoiding cross-polarization and greatly lowering feed obstruction. The planar reflector, the microstrip reflectarray, the use of phased arrays and clustered feeds as reflector feeds to create limited scan arrays, and other reflector concepts are among them.

(i) Microstrip Antennas

The benefits of microstrip antennas are low profile, light weight, and ease of production with modern printed circuit board materials and fabrication methods. The cellular phone satellite and wireless communication networks and the Global Positioning System have all made substantial use of this technology. Narrowbandwidth and substantial loss at higher microwave and millimeter-wave frequencies are generally their drawbacks.

Single feeds are used to achieve linear polarized designs. Microstrip transmission line feeds are placed on the same layer as the radiating element, whereas coaxial probe feedings are used alternatively. Due to its single-layer printed circuit board architecture, the microstrip transmission line feed has the benefit of being simple to fabricate; however, spurious radiation from the feed lines can be problematic, especially in array applications. Though feed probe inductance needs to be considered, the coaxial probe feeding approach functions effectively. Although aperture coupling provides a wider range of band excitation, the requirement for multi-layer printed circuit boards raises the difficulty of production.

Rectangular patch geometry allows for the generation of circular polarization from a single feed. The resulting radiation is a highly narrowband and circularly polarized. When dual feeds are provided in phase quadrature, square and circular patches can produce circular polarization. The resulting circular polarization signal has a bandwidth of more than 5% at a 6.0 dB axial ratio, making it substantially broader.

Both the transmission line model and the more precise cavity model allow for first-order easy patch design [13]. For thorough design, the finite element method (FEM), the method-of-moment (MOM), and the finite difference time domain (FDTD) approaches are employed.

Considerable work is still being done in the field of bandwidth optimization. The most recent cutting-edge microstrip-radiating element broad banding methods are reviewed by Kumar and Ray [13]. Patch shape optimization, substrate parameter optimization (lower dielectric constant and increasing thickness), multiple resonator planar configurations, stacked patch multiple resonator configurations, parasitically coupled top-hat configurations, aperture and electromagnetically coupled feed configurations, slot-loading techniques, travelling wave circularly polarized patches, quadrature-fed patch for circular polarization, broadband impedance-matching techniques, tunable ferrite material substrates, multi-mode excitation, and linear and planar log-periodic array embodiments are some examples of broadband techniques [13].

Modern phased array systems frequently incorporate microstrip-radiating elements. A big dielectric slab can support a surface wave mode of guided wave propagation, which deteriorates the array radiation and causes scan blindness, as was covered in the section on phased array antennas. The phased array design challenge gets more complicated when there is mutual coupling. For in-depth discussions, the reader is directed to the literature.

(j) Millimeter-Wave Antennas

The reflector, lens, and horn antennas are now the most popular millimeter-wave antennas in use. Apart from these configurations, a few lower frequency antennas, such as slot, dipole, biconical dipole, and monopoles, can be modified to operate at millimeter-wave frequencies. Up to 94 GHz, slotted waveguide arrays have been demonstrated. Increased slot diameters and loosened mechanical tolerance have rekindled interest in the Purcell slot array, making it appealing in the millimeter-wave regime [13].

Up to 100 GHz, printed microstrip patch arrays on substrates with low permittivity have been demonstrated on millimeter-wave frequency bands, feed loss on typical microstrip lines is particularly problematic. The feeding of a microstrip-radiating element with the fringing fields connected to a dielectric image guide and the E field probe connection of the microstrip patches to a feed waveguide are two techniques that have been researched to minimize feed loss. A particular focus on assembly methods and mechanical tolerances is required for implementing these structures at the millimeter-wave scale. Given that most millimeter-wave transmission lines have open constructions, open dielectric structure antennas are appealing. These constructions contain strategically positioned discontinuities that cause a surface wave phenomenon that in turn causes radiation. The periodic dielectric antenna and the tapered dielectric rod are two examples [13].

The long slot in a waveguide, the regularly loaded waveguide, and the trough waveguide structures are examples of leaky wave structures. When an open or closed waveguide experiences continuous or periodic perturbation, leaky waves are set off. Typically, the leakage per waveguide length is kept low to reduce the antenna feed's impedance mismatch. The purpose of the electrically long antennas is to allow most of the energy to radiate out before the guided wave reaches the antenna's end. This kind of antenna allows beam scanning at the input frequency.

The periodically loaded waveguide antenna starts the surface wave mode with a periodic surface perturbation, typically a metal grating or a corrugation.

(k) Dielectric Resonator Antennas

Dielectric resonator antennas, or DRAs, are becoming more and more common because of their small size, more design flexibility, and low mutual coupling in an array setting. An organized review of this technique is given by Luk and Leung [13]. Similar to a microstrip patch, a DRA is simply a resonant structure intended to emit electromagnetic radiation. The resonance and bandwidth of the device are mostly determined by the resonator's size (dimensions) and the characteristics of the material used. Rectangular and body of revolution (BOR) configurations are common shapes. Another way to think of hemispherical DRA antennas is as material-loaded monopoles. DRAs have mode-dependent radiation properties; multi-port feeds can be used to stimulate multiple modes. Different mode excitations enable the realization of both cardioid and monopole radiation patterns. Instantaneous bandwidths up to 25% have been proven and both linear and circular polarizations are feasible; however, the standard trade-offs between dielectric loading, operating frequency, size, and bandwidth still hold true. Aperture-coupled microstrip and coaxial probe are common feed setups.

Since lens antennas are also used to collimate spherical waves into plane waves, they are comparable to reflector antennas in many ways. In general, reflector antennas are more adaptable than lens antennas. The following phenomena affect the lens technology: (a) bulkiness and weight; (b) need for edge support; and (c) surface loss from internal and external reflections, spillover loss, lens zone obstruction, dielectric material loss, and tangent losses. On the other hand, lenses work better in specific situations, including wide-angle scanning [13].

There are significant optical similarities with lenses. The lenses may become quite heavy and thick if there is a significant subtended feed angle between the feed and the lens extremities. Bulk can be decreased by arranging steps in the lens to form distinct zones. For a thorough explanation, the reader is directed to Elliot [13]. Additionally, parallel metal plates, waveguide technologies, and artificial dielectrics can be used to create lenses [13].

A type of lens known as Fresnel zone plate antennas uses diffraction rather than refraction to collimate spherical electromagnetic waves that are impinging on it [13]. Compared to traditional lens technology, the zone plate architecture has several benefits, such as a simpler, lighter, and planar construction with less material-induced insertion loss. In [13], Wiltse provides an overview of the development of Fresnel zone plate antennas. A Fresnel Zone plate lens with improved focusing capabilities is described by Hristov and Herben [15]. In [13], dual-band Fresnel zone plate antennas are covered. An integrated circuit Fresnel zone plate antenna operating at 230 GHz was described by Gouker and Smith [13].

(l) The Fractal Antenna

Benoit Mandelbrot and others popularized fractal geometric curves, which served as the basis for the physical forms of fractal antenna radiating elements [13]. Werner [13] provides a summary of fractal electromagnetic research. Fractal curves can be classified as either random or deterministic. A recurring motif or pattern is applied to progressively larger size scales in deterministic fractals. In mathematics, fractals are expressed in terms of "*n*" iterations of the theme, where *n* is an infinite number. Fractal structures are, in practice, based on a finite number of repetitions. The basic idea behind a fractal-based radiating element is to reduce the element's physical size for a particular frequency range by introducing fractal-based impedance loading into the radiating element construction.

Instead of low VSWR throughout a continuous bandwidth, standard fractal antennas typically show non-harmonically related numerous sub-band resonances (segregated sub-band regions with lower VSWR). For genuine wide-band operation, fractal antennas usually require additional impedance-matching networks. Monofilar helical antennas and fractal loop arrays have been demonstrated by Cohen [13]. Over a limited bandwidth, the real part of the impedance of an electrically tiny loop antenna has been successfully raised using fractal approaches [14]. The use of printed fractal technology in multi-band radio and television service antennas is discussed by Breden and Langley [13]. Sindduo and Sourdois [15] discuss printed fractal branching antennas that are both wide-band and multi-band.

The Chu-Wheeler electrically tiny antenna principles [13] promote the use of fractal antennas as one method of fulfilling the available space (volume) for an electrically small antenna in order to optimize bandwidth. It has been demonstrated by Best that fractal antenna geometries are not much better than alternative space-filling antenna geometries, such as spherical monopoles, helical monopoles, structures generated by evolutionary algorithms, and randomly designed three-dimensional wire radiators [13]. In [13] and [15], it is explained how fractal antennas relate to log-periodic antennas, frequency-independent antennas, and genetic algorithm-based wire antenna design.

(m) Frequency Selective Surfaces

In essence, frequency selective surfaces (FSS) structures are plane wave "filters" having band-pass or band-stop characteristics that rely on both frequency and angular parameters. They find utility in wide-band phased arrays, multi-band applications, and situations where it is preferable to obstruct one frequency band's signal transmission while permitting another [13]. Another FSS use is dichroic reflector technology [13].

(n) The Genetic Algorithm

Genetic Algorithm-Based Antenna Design: Because of the unusual class of antennas that arise from these design techniques, genetic algorithm-based optimization techniques are briefly discussed here even though they technically belong to the computational electromagnetic techniques area. Haupt and Haupt [13] provide an overview of genetic algorithm optimization strategies. Haupt [13], and Johnson and Rahmat-Samii [13] provide a great summary of genetic algorithms for electromagnetic applications and provide comparisons with more conventional local optimization methods. Cuthbert provides a detailed description of classic gradient optimization methods based on local minima [13]. In [13] a selection of genetic algorithm-based wire antenna designs are covered.

(o) Micromachined Electromechanical Switches

Miniature semiconductor switches known as micro-machined electromechanical system (MEMS) switches are employed in a range of antenna applications. Brown gives a great summary of MEMS technology for applications using radio frequencies [13]. Compared to ferrite waveguide and semiconductor switching elements like PIN diode and FET transistor switches, the MEMS switch is physically smaller, has a very low on-state loss, strong off-state isolation, and a low average DC power consumption.

Cantilever, or type 1 switch, and membrane, or type 2 switch, are the two most popular switch topologies. There have been reports of phase shifters and real time-delay devices operating at the C through W bands utilizing MEMS type 2 switches. Because of its excellent off-state isolation, the type 1 switch architecture is better suited for lower (below X Band) applications.

Reconfigurable antenna technology is an interesting use of MEMS device technology [13]. The goal of developing shared asset apertures with many functions and bands for sophisticated radar and communications systems has prompted scientists to investigate radiating elements with dynamically adjustable physical characteristics. Reactive loads spaced along wire antenna lengths, like dipoles, to provide multi-band performance, have already, to a lesser extent, proved this [13]. Using a switching matrix, the MEMS switch provides a way to extend this idea to two dimensions. MEMS technology has been successfully shown in the scientific community, and initiatives to commercialize the technology are being undertaken. These initiatives include environmental packaging approaches, RF compatibility, and lifetime reliability studies [13].

Because MEMS switch electromechanical devices require free mechanical motion within the package, device packaging becomes extremely difficult. For a microstrip antenna substrate, a low dielectric constant artificial dielectric material can be synthesized locally from a high dielectric substrate using micromachined antenna technology. For a given substrate thickness, the lower effective dielectric constant lowers surface waves and increases radiation efficiency. To lower the substrate's local dielectric constant, Gauthier et al. [13] present a technique that involves drilling a tightly spaced two-dimensional grid of holes into the substrate.

It was reported that the intrinsic substrate dielectric constant of 10.8 was reduced to 2.3. The authors Nguyen et al. [13] report the utilization of micromachining techniques to achieve a local substrate dielectric constant of around 1.0 by suspending dipole or slot radiators on a very thin membrane. A micromachined diagonal horn operating at 110 GHz is described by Veidt et al. [13].

5.3.2.2 Antenna testing and measurements

Only a handful of the fundamental ideas in antenna measurements are briefly covered here because there are many great publications on the subject. For a more thorough explanation, the reader is referred to the relevant literature [15]. The radiation parameters that are most frequently measured are impedance, VSWR, return loss, radiation patterns, cross-polarization, axial ratio, and the polarization ellipse tilt angle. With the antenna radiating into an anechoic environment, impedance and VSWR are measured using standard network analyzer techniques [13]. Generally, sweeping frequency measurements are performed in an open, reflectionless environment, in a full-scale anechoic chamber used for radiation pattern measurements, or in compact anechoic chambers specifically developed for impedance and VSWR testing.

When the source antenna is illuminated, radiation pattern cuts are measured by carefully rotating the AUT and logging the ensuing phase and amplitude response as a function of position. The response in terms of both phase and amplitude is measured in relation to two orthogonal spherical coordinates. The substitution approach is the most common way to quantify gain. A calibrated gain standard and the source antenna are used to record the signal level. Usually, the two antennas are oriented axially. Next, the same source antenna and power levels are used to measure the AUT.

To find the AUT's gain above isotropic at the measurement frequency, the difference in the receive levels of the two measurements is computed and added to the calibrated gain number of the gain standard. In the VHF and UHF frequency ranges, gain standards are made using dipoles and dipole arrays. Horns with pyramidal and conical shapes are employed in frequencies over 800 MHz and up to millimeter waves. Circularly polarized gain standards can be immediately applied to axial mode helical antennas, dual mode ridged square aperture/OMT-fed waveguide horns, and OMT-fed circular horns of different kinds.

By rolling the source antenna to align its polarization at a 90° angle from the AUT, one can effortlessly measure the cross-polarization of a linearly polarized antenna. Cross-polarization is commonly expressed in terms of the copolarized beam signal peak. By comparing the received signal from the AUT under copolarized stimulation with a cross-polarized signal using a similar source antenna of opposite polarization, one can directly quantify the cross-polarization of a circularly polarized antenna. Furthermore, the polarization tilt angles and axial ratios of the source antenna and AUT can be used to compute cross-polarization rejection. Typically, the spinning linear source antenna approach is used to measure the axial ratio.

While the AUT's positioner is recording a pattern at a considerably slower rotation rate, a linear source antenna with low cross-polarization is rapidly rotated about its roll axis. The final pattern consists of the original antenna design with a high-frequency sinusoidal-like pattern placed on it. The response of the AUT will change significantly when the source antenna is rolled if it is highly linear. However, as the circular antenna is equally sensitive to all orientations of linear polarization, if the AUT is purely circular, the signal will not undulate. It is demonstrated that the axial ratio of the AUT for that specific direction is equal to the difference in amplitude between neighboring maxima and minima in the spinning linear plot.

Therefore, a simple method for figuring out the axial ratio of the AUT throughout a pattern cut is the spinning linear measurement. A polarization ellipse measurement can be used to find the axial ratio and polarization tilt angle for a single spot on the antenna radiation pattern. The source antenna is wheeled one full revolution while the AUT is stationary in the test position. The polarization ellipse is the recorded contour, and the axial ratio is the major-to-minor diameter ratio. The polarization tilt angle and the main diameter's angle with respect to a certain reference are connected.

For phased array antenna design and diagnostics, near-field measurements are becoming more and more common [13]. The equivalency theorem, an electromagnetic theorem, is connected to the fundamental idea of near-field metrology [15]. The following is a summary of the theorem: Any other observation point's electromagnetic field can be determined if the two orthogonal directions of the electric field are known everywhere on a closed surface. In the near-field measurement scenario, two orthogonal polarizations of the electrical field are sampled along the closed surface using a field probe. Then, a correction algorithm is applied to eliminate the probe's perturbation of the measured field, and the far-field radiation pattern is computed using Fourier transform techniques.

The planar, cylindrical, and spherical near fields—named for the closed surfaces in which the data is sampled—are the three basic types of near-field ranges. The enclosed measurement surface in the planar near-field setup is roughly represented by a plane. This is suitable for directional antennas like reflector antennas and planar arrays. Fan beam antennas like linear arrays are suitable for the cylindrical near-field range. Any kind of antenna, including broad-beam antennas, can use the spherical near field. The least computationally and data acquisition inefficient near-field technique is the spherical near-field algorithm. The most computationally efficient near-field approach is the planar one. Though they need substantially more time for data processing and acquisition, near-field measurement techniques have a number of advantages over typical far-field pattern cut approaches.

The following is a list of the benefits:

(a) The analysis can use the entire data set at a later time.
(b) Phased array measurements allow for the time multiplexing of the measurement frequency and the beam location, allowing for the extraction of many data sets from a single data acquisition session that span a variety of beam positions and operating frequencies.
(c) The aperture excitation can be holographically imaged using techniques for inverse Fourier transform.

A very useful feature in the array design process is the ability to directly generate an image of the array aperture excitation. Along the array face, it can also be utilized to diagnose failures of phase shifters, elements, and other components.

5.3.2.3 Emerging antenna technologies

Antenna technology has made major strides in the last five years, which are briefly described in this section. One of the biggest developments in antenna technology in recent years, meta-materials have the potential to have a huge influence on the fields of microwave and millimeter-wave antenna and circuit design [13]. The extensive corpus of literature on this subject is recommended reading for the curious reader. Kildal et al., Yeo and Mittra, and Eleftheriades et al. offer excellent summaries [13]. Engineered composites with constituent characteristics that display properties not seen in nature are known as metamaterials. This gives designers more options than just the conventional magnetic, dielectric, and metallic materials.

Double negative materials (DNM) or electromagnetic band gap (EBG) materials are two general categories for metamaterials [15]. DNMs can be engineered to have both negative magnetic permeability and negative apparent (macroscopic) dielectric, which permits backward wave propagation. Their feature sizes are significantly smaller than a wavelength. The feature sizes of EBGs range from one half wavelength to several wavelengths. Artificial magnetic conductors (AMC) and tunable impedance surfaces (TIS) are two EBG variants that can be implemented for narrow-to-moderate bandwidths.

Ground plane impedance tailoring (AMC), directivity enhancement of microstrip patch radiators, multi-band/wide-band enhancement, radar cross section reduction (FSS), applications of electromagnetic "hard" and "soft" surfaces to waveguides, horns, and reflector antennas, series feed antenna array beam squint mitigation, double negative index lens antennas, EBG analog waveguide and phase shifters and reflect-arrays, and FSS-based leaky wave antennas are just a few ways that metamaterials can be used in antenna technology. Perfect "*H*," or Neumann boundary condition, is synthesized by AMC materials over a finite bandwidth, where reflection at surface interfaces is unity with 0° phase shift. At frequencies where natural ferrite materials perform poorly, a manufactured "magnetic" substance can be employed.

For conformal applications, this feature can be used to create low-profile antennas, such as horizontal dipoles, very close to artificial AMC surfaces. Low-profile structures can have their input impedance improved by utilizing AMC characteristics. Furthermore, for vertically polarized monopoles, AMC can reduce the parasitic low elevation angle scattering that is typically connected to metallic ground planes [14].

Widespread use has been made of electromagnetic "hard" and "soft" surfaces, which are analogous to the acoustics field. Hard waveguides that are physically compact and have uniform TEM excitation can be achieved without the limitations of metallic waveguides. In reflector antenna applications, hard horn antennas can be customized for lowest aperture obstruction, low cross-polarization, and pattern symmetry. A popular element in reflector antenna art, the soft horn is often referred to as the conventional corrugated horn. The purpose of hard/soft reflector feed strut assemblies is to reduce parasitic diffraction and aperture shadowing.

An additional degree of freedom in quasioptic lens antenna design is provided by DNM lens technology. Its permittivity and permeability are directly proportional to its optical index of refraction. For a limited bandwidth, DNM materials can thus achieve negative index of refraction. As a result, the designer is no longer limited to materials with positive index of refraction. Physical compactness, impedance matching to open space, and improved focusing capabilities are all possible with DNM spherical lenses.

5.3.3 Phased Arrays

An array antenna is a device made up of several separate radiation elements that are spatially combined to produce a particular radiation pattern, typically in the far field. Each element is excited with a different amplitude and phase. Another way to conceptualize an array antenna is as a continuous aperture distribution that is sampled at specific points inside the aperture. An array whose primary beam can be electrically directed in one or more dimensions by varying the excitation phases on each of the array's individual radiation elements is known as a phased array antenna [13]. Certain array topologies can achieve beam scanning rates of approximately 1 μs, which is far faster than the current state-of-the-art mechanical scanning with motor technology. In fire control and other radar systems where large scan volumes need to be quickly analyzed and multiple target tracking is necessary, rapid beam scanning is essential.

There are four types of array antennas: (a) a linear array, in which the radiating elements are arranged in a single line; (b) a circular array, in which the radiating elements are arranged in a circle with respect to a center reference location; (c) a planar array, in which the radiating elements are arranged in a two-dimensional grid; and (d) a conformal array, in which the locations of the radiating elements are described by a three-dimensional surface.

One may see the different array architectures in Fig. 5.38. To keep things brief, we will be talking about linear phased array technology here. Two-dimensional and conformal arrays can benefit from the basic phased array concepts that were created for linear arrays.

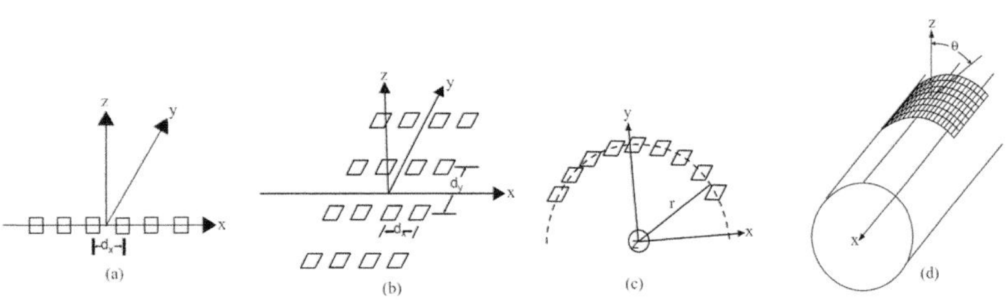

Figure 5.38 Different phase array architectures: (a) linear array; (b) planar array; (c) circular and planar array; (d) conformal array.

5.3.3.1 Linear array technology

Figure 5.39 conceptually illustrates a linear array. This structure is made up of several radiating components arranged in a straight line to create the transition from the array feed system to free space. A number of the antennas that are included in the antenna catalog are suitable options for phased array architecture. TEM horns, microstrip patches, printed dipoles, monopoles, crossed dipoles, bow tie monopoles, open sleeve dipoles, printed Vivaldi notches, stripline- or microstrip-fed linear and crossed printed slots, circular, rectangular, and ridged open-end waveguide, and printed dipoles are among the frequently utilized devices.

Along with its element amplitude and phase excitation, each radiating element has a unique radiation pattern that adds to the overall radiation pattern. The device that transfers microwave energy between the radiating devices and the transmitter/receiver is called the feed system. The feed systems can be designed as radiation-based space feeds or as guided wave transmission lines. A microwave/millimeter-wave circuit component called a phase shifter produces variable insertion phase, or time delay, in relation to its input for a broadband array system. The appropriate relative phase shift across the radiating elements is achieved for beam scanning by varying the insertion phase of each phase shifter.

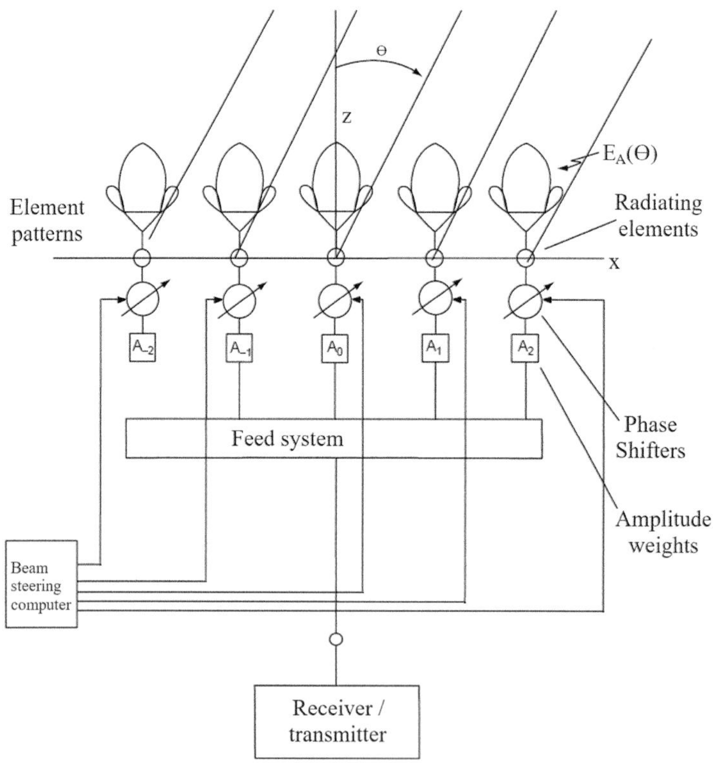

Figure 5.39 Linear phased array's controller section.

The non-uniform distribution of power across the array, known as amplitude weighting, is what creates the low side lobe far-field radiation patterns. Although the amplitude weighting is often applied in the feed system, it is conceptually represented in the picture as blocks A_{-N} through A_N. For a particular beam pointing direction, the beam steering computer (BSC) calculates the correct insertion phase value and instructs each phase shifter accordingly.

It is assumed in the first-order array analysis that there is no mutual coupling and that the radiating elements are isotropic and separated from one another. Due to mutual coupling, the amplitude and phase excitations of surrounding radiating elements, as well as the radiating element itself, can affect each other's impedance characteristics and radiation pattern. The paragraphs that follow will go into additional depth about this effect.

Consider a linear array in the transmit mode. In the far field, the observation point is at infinity, which causes the line-of-sight rays from each radiating element and the observation point to be parallel. This approximation leads to the following expression for the far-field electric field patterns as a function of θ:

$$E_A(\theta) \equiv \underbrace{\left[\frac{\exp\left(-j\frac{2\pi r_0}{\lambda}\right)}{r_0} \right]}_{\text{Isotropic element pattern}} \times \underbrace{\sum_N A_n \exp\left[-j\frac{2\pi}{\lambda} n\Delta x(\sin\theta - \sin\theta_0)\right]}_{\text{Array factor}}, \quad (5.16)$$

where N = number of radiating elements, r_0 = distance from the center of the array to the far field observation point, A_n = array element amplitude weighting coefficients, θ = beam pointing, referenced to broadside (θ = 0), x = the array element spacing, λ = wavelength, θ_0 = the desired beam scan angle, and $n \equiv 2\pi$.

$\Phi_n \equiv \frac{2\pi}{\lambda} n\Delta x \sin\theta_0$ the phase shifter insertion needed to scan the beam to θ_0 is the nth phase shifter. The isotropic pattern is usually removed from the formula since it has no effect on the array's far-field amplitude pattern. It is evident that the radiation pattern is influenced by the number of radiating elements, their relative spacing, the array's operating frequency (wavelength), and the phase shifter settings.

In many system applications, just one primary beam is required. Grating lobes, also referred to as secondary or false main beams, depend on two factors: (a) the array element spacing; and (b) the phased array's scan volume. Examine Eq. (5.16) and every radiation element contributes positively in phase when the term

$$\frac{1}{\lambda} n\Delta x(\sin\theta - \sin\theta_0) \equiv m, \qquad \text{where } m = 0 \pm 1,2,3\cdots \tag{5.17}$$

The major beam is represented by m = 0, and the other beams are grating lobes. The following expression predicts grating lobes as a function of beam scanning and element spacing:

$$\sin\theta_{gl1} \equiv \sin\theta_0 + \frac{\lambda}{\Delta x}, \tag{5.18}$$

where $\sin\theta_{gl1}$ is the first grating lobe in visible space.

The necessary spacing is if the array must scan over the $\pm\theta_m$ scan volume.

$$\Delta x \leq \frac{\lambda}{\sin\theta_m} \tag{5.19}$$

The array can scan ±90° without grating lobes if the interelement distance is one half of open space wavelengths. For a given array element spacing, Fig. 5.40 shows the relationship between the viewable space, grating lobes, and the main beam. Low side lobe patterns are produced by amplitude weighting. For a uniformly stimulated linear array, the theoretical side lobe level is –13.5 dB with respect to the main beam. For a specific number of array elements and interelement spacing, the uniformly lit array also has maximum gain and minimum beam width. The light across the array must have a non-uniform amplitude for designs involving lower side lobe levels.

Gain, beam width, and side lobe level are all trade-offs. When compared to uniform illumination, low side lobe level arrays have greater beam width and lower gain. One way to express the directivity of a linear array of perfect isotropic architecture parts is as

$$D_0 \equiv \frac{\left|\sum_N A_n\right|^2}{\sum_N \left|A_n\right|^2} \tag{5.20}$$

where A_n = the element excitation coefficients.

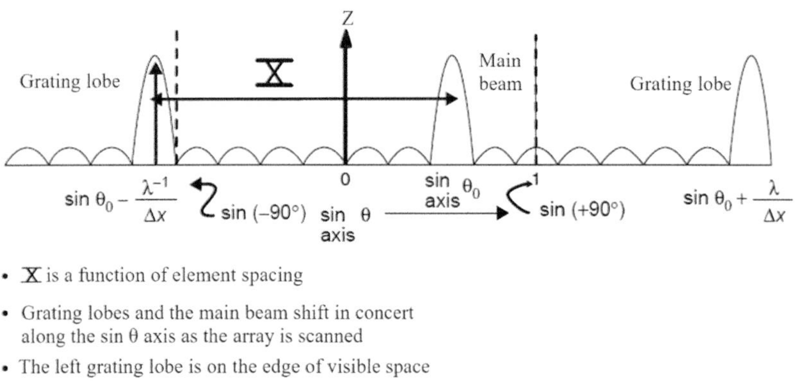

Figure 5.40 Understanding the grating lobe and the main beam.

The decrease in gain brought on by amplitude weighting is known as aperture efficiency, and it can be expressed using the following formula:

$$\eta_{\mathrm{ap}} \equiv \frac{\left|\sum_N A_n\right|^2}{N\sum_N \left|A_n\right|^2}, \tag{5.21}$$

where A_n is the element excitation coefficient.

At the low side lobe level, several methods of pattern synthesis have been established for two-dimensional apertures and continuous line sources. These methods have been extended to array theory by employing continuous aperture sampling at the array elements' positions or null matching approaches [13]. The approach that is most frequently used is the Taylor weighting method. The optimal compromise between beam width and side lobe level is achieved in the far-field radiation pattern created by the Taylor design. For additional information on the creation of the Taylor coefficients, which is outside the scope of this literature, the reader is referred to the literature [13].

Realistic radiation elements have an element pattern that weighs the array factor; as Eq. (5.22) shows, they are not isotropic.

$$E_A(\theta) \equiv \underbrace{[E_e(\theta)]}_{\text{Element pattern}} \times \underbrace{\sum_N A_n \exp\left[-j\frac{2\pi}{\lambda} n\Delta x(\sin\theta - \sin\theta_0)\right]}_{\text{Array factor}} \quad (5.22)$$

For the sake of clarity, the isotropic element pattern is suppressed in this equation. An appropriate estimate for an ideal element is

$$E_e(\theta) = \cos(\theta), \quad (5.23)$$

where the Z axis is used as a reference for θ.

Figure 5.41 illustrates the ideal element layout, which accounts for the aperture projection loss when the array scans off bore sight, but not for the effects of mutual coupling between the radiating components. Every linear- and planar-phased array experiences a scan loss of at least cos(θ). When mutual coupling occurs, this loss can potentially be higher, that is, cos$x$ (θ), where x is a real integer greater than 1.0.

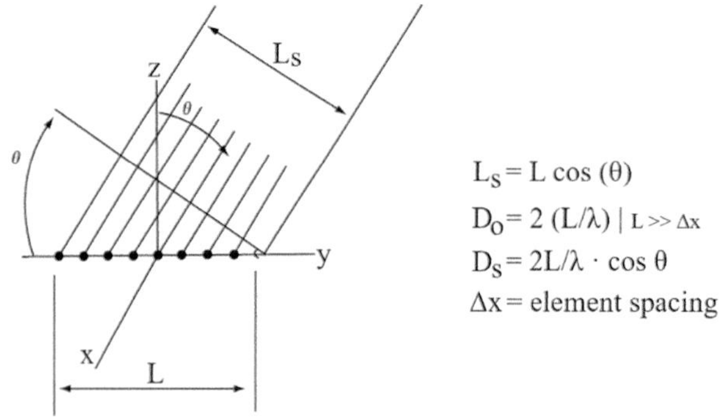

Figure 5.41 Loss determination of aperture projection.

We shall talk about this effect in the following paragraphs. The impact of element pattern weighing on the array space factor's side lobes and main beam is shown in Fig. 5.42. This figure makes several things clear:

(a) With scan angle, the main beam gain decreases by at least $\cos(\theta)$.
(b) High sidelobes are the manifestation of grating lobes that are only beginning to be seen on scan.
(c) Because $\cos(90°) = 0$, a broad side array cannot scan to end-fire. It is possible to construct arrays to function in this "end-fire" scenario. The reader is referred to the literature [15].

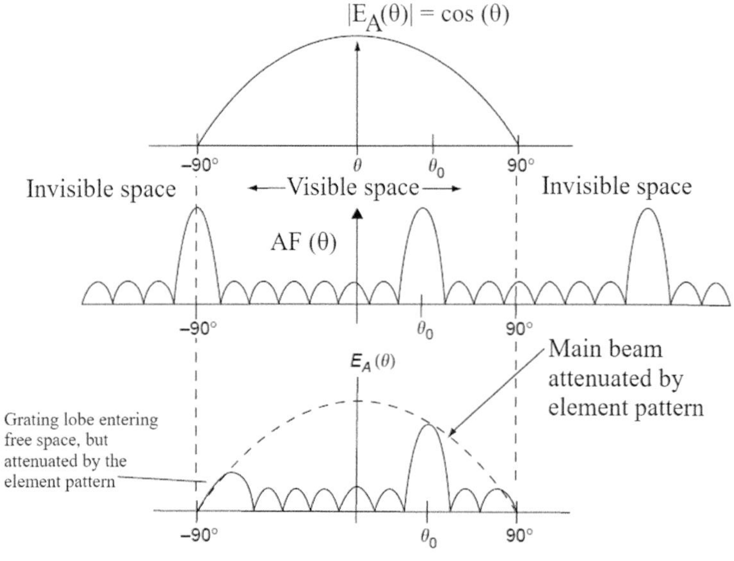

Figure 5.42 Weighting of the phased array elements.

A number of factors influence array bandwidth, such as mutual coupling between the radiation elements and the radiating elements' impedance, polarization, and radiation pattern bandwidth. The phase shifters' frequency response causes a beam squint in the array factor as a function of frequency. The following expression can be used to quantify, in percentage terms, the frequency change for a linear array that results in a shift of ±1/2 of the nominal 3 dB beam width at the beam position:

$$\text{Fractional band width} = \frac{\Delta f}{f} = \frac{\text{BW}_0}{\sin\theta_s}, \tag{5.24}$$

where θ_s = the scan angle off bore-sight, in radians, and BW_0 = the beam width of the array at bore-sight.

When true time delay (TTD) devices are used in place of the array phase shifters, broad bandwidth can be achieved. In order to compensate for the variations in propagation times between each radiating element and any random observation point in the far field, beam scanning is accomplished by programming a time delay into each radiating element. Since there is no beam squint with frequency in this scenario, the bandwidth is not constrained by the array factor. In the end, the radiating element bandwidth in the array environment and the feed network bandwidth decide the array bandwidth.

Sub-arrays and steering with TTD phased arrays use programmable time delay units at each radiating element in place of the phase shifter devices found in conventional devices. Time delay is used in scanning to make up for variations in the propagation times between each radiating element and the target. As the array is scanned off bore-sight, the propagation delay across the phased array aperture increases significantly. There is no beam squint with frequency, meaning that the instantaneous bandwidth is not constrained by the antenna array factor.

Although TTD devices have always been large, heavy, bulky, and costly due to their lossy nature, new developments in MEMS offer the possibility of great performance in small monolithic semiconductor structures at modest input power levels. In Fig. 5.43, a time delay-steered phased array is shown. Following a sequence of mathematical reductions, the array factor for a phased array steered by a temporal delay can be written as

$$E(\theta) = \sum_{n=1}^{M} A_n \exp\left[\left(\left(j \times \frac{2\pi}{\lambda}\right) \times n \times \Delta x\right) \cdot (\sin\theta - \sin\theta_0)\right]. \tag{5.25}$$

There is no beam squint since the electric field's maximum occurs at θ_0 regardless of the wavelength (or operating frequency). The array beam width and gain are still functions, to be considered.

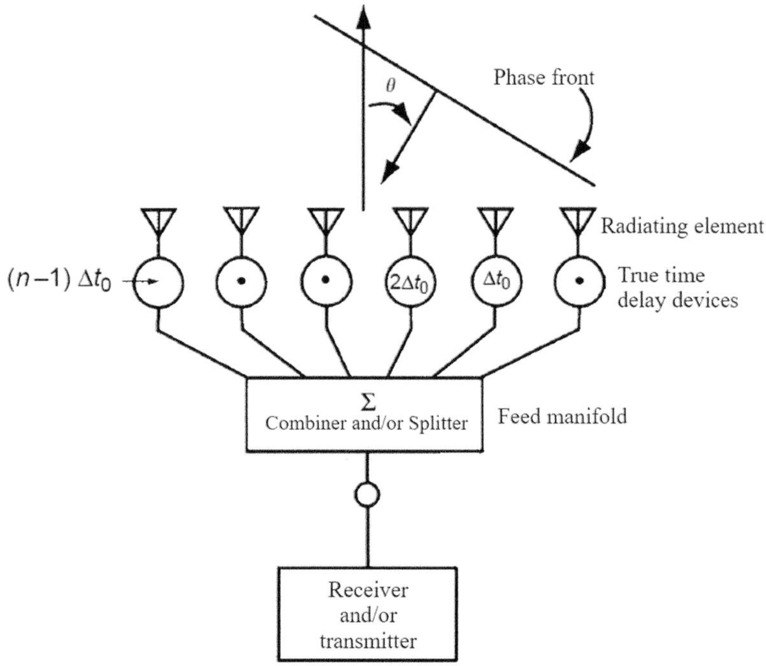

Figure 5.43 Phased array topology with true time delay.

5.3.3.2 Phased array component technologies

Basic linear phased array theoretical ideas were covered in the preceding sub-section; these ideas can be easily extended to planar and broadly conformal array apertures. The industry's present phased array hardware situation is highlighted in the discussion that follows.

The importance of monolithic microstrip antenna-phased arrays is growing as a result of recent advancements in integrated circuit technology, test procedures, and cost-effective semi-conductor production, as well as stricter requirements for systems. Monolithic integration is conceivable because the requisite dimensions of the microstrip-radiating elements at millimeter-wave frequencies are on the same order as the active circuit devices. The T/R module sub-circuits, which include power amplifiers, LNAs, phase shifters, attenuators, and T/R switches, are integrated with the radiating elements.

Industry research is currently focused on packaging that shields semiconductor devices from severe environments and mechanical damage [13]. The antenna designer faces additional issues related to radiation efficiency, bandwidth, and surface wave/scan blindness because of the electrical parameter limits imposed by semiconductor substrate technology for silicon (Si) and gallium arsenide (GaAs).

In broad-beam scanned microstrip-phased array applications, micromachining technology is employed to reduce mutual coupling caused by grounded dielectric slab substrate surface wave modes. This issue is especially problematic for high relative dielectric constant substrates, like Si and GaAs, which are commonly used semiconductor substrates for integrated radiating element T/R module topologies. A low dielectric constant artificial dielectric material can be synthesized locally from a high dielectric substrate for use as a microstrip antenna substrate through the application of micromachining techniques.

For a given substrate thickness, the lower effective dielectric constant lowers surface waves and increases radiation efficiency. A closely spaced two-dimensional grid of holes is drilled into the substrate by Gauthier et al. [13] to lower the substrate's local dielectric constant. It was reported that the intrinsic substrate dielectric constant of 10.8 was reduced to 2.3. Nguyen et al.'s [13] description of micromachining methods allows one to achieve a local substrate dielectric constant of roughly 1.0 by suspending dipole or slot radiators on an extremely thin membrane—a useful property for millimeter-wave applications. For substrate materials with dielectric constants similar to Si, Papapolymerou et al. [13] have shown a 64% bandwidth enhancement and a 28% radiation efficiency improvement using the membrane approach.

MEMS, or micro-mechanical element switches, are employed in a range of antenna applications. For antenna RF applications, Rebiez offers a comprehensive review of MEMS technology that includes a state-of-the-art evaluation of switch lifetime and packaging technique [13]. Comparing the MEMS switch to ferrite waveguide and semiconductor switching elements like PIN diode and FET transistor switches, the MEMS switch features low average DC power consumption, strong off-state isolation, very low on-state loss, and a compact physical footprint.

Due to these characteristics, some researchers have suggested that MEMS phase shifter-based passive electronically scanned array (PESA) architectures for very large space-base military radar and communications systems be used instead of monolithic T/R module-based active electronically scanned array (AESA) architectures [13]. In terms of weight, DC power consumption, and necessary system effective isotropic radiated power (EIRP), the MEMS topologies compare favorably. Reconfigurable antenna technology is an interesting use of MEMS device technology.

The goal of developing shared asset apertures with many functions and bands for sophisticated radar and communications systems has prompted scientists to investigate radiating elements with dynamically adjustable physical characteristics. Using a switching matrix, the MEMS switch provides a way to extend this idea to two dimensions.

Impedance-matching phased array radiating elements dynamically as a function of beam scan is another possible use case for MEMS switches. The scan impedance of the radiating element varies with beam scan location due to mutual interaction between radiating elements within the array environment. The radiating element can be dynamically tuned using a dynamic impedance-matching network made up of planar transmission lines and switchable banks of reactive elements. With the help of environmental, RF-compatible packaging techniques and life-cycle reliability studies, significant progress has been made in commercializing the technology. Once more, Rebiez is cited for the reader [13].

Fractal antenna array technology: The radiating element shape and array element spacing of fractal array antennas can be based on fractals. Numerous academics have examined fractal-based arrays [13]. Werner, Haupt, and Werner provide a thorough summary of fractal antenna array theory [13]. With fewer elements, fractal arrays can combine the efficiency of a periodically spaced array and the robustness of a random array.

drawback of Vivaldi arrays in dual polarized apertures is that they need contiguous grounds across the neighboring radiating elements. For example, solder metallization is very hard to achieve and makes actual field repair very challenging. Scholars have tackled these problems in a number of ways.

A linearly polarized balanced antipodal Vivaldi antenna (BAVA) element with a 3:1 bandwidth and good scan impedance performance up to 45° off bore-sight has been demonstrated by ElSallal and Schaubert [13]. Contiguous mechanical and electrical element-to-element contacts are not necessary for this element. Consequently, this component can serve as the foundation for an easily field-repairable integrated "plug and play" radiating element/TTD or T/R module. In an infinite array context, Holter has proven a dielectric-free, body of revolution "ogive shaped" radiating element with good scan impedance performance across a decade of bandwidth [13].

Typically, maximal aperture efficiency throughout a bandwidth is required for broadband arrays. Because the overall array lattice spacing is limited to be one half wavelength at the highest operating frequency for grating-lobe-free operation ($\lambda_{o\text{high}}/2$) this leads to an element "compaction" problem, similar to multi-band array architectures, where the low-frequency band is oversampled (too many elements) for grating-lobe-free operation.

The elements closest to the array center excite the highest frequency portion of the array, whereas elements farther from the array center excite progressively lower frequencies. With careful radiating element design, the array's active component can grow and decrease with frequency, and its lattice effectively operates at any band of operation with a $\lambda_o/2$ lattice spacing. Thus, this architecture achieves the best possible balance between element count and lattice density.

Beam Formers Analogous: Modern radar and telecommunication systems are currently interested in simultaneous multiple beam operation. Both analog and digital beam former topologies allow for the realization of simultaneous multi-beam operations.

The briefing on digital beam forming (DBF) is covered in the section on developing technologies that follows. The Butler and Blass matrices, variants on the Rotman and Luneburg lens feeds, passive, cross-bar, and active vector modulation methods are examples of classic simultaneous beam feed networks. A set of orthogonal, equiangular gazing beams is produced by the Butler and Blass matrices. To individually and electronically steer these beams, a time delay or phase shift at the element level is necessary. For a thorough explanation, please refer to Fourikis' text [13]. Smart Antennas: Phased array antenna, DBF, and digital signal processing (DSP) technologies are used to create smart systems.

Communications systems that use spatial signal processing techniques to sense, adapt, and enhance performance are utilizing smart antennas. Adaptive, or agile, directed radiation patterns are utilized by smart systems. In order to provide narrow beam width pattern coverage across a large field of view, switched beam antennas take advantage of a group of common beam width antennas that are commutated. These technologies are being used at frequencies lower than the S Band by contemporary wireless communications systems and cellular telephone systems.

The following functions of smart antenna systems include increasing system gain, reducing or nulling parasitic interference signals, compensating for multi-path-induced signal fading in electromagnetically complicated settings, and enabling simultaneous multiple beam operation. The direction of arrival (DOA) and adaptive beam forming (ABF) algorithms, which are incorporated in DSP technology, are the brains behind a smart antenna system. Because it may simultaneously maximize the pattern gain for a signal of interest (SOI) and minimize signals of non-interest (SONI), adaptive nulling is incredibly strong. An outstanding high-level presentation of smart antenna systems is given by Balanis [15].

5.3.3.3 Emerging phased array antenna technologies

Phased array technology is an active field of study, especially in the DOD, where several novel technologies are being explored recently. A good and concise summary of the state of the art for phased arrays is given by Brookner [13]. We will touch on a few of the significant ongoing studies in this section. For a more thorough explanation, the reader is referred to the relevant literature. Phased array applications can benefit from photonics systems in several ways, including as immunity to electromagnetic interference (EMI), signal bandwidths greater than one octave, low radar cross section, and very low loss for RF signal interconnections and digital control.

Within phased array architectures, the photonic system can execute one or more of the following system functions: (a) distribution of digital control signals; (b) distribution of RF energy to and from the array radiating elements; and (c) optical signal processing of the array's RF signals. To modify the phase weighting and amplitude of each radiating element in the phased array antenna architecture, digital control signals must be transmitted. For phased arrays with hundreds or thousands of radiating elements, traditional connectivity systems including wire bundles, printed circuit boards, and flexible circuit board technology are exceedingly inefficient.

An optically controlled Ka Band phased array antenna with an opto-electronic interface circuit (OEIC) that enables control signals to be distributed to individual antenna elements over a single fiber optic connection is described by Kunath et al. [13].

By modulating the RF signal to optical frequencies for the distribution of RF energy to and from the phased array's radiating elements, microwave/millimeter-wave feed waveguide manifold technology can be substituted. In comparison to future optical cable technology, traditional microwave/millimeter-wave-guided wave transmission line technology is very heavy, thick, and narrowband by nature. After the RF signal is converted to the optical domain, optically based space-time signal processing becomes possible.

Large DBF phased arrays demand a high processing burden, which the optical channel's huge bandwidth would provide. In the optical domain, phase shifting and TTD—which are necessary for array beam scanning—can be completed. Phased arrays have a lot of potential thanks to optical technology, but significant technological advancements are still needed for embedded fiber interconnect technologies, physically compact and more mature electrical-optical integrated circuits (OEIC) device packaging and improved optical-to-microwave frequency conversion efficiency (DC power consumption).

5.3.4 Beamforming and Beam Steering

In terms of flexibility, low probability of detection (LPD), low probability of intercept (LPI), interference suppression, direction finding, real-time pattern synthesis, radiating element parasitic mutual coupling mitigation, and multi-beam generation, affordable and practical DBF systems are the best phased array solution. Radar community research on space temporal adaptive processing (STAP) algorithms is currently flourishing [13]. All of the features of the traditional analog beam former can be achieved with a digital beam former. DBF generates and steers numerous beams by utilizing the mathematical power of DSP.

In order to convert analog RF array signals into digital data streams, the ultimate DBF systems would integrate LNAs at the individual element level and then use analog-to-digital (ADC) and digital-to-analog (DAC) converters. Since conformal array applications require beam steering algorithms, polarization control, amplitude and phase control, array sector switching, element pattern compensation, built-in-test (BIT), and in situ calibration, DBF technology is especially well suited for these types of applications.

Lossless beamforming is achieved because, in contrast to analog beam forming, the LNA determines the signal-to-noise ratio of the receiving system and is unaffected by additional downstream processing. At carrier frequencies, when ADC/DAC technology is unable to directly sample and rebuild the analog waveforms, hybrid systems where there is a first up and down-conversion in analog RF are employed. Extreme demands are placed on the ADC/DAC technology as well as the processing, storing, and transmission of massive volumes of digital data especially for large arrays by this digital solution. It is hoped that the tremendous pace at which computer technology and software are developing will continue in the future.

The DBF systems in use today are made up of many sub-systems and component combinations. Problems may arise from these constitutive sub-systems' imperfections and non-uniformity. Significant issues include receiver channel imbalance, amplitude and phase errors, A/D offset errors, frequency errors, and so on.

An essential component of DBF system design is array calibration, which includes mutual coupling mitigation. Furthermore, for actual systems operation, an array's capacity to self-heal following channel failures is essential.

The literature has several DBF systems that are demonstrated. Ref. [13] describes an experimental X band system with 570 active units that conforms to an airplane wing's leading edge. It was proved how to synthesize sectored beams with low side lobe levels (−40 dB).

Since the unit cell for any scanning array is typically chosen as $\lambda_o/2 \times \lambda_o/2$, it is difficult to place the DBF circuitry and DSP computer interconnect directly behind the array aperture. This is especially problematic for electrically large arrays of hundreds or thousands of elements, and especially for higher microwave and millimeter-wave frequencies, where the $\lambda_o/2 \times \lambda_o/2$ unit cell is very physically small. Ultimately, DBF technologies must ride the wave of miniaturization as driven by the commercial wireless industry.

The literature has reports on recent studies in electronics packaging and DBF circuit designs. In a device of just 50 mm on one side, an 8-channel, 4-beam 10 MHz bandwidth beamformer was realized using hybrid wafer-scale integrated circuits technologies [13]. In order to achieve economies of scale in high-volume production, Curns et al. talk about a modular, scalable architecture [13]. They created a 512-element X band receive-only system that uses 32 columns of 16-channel beamformers to do digital beamforming in azimuth.

A *n*-channel real-time parallel processor board is connected to each columnar digital beam former as shown in Fig. 5.44. Each DBF transmitter consists of a digital back-end and an RF front end that communicate to the antenna arrays. The RF modules will receive the digital data that can set the beam angle and switch from the first beam as shown in Fig. 5.44(a) to the last beams as in Fig. 5.44(d). The transition and the resolution of the angle switching of the beams is determined by the number of antennas in the array and the digital control bits. Hence enabling the feature as a scalable multi-beam DSP processor design.

In order to keep costs down, the researchers took advantage of commercial off-the-shelf (COTS) components like field-programmable gate array (FPGA) technologies. It is imperative that these fields continue to progress in order to render DBF solutions for electrically massive systems feasible.

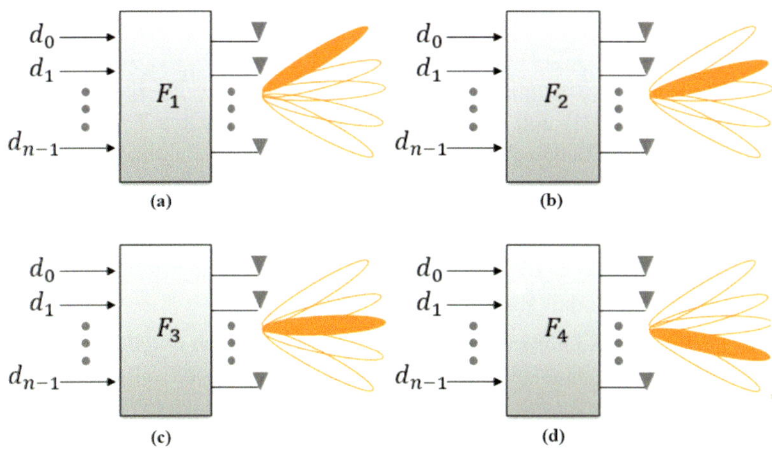

Figure 5.44 Digital beamforming example.

5.4 Conclusion

This chapter provides the RF transceiver integration with the consideration of interfaces to the external world for measurements and applications. The details of the overall system integration taking into consideration all the sub-systems and the ESD protection bondpads are briefed in this chapter. The various prototyping methods and technology along with the post-semiconductor IC fabrication processing steps are explained that will enable the RF transceiver IC to be accessible for measurements and overall system integration for usage in practical applications. This chapter culminates in the discussion about the antennas, phased arrays and beamforming which will complete the whole RF transceiver system to be able to function as wireless communication system-on-chip solution.

References

1. Serial Peripheral Interface (SPI), User Guide, Texas Instruments, Literature Number: SPRUGP2A, Mar. 2012, https://www.ti.com/lit/ug/sprugp2a/sprugp2a.pdf?ts=1719315728316.
2. S. Dabral and T. J. Maloney, *Basic ESD and I/O Design*, John Wiley and Sons, Inc., 1998.
3. ANSI-ESD SP5.1.2–2006 For the Protection of Electrostatic Discharge Susceptible Items – Standard Practice for Human Body Model (HBM) and Machine Model (MM) Alternative Test Method: Supply Pin Ganging Component Level.
4. ANSI-ESD S5.3.1–2009 For Electrostatic Discharge Sensitivity Testing – Charged Device Model (CDM) – Component Level.
5. ANSI-ESD S5.2–2009 For Electrostatic Discharge Sensitivity Testing – Machine Model (MM) – Component Level.
6. S. G. Beebe, *Characterization, Modeling, and Design of ESD Protection Circuits*, Technical Report, Stanford University, 1998.
7. M.-D. Ker and C.-C. Yen, Investigation and design of on-chip power-rail ESD clamp circuits without suffering latchup-like failure during system-level ESD test, *IEEE Journal of Solid-State Circuits*, vol. 43, no. 11, pp. 2533–2545, Nov. 2008.
8. J. E. Vinson and J. J. Liou, Electrostatic discharge in semiconductor devices: an overview, *Proceedings of the IEEE*, vol. 86, no. 2, pp. 399–420, Feb 1998.
9. *International Technology Roadmap for Semiconductors, Assembly and Packaging*, 2009 ed., https://www.semiconductors.org/wp-content/uploads/2018/09/Assembly-Packaging.pdf.
10. A. Andrae, *Usefulness and Precision of Life Cycle Inventory in Microelectronics Packaging*, 2009.
11. E. E. Times, 3D IC design, Sep. 2011 https://www.eetimes.com/3d-ic-design.
12. O. Burkacky, T. Kim, and I. Yeom, *Advanced Chip Packaging: How Manufacturers Can Play to Win*, McKinsey & Company, May 2009, https://www.mckinsey.com/industries/semiconductors/our-insights/advanced-chip-packaging-how-manufacturers-can-play-to-win.
13. M. Golio and J. Golio, RF and microwave passive and active technologies, in *The RF and Microwave Handbook*, 2nd ed., CRC Press, Taylor & Francis Group, Boca Raton, London, New York, 2008.

14. IEEE Standard Definitions of Terms for Antennas, IEEE Standard 145–1983, Institute of Electrical and Electronics Engineers, 1983.
15. C. A. Balanas, *Antenna Theory, Analysis, and Design*, 2nd ed., John Wiley and Sons, New York, 1997.

If you don’t have the heart, don’t teach.
Teaching without love is meaningless.
If you don’t have the passion, don’t do research.
Research without affection is impactless.

—Kiat Seng Yeo

In a VUCA (Volatility, Uncertainty, Complexity and Ambiguity) world, universities should prepare Global-ready, Industry-ready, Future-ready, Technology-ready, and Social-ready (GIFTS) graduates.
“GIFTS” bring significant economic value to the country and improve the well-being and culture of the nation.

—Kiat Seng Yeo

We should not be just markers of our own destiny but makers of our society.

—Kiat Seng Yeo

$$\frac{Y(s)}{X(s)} = \frac{G(s)}{1 - G(s)H(s)}. \quad (6.1)$$

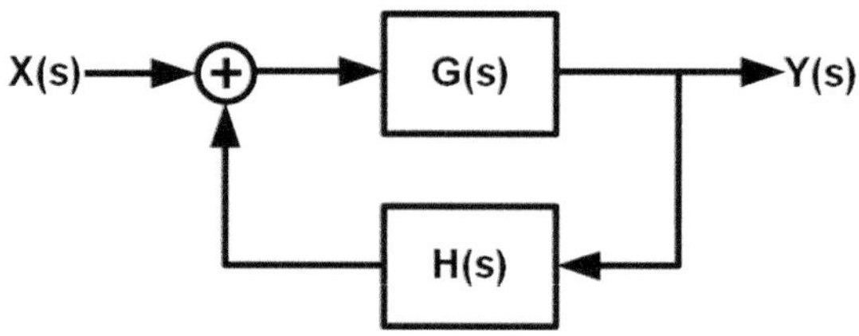

Figure 6.1 Oscillator feedback model.

For the system to oscillate, the closed loop gain should be infinity and the root of the denominator should be in the right half of the complex axis. The necessary condition for oscillation is defined at the frequency of oscillation where $s = j\omega_0$ and satisfies both the gain and phase condition known as the Barkhausen criteria given as follows:

1. The loop gain is equal to unity at the frequency of oscillation i.e. $G(s) \cdot H(s) = 1$.
2. The total phase shift around the loop is 0° or integer multiples of 360°.

Though the Barkhausen criteria explains the steady-state condition of the oscillator, in most practical cases, it only defines the necessary condition and is not sufficient for the circuit to oscillate. The closed loop gain required is larger than unity for oscillator start-up. To ensure proper oscillator start-up in the presence of noise, the loop gain $|G(j\omega_0).\ H(j\omega_0)|$ is made 2–3 times larger. As the oscillation is properly started, the noise in the circuit gets amplified leading to an increase in amplitude until the amplifying device is saturated and the amplitude of the steady-state output is limited by the non-linearity in the oscillator. In RFIC applications, it is required that the oscillator outputs be made tunable, where the output frequency is a function of a voltage or a current. In most cases, the oscillator output is varied by an input control voltage and these oscillators are commonly known as voltage-controlled oscillator (VCO).

6.2 VCO Architectures

6.2.1 Resonator-Less Oscillators

The simplest resonator-less oscillator is the ring oscillator as shown in Fig. 6.2. The ring oscillator is formed by cascading n-stages of inverting gain stages with the output of the last stage connected to the input of the first stage forming the feedback loop. The oscillation frequency of the ring oscillators is given by

$$f_{\text{osc}} = \frac{1}{2n\tau}, \tag{6.2}$$

where n is the number of stages and τ is the delay in each stage. In general, odd number of stages is cascaded in single-ended implementation of a ring oscillator, while for differential implementation even number of gain stages can be used. For the differential case, the output of the last gain stage is cross-coupled and fed back to the input.

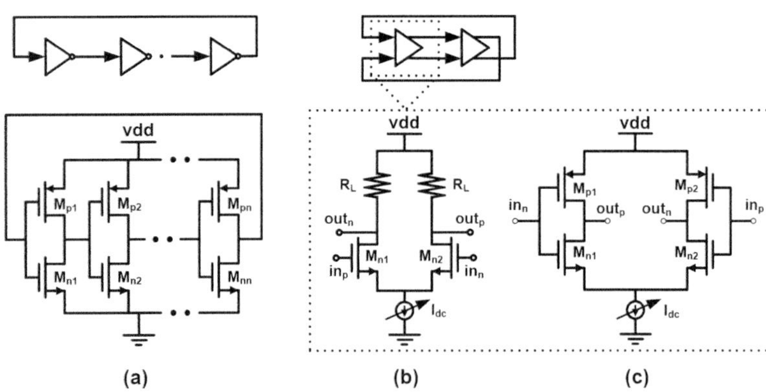

Figure 6.2 Ring oscillator: (a) single-ended, (b) differential with passive load, and (c) differential with active load.

The ring oscillator can be made tuneable by varying the transconductance of the gain stages. In comparison with the resonator-based oscillator, the ring oscillator can achieve a very wide tuning range occupying only a fraction of the die-area, but the spectral purity of the output signal is poor. The ring oscillators achieve poor phase noise due to the presence of noise

6.2.3.1 Flicker noise in LC oscillators

With the continuous scaling of semiconductor technology, CMOS oscillators have been preferred over their bipolar counterparts. However, CMOS technologies suffer from higher flicker noise corner frequency [3]. In the phase noise profile of LC oscillators as shown in Fig. 6.6, the noise close to the carrier frequency is dominated by the flicker noise and far away noise is dominated by thermal noise. In the VCO, this flicker noise is up-converted to $1/f^3$ phase noise close to the carrier frequency with a slope of –30 dB/decade. Several works have been reported in the literature to predict, characterize and reduce the phase noise and flicker noise in oscillators [4–28]. The major contributors for flicker noise in oscillators are:

i. Amplitude modulation (AM) – frequency modulation (PM) noise conversion due to varactor non-linearity [5–8],
ii. Tail current noise [9–12], and
iii. Modulation of parasitic capacitances [11, 13, 14] and output voltage waveform [10, 15–20].

(a) AM–FM noise conversion in oscillators

In the typical high-frequency oscillators, the frequency fine-tuning is achieved using varactors where the variation in voltage across the varactor terminals results in a small variation in capacitance thus resulting in frequency tuning. However, one of the varactor terminals is directly connected to the oscillator output voltage which typically has a large voltage swing.

Therefore, the oscillation amplitude also influences the effective capacitance of the varactor as shown in Fig. 6.7 and the effective capacitance is calculated using large-signal analysis [29]. Therefore, the AM noise converts to FM noise and the AM–FM sensitivity can be given as [8]

$$K_{\mathrm{AM-FM}} = \frac{\partial \omega_0}{\partial A} = \frac{\partial}{\partial A}\left(\frac{1}{\sqrt{\mathrm{LC}_{\mathrm{eff}}}}\right) = -\frac{1}{2}\frac{\omega_0}{C_{\mathrm{eff}}}\frac{\partial C_{\mathrm{eff}}}{\partial A}, \tag{6.8}$$

where ω_0 is the oscillation frequency equal to $(LC_{eff})^{-0.5}$ and A is the oscillation amplitude. The sensitivity of effective capacitance in terms of oscillation amplitude is expressed as

$$\frac{\partial C_{eff}}{\partial A} = \frac{C_{max} - C_{min}}{\pi} \frac{2V_{eff}}{A^2} \sqrt{1 - \left(\frac{V_{eff}}{A}\right)^2}, \tag{6.9}$$

where V_{eff} is the effective voltage across the varactor terminals and C_{max} and C_{min} are the maximum and minimum capacitance of the varactor. The spectral density of the FM sideband at a given offset frequency ω_m can be given as [8]

$$L(\omega_m) \approx \left(\frac{K_{AM\text{-}FM}}{2\omega_m}\right)^2 V_n^2, \tag{6.10}$$

where V_n is the amplitude-modulated noise at the output of the oscillator. From Eq. (6.8)–(6.10), it can be observed that AM-FM noise conversion can be improved by reducing the varactor capacitance variation C_{max}–C_{min}.

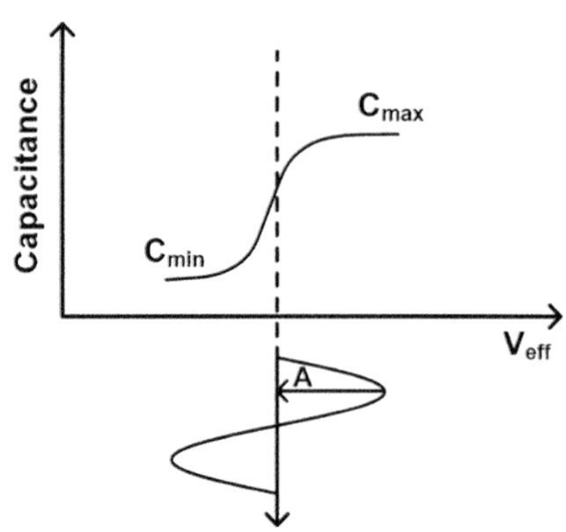

Figure 6.7 Typical varactor capacitance curve over oscillation period.

(b) Groszkowski effect–induced $1/f^3$ noise

The Groszkowski effect describes a phenomenon in LC oscillators where the harmonic current arising due to the non-linearity in actives devices flows in the LC tank. When these current signals with higher harmonics reach the resonant tank, a frequency and phase shift in the oscillator is observed and was first quantified by Groszkowski in 1933 [30]. In the oscillators, this noise current can be either from the tail current transistors or the cross-coupled pair and is a dominant mechanism in the flicker noise up-conversion to $1/f^3$ phase noise [31, 32].

The schematic of cross-coupled VCO with tail current is shown in Fig. 6.8. The cross-coupled transistors M_1 and M_2, switch the current signal of the tail current source. The output current can be regarded as a square wave and contains odd harmonic components as shown in Figs. 6.8(b) and 6.8(c). The phase noise of the VCO due to the tail current can be given as [33]

$$L(\Delta\omega) = \frac{\sigma^2}{2Q^4}\left(\frac{\omega_0}{\Delta\omega}\right)^2 \frac{S_{\Delta I}(\Delta\omega)}{I_B^2}, \tag{6.11}$$

where σ is related to the conduction angle of the oscillator. Therefore, the phase noise due to the Groszkowski effect will be more dependent on the tank quality factor and a higher tank Q helps to improve the flicker noise–induced phase noise. At the same time, since flicker noise is directly proportional to the tail current and phase noise is inversely proportional, increasing the tail current will improve the flicker noise.

Several other works are reported in the literature for the reduction of $1/f^3$ phase noise as shown in Fig. 6.9. Linearization of transconductance with source degeneration resistors to cross-coupled transistor is proposed in [22] as shown in Fig. 6.9(a). In comparison to the conventional oscillator, the addition of source degeneration resistors presents a low transconductance variation over a large voltage swing and an optimum value for resistors effectively suppresses the $1/f$ noise of the cross-coupled pair [22]. Also, as shown in Fig. 6.9(b), [34] and [21] proposed the addition of a series inductor at the source of the switching transistors resonating with the parasitic capacitance at twice the oscillation frequency to suppress the $1/f$ noise.

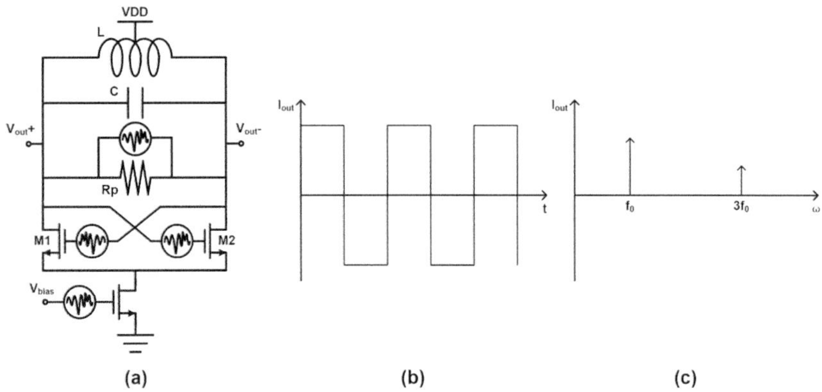

Figure 6.8 Typical VCO noise: (a) noise source, (b) output current (time-domain), and (c) output current (frequency domain).

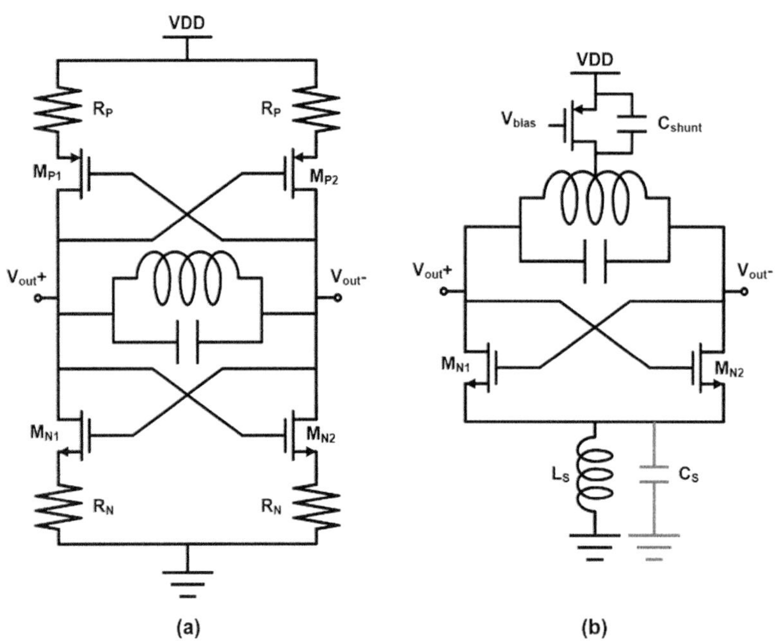

Figure 6.9 Typical VCO topologies to reduce flicker noise: (a) source degeneration resistor and (b) source degeneration inductor.

(c) Modulation of parasitic capacitance

In the LC oscillator, the parasitic capacitance of the active devices such as gate-source (C_{gs}), gate-drain (C_{gd}) and drain-bulk (C_{db}) exhibit a highly non-linear behavior due to the large swing at the oscillator output. This causes a reactive power imbalance effect in the conventional tail current topology [14]. To minimize the effect, [14] proposes a VCO topology with split tail current as shown in Fig. 6.10 [14]. In the proposed topology, the parasitic capacitance C_1 provides an additional signal path for the harmonics to flow. Due to the current flowing through the capacitance C_1, this effectively creates a quadratic voltage at the source of the switching transistors to balance the tank quadratic current responsible for frequency variation [14].

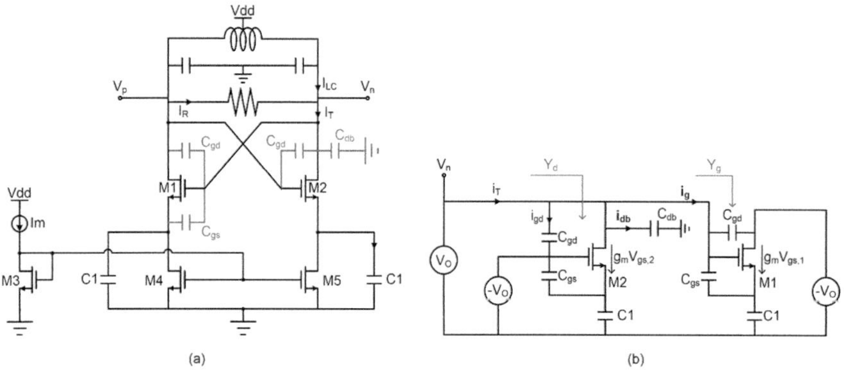

Figure 6.10 VCO with reactive power cancellation (a) schematic, and (b) one-port representation.

6.2.3.2 Review of oscillator noise models

(a) Time-invariant model

The phase noise in oscillators has been a topic of widespread research due to the complex nature and process of how the noise is generated in the oscillators. One of the earliest phase noise models still widely used by circuit designers was proposed by D.B. Leeson in 1966 [35]. The time-invariant Leeson's phase noise model for resonator-based oscillators is

$$L\{\Delta\omega\} = 10\log\left[\frac{2FKT}{P_s}\left\{1+\left(\frac{\omega_0}{2Q\Delta\omega}\right)^2\right\}\left\{1+\frac{\omega_{1/f^3}}{|\Delta\omega|}\right\}\right], \tag{6.12}$$

where F is the excess device noise factor, k is the Boltzmann constant, T is the absolute temperature, P_S is the power of the output signal, Q is the quality of the tank circuit, ω_o is the frequency of oscillation, $\Delta\omega$ is the frequency offset from the carrier frequency and ω_{1/f^3} is the corner frequency between the $1/f^3$ and $1/f^2$ region as shown in Fig. 6.11. In summary, according to Leeson's model, the phase noise in the oscillator can be improved with higher signal power or high quality in the tank circuit.

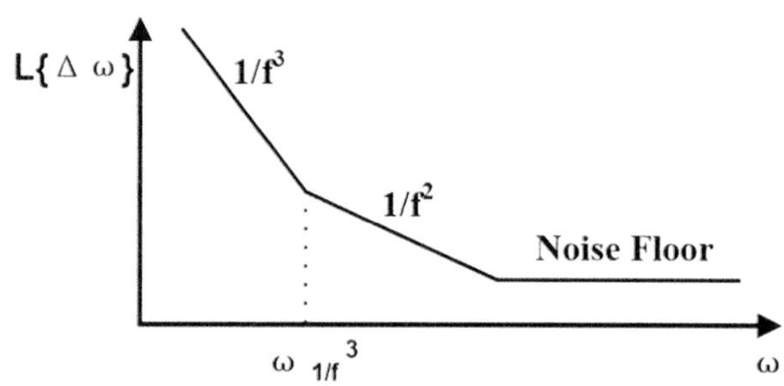

Figure 6.11 Typical oscillator phase noise spectra.

Though simple and advantageous for relatively accurate prediction of VCO phase noise, one of the main drawbacks of the Leeson phase noise model is the estimation of device noise factor F. The empirical parameter F cannot be obtained analytically and depends on the oscillator topology. Also, ω_{1/f^3} in Eq. (6.12) is not the typical device $1/f$ noise corner and is not known to the circuit designer without measurements. In addition, the model assumes the VCO is linear, but oscillators are generally non-linear time-varying systems. However, several attempts have been made in recent literature to improve the phase noise prediction based on the Leeson's model [21, 36, 37].

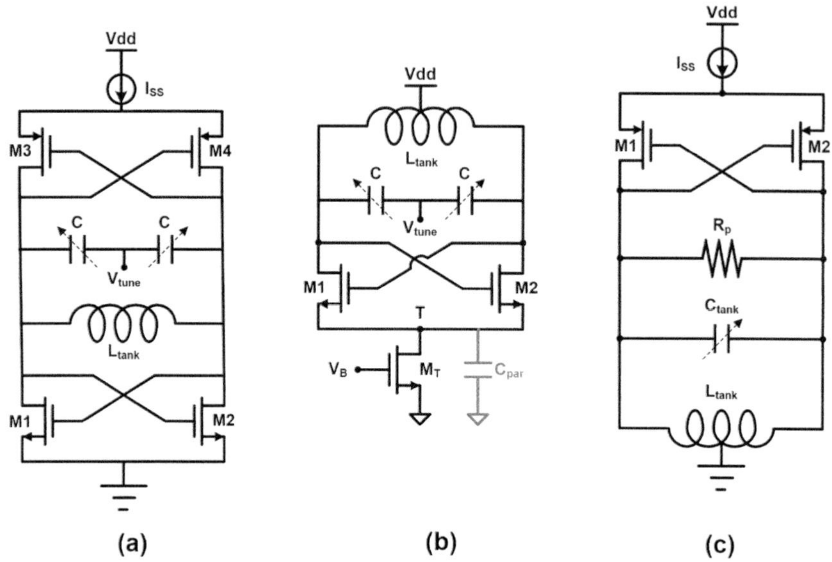

Figure 6.13 Class-B oscillators: (a) CMOS, (b) NMOS and (c) PMOS.

The oscillator start-up is shown in Fig. 6.14(a). As the oscillator starts, the output voltage increases linearly and reaches a steady state. As the output voltage increases, the cross-coupled pair act as hard switches alternating between ON and OFF state as shown in Figs. 6.14(b) and (c). As they enter the triode region, the transconductance decreases and the output voltage stops growing reaching a steady state.

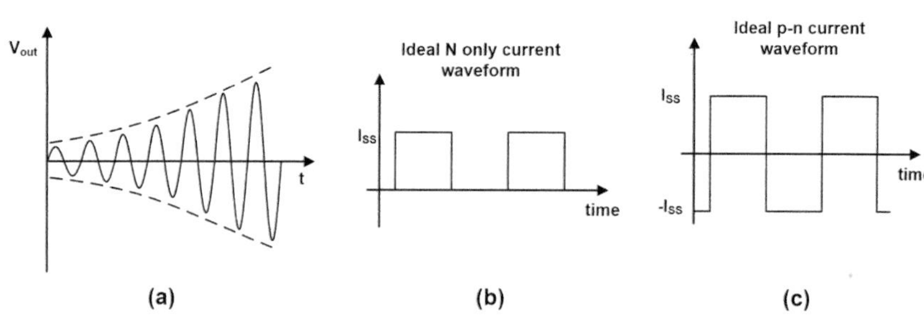

Figure 6.14 Class-B oscillator output: (a) start-up voltage, (b) ideal current (NMOS) and (c) ideal current (CMOS).

In the steady-state conditions, the oscillator operation can be classified into voltage and current-limited regimes [40] as shown in Fig. 6.15. In the current-limited regime, the oscillator output swing is directly linear to the bias current and when the oscillator output voltage saturates due to finite voltage headroom, the oscillator operates in the voltage-limited regime. In the voltage-limited regime, increasing the bias current of the oscillator only increases the power consumption and reduces the oscillator figure-of-merit (FOM).

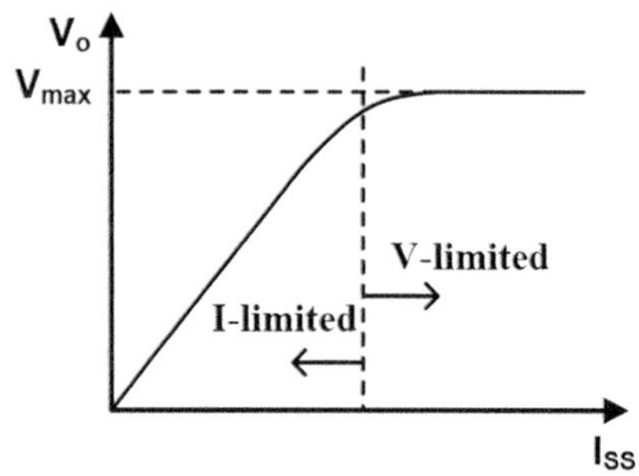

Figure 6.15 LC VCO operation regions at steady state.

In the current-limited regime, the negative resistance transistors fully conduct the bias current due to the large output voltage. The current exhibits an almost square waveform similar to an ideal current source. The square current waveforms produce a voltage swing across the LC tank and the output voltage is given as $V_{\mathrm{od}} = 2 \cdot I_{\mathrm{SS}} \cdot R_T/\pi$, where $2 \cdot I_{\mathrm{SS}}/\pi$ is the fundamental harmonic component and R_T is the equivalent parallel resistance of the tank circuit. For the CMOS implementation, due to the current-reuse in the PMOS and NMOS transistors, the voltage swing is twice the NMOS/PMOS implementation as shown in Figs. 6.14(b) and 6.14(c), respectively, and the output voltage is given as $V_{\mathrm{od}} = 4 \cdot I_{\mathrm{SS}} \cdot R_T/\pi$.

In the realistic implementation, the ideal current source is replaced with a current bias transistor and there exists a minimum voltage V_{tail} to keep the tail transistor in saturation. As the bias current I_{ss} is increased, the V_{ds} of M_{tail} decreases due to the increase in V_{gs} of $M_{1,2}$ (Fig. 6.13(b)) and M_{tail} enters the triode region. The output voltage of the oscillator in the voltage-limited regime is given as $V_{m,\mathrm{out}} = I_{\max} \cdot R_T$ where $I_{\max}$ is the maximum current as the current source transistor enters the triode region.

In comparison to the NMOS / PMOS VCO, the CMOS-based VCO saturates earlier due to the additional V_{gs} drop though the output swing is high. The best performance of the class-B oscillator is achieved when the output oscillation amplitude is around VDD [40, 41].

Phase noise is one of the important performance parameters of the VCO. The phase noise contribution of the popular class-B topology is based on the periodic steady-state model and is proposed in [42]. The noise figure contributions from different noise source in the LC oscillator consisting of tail current source, cross-coupled transistor and the resonator can be given as [21]

$$F = 1 + \frac{4\gamma I_{SS} R_T}{\pi V_{m,\text{out}}} + \frac{4}{9}\gamma g_{m,\text{tail}} R_T, \tag{6.14}$$

where γ is the channel noise coefficient of FET (2/3 for long channel and higher for short channel devices), I_{SS} is the tail current, $g_{m,\text{tail}}$ is the transconductance of tail current transistor and R_T is the equivalent impedance of the resonator. For the ideal class-B oscillator, as given in Eq. (6.14), $F = 1 + \gamma$ as the tail transistor is ideal and provides a high impedance for the whole oscillation period. However, in real implementations, the noise contribution from the tail transistor affects the oscillator's maximum achievable amplitude and phase noise performance as shown in Fig. 6.16 [43].

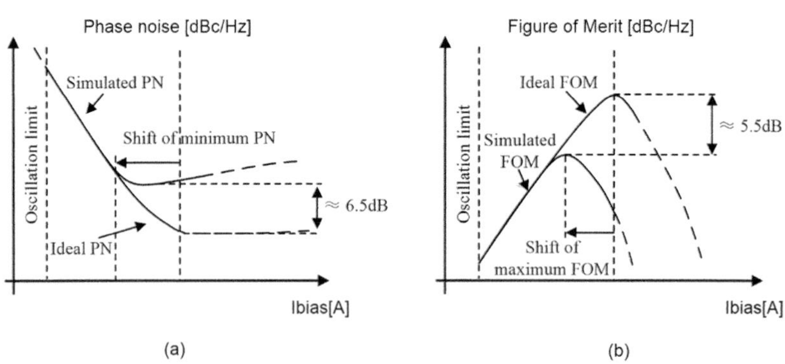

Figure 6.16 Effect of a tail transistor in real class-B VCO: (a) phase noise, and (b) FOM [43].

In the class-B topology with a current source, the tail transistor set the bias currents and also inserts a high impedance path in series with the negative resistance transistors. Therefore, the implementation of a tail current transistor in the class-B topology is also of particular interest. The current source can be implemented in 3 ways: (1) as a single transistor, (2) in cascode configuration, or (3) using a resistor. In advanced process technology, due to lower supply voltages and additional high-frequency noise from the cascode transistors, a single transistor implementation is preferred for RF VCOs. In the single transistor current source implementation, the transistor is typically sized relatively large to reduce the transistor flicker noise. However, this introduces a large parasitic capacitance (C_{par}) at the source of $M_{1,2}$ (as shown in Fig. 6.17) and this C_{par} tends to maintain a common source voltage and the current experiences a dimple reducing the current efficiency as $M_{1,2}$ enters the triode region [43]. As $M_{1,2}$ enters deep-triode region (voltage-limited regime), C_{par} creates a discharge path to ground degrading the tank Q and phase noise. In some implementations, to avoid the tail transistor flicker noise up-conversion, resistive bias is used [9]. The thermal noise due to resistors can be reduced with a large tail capacitance to provide a low-frequency path to filter the thermal noise.

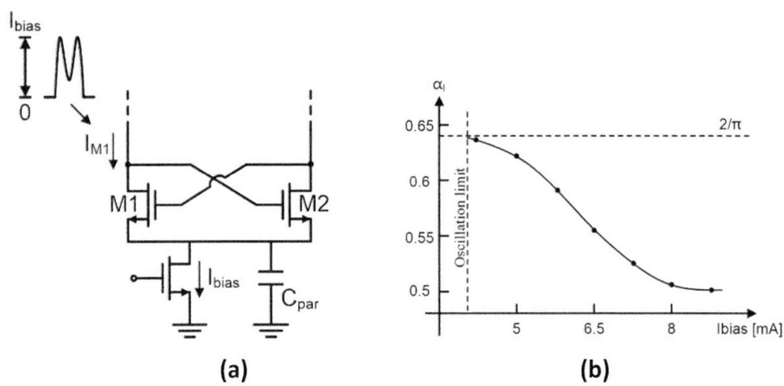

Figure 6.17 Class-B VCO with tail transistor parasitic capacitance: (a) schematic and (b) current efficiency [43].

Additionally, other solutions are proposed in the literature to reduce the flicker noise up-conversion in a class-B oscillator with a tail current source. Noise filtering as proposed in [21] is one of the most popular techniques to considerably reduce the flicker noise up-conversion of the tail current transistor. In this technique, a large capacitor along with the inductor forms a tuned circuit at twice the harmonic frequency to provide a high impedance path for the even harmonics flowing in the common-mode path and to prevent the discharge to the tank circuit. However, it is only effective for narrow-band applications, and it increases the die-area due to the presence of an additional resonator.

6.3.2 Class-C Oscillator

The oscillator topologies have evolved over the years to improve the phase noise and power consumption trade-off and oscillator FOM. The tail current shaping was introduced in [44] to obtain a high output swing and better phase noise. The basic idea was to make the tail current large when the oscillator output is at its maximum or minimum and to turn off the current during the zero-crossing resulting in an improved DC to RF efficiency. The improved version of the class-C oscillator is shown in Fig. 6.18(a). In comparison to the class-B topology, class-C oscillator has their negative transconductance transistors $M_{1,2}$ AC coupled to the output voltage and biased separately at a voltage lower than the supply voltage. This ensures $M_{1,2}$ remain in saturation over the whole oscillation period. The drain currents of $M_{1,2}$ for the class-C are shown in Fig. 6.18(b). With a large C_{tail} similar to class-B, the conduction angle is lower than half a period and the current waveforms are tall and narrow pulses resulting in a better phase noise. For the same bias current, class-C oscillator achieves almost twice the excess noise factor delivering a better performance in comparison to class-B oscillators [33]. However, at large oscillation amplitudes, due to the large C_{tail} and class-C operation, when $M_{1,2}$ enters the triode region, they load the tank circuit and the drain current no longer exhibits the tall narrow pulses limiting the phase noise performance. The maximum swing the transistors $M_{1,2}$ is still at saturation is $(V_{\text{dd}} - V_B)$ assuming the $M_{1,2}$ are biased close to the threshold voltage.

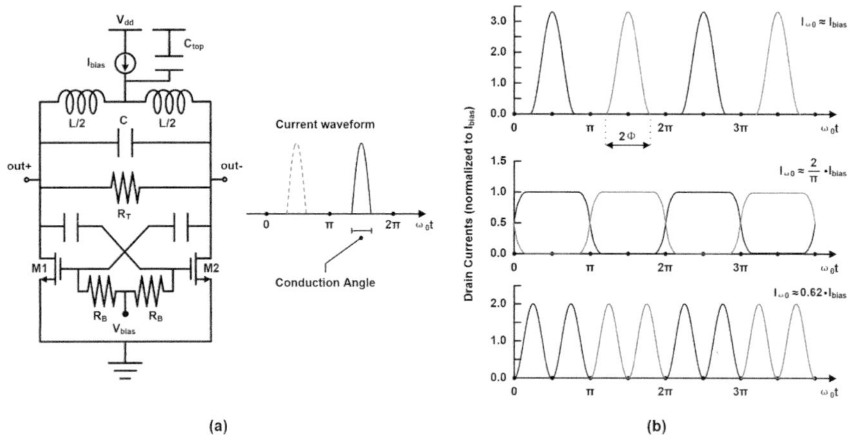

Figure 6.18 Class-C VCO: (a) NMOS only schematic and (b) drain current waveform [33].

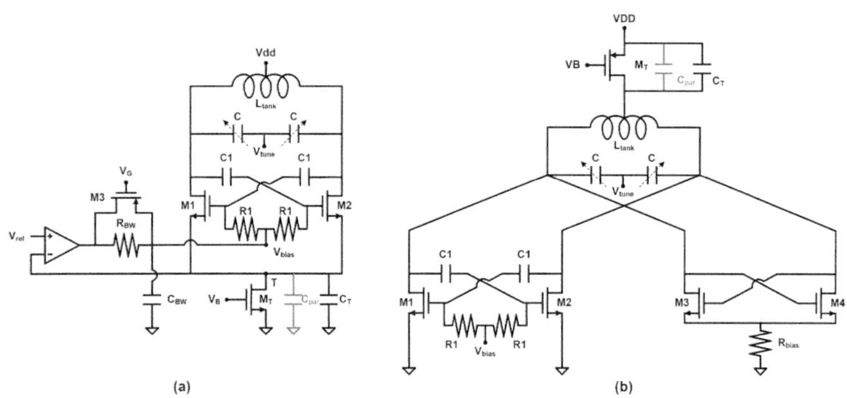

Figure 6.19 Class-C VCO: (a) with dynamic V_{bias} [43] and (b) hybrid class-B/class-C [48].

One of the important drawbacks of the class-C topology is its high V_{bias} requirement for robust start-up. In practical designs, the best performance is achieved when the V_{bias} is above the threshold voltage. However, this limits the start-up of the class-C oscillators. Several works have been reported in the literature to improve the performance of class-C oscillators. In [43], a dynamic V_{bias} class-C oscillator is proposed for an optimum trade-off between start-up and steady-state conditions. V_{bias} is set at a large value to facilitate start-up and dynamically adjusted in steady-state conditions [43, 45, 46]. In [47] and [48], a hybrid class-B topology in parallel to class-C is proposed as shown in Fig. 6.19. The class-B pair ensures the start-up robustness while the class-C pairs are applied a low V_{bias}. Though the hybrid class-B/class-C has less efficiency in comparison to pure class-C topology, it still achieves 25% higher efficiency than pure class-B oscillators for the same bias current [48].

6.3.3 Class-D Oscillator

Class-D oscillator topology is an extension of class-B topology with the removal of the current source transistor as shown in Fig. 6.20. With the removal of the current source, the class-D oscillator is best suited for low supply voltage conditions and the output oscillation amplitude is also maximized. However, class-D oscillators are very sensitive to supply variations in comparison to class-B and class-C oscillators and require voltage regulators and on-chip LDO to stabilize the supply voltage in real-life applications [49].

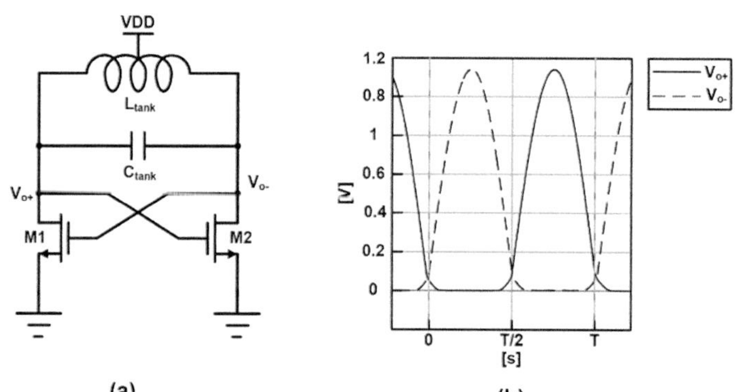

Figure 6.20 Class-D VCO: (a) schematic and (b) output waveform.

One of the attractive features of the class-D oscillator is its high oscillation amplitude $\sim 3V_{DD}$ and achieving a low phase noise with low supply voltage [50, 51]. Also, for a good class-D operation, the transistors $M_{1,2}$ are chosen to be large, so that they act as ideal switches and their equivalent resistance is less in comparison to the tank equivalent resistance. Due to ideal switching, in comparison to class-B, $M_{1,2}$ are in deep-triode region and are in voltage-limited regime. However, the oscillation voltage at the output is forced to ground for at least half the period as either M_1 or M_2 is in triode and thus preventing the $M_{1,2}$ noise to be up-converted to the phase noise due to almost zero ISF during the half-period.

Also, the oscillation frequency and phase noise of class-D oscillators are different in comparison to class-B oscillators. The class-D oscillators exhibit different frequency and phase noise characteristics based on the capacitance of the tank circuit as shown in Fig. 6.21 [52]. During one half of the oscillation period, T_1 $(0 < t < T_{osc/2})$, M_1 shorts the inductor to ground.

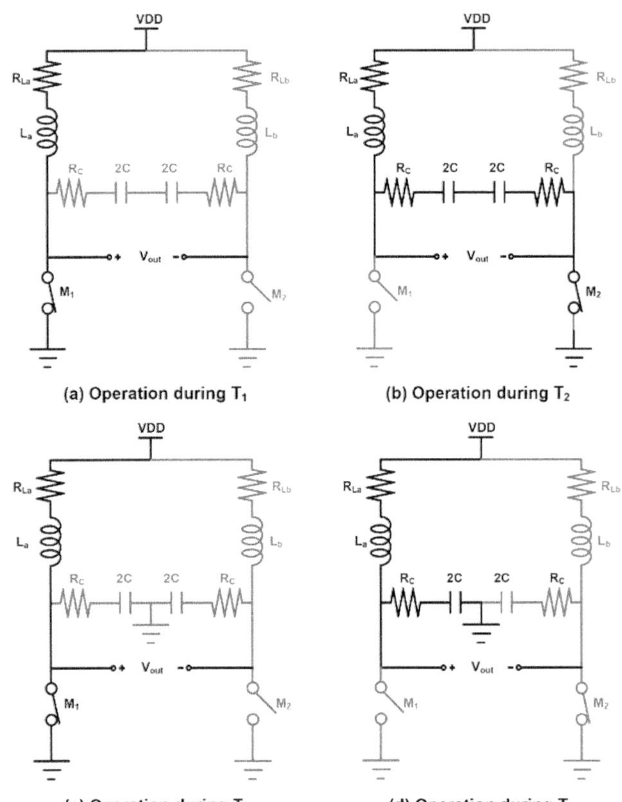

Figure 6.21 Class-D VCO operation: (a) floating capacitance (T_1), (b) floating capacitance (T_2), (c) single-ended capacitance (T_1), and (d) single-ended capacitance (T_2) [52].

In the second half of the oscillation period, T_2 ($T_{osc/2} < t < T_{osc}$), $M_1(M_2)$ is OFF and the inductor resonates with the equivalent tank capacitance (C_{eq}). However, C_{eq} equals $2C$ in floating configuration and C in single-ended configuration.

Therefore, the frequency of oscillation and phase noise for floating and single-ended capacitors in class-D is given as [52]

$$\omega_{0,float} = \frac{\sqrt{2}}{\alpha}\sqrt{\frac{1}{LC}} \tag{6.15}$$

$$\omega_{0,se} = \frac{1}{\alpha}\sqrt{\frac{1}{LC}} \tag{6.16}$$

and

$$L_{float}(\Delta\omega) = 10\log_{10}\left[\frac{\omega_{0,float}^2}{\Delta\omega^2}\frac{k_B T}{V_{DD}^2}(0.104R_L + 0.141R_C)(1+n_{MOS})\right] \tag{6.17}$$

$$L_{se}(\Delta\omega) = 10\log_{10}\left[\frac{\omega_{0,se}^2}{\Delta\omega^2}\frac{k_B T}{V_{DD}^2}\left(0.104R_L + \frac{0.141}{2}R_C\right)(1+n_{MOS})\right], \tag{6.18}$$

where $\alpha \approx 1.3$, R_L and R_C are inductor and capacitance losses, respectively, k_B is the Boltzmann constant, T is the temperature and n_{MOS} is the noise generated by switch transistors normalized by tank noise. Therefore from Eqs. (6.15) and (6.16), it can be observed that for the same values of L and C, floating (differential) and single-ended capacitor implementation achieves a higher and lower resonant frequency, respectively, in comparison to a fixed resonant frequency in class-B oscillators. Also, from Eqs. (6.17) and (6.18), it can be observed that the best phase noise in class-D is achieved with differential capacitance and tank loss dominated by the inductor. Also the $1/f$ noise up-conversion of $M_{1,2}$ to the overall phase noise can be reduced with large switch transistors resulting in reduced on-resistance. However, in practical designs, the large size of $M_{1,2}$ leads to higher parasitic capacitance limiting the oscillation frequency.

6.3.4 Class-F Oscillator

Class-F oscillator acts on the resonator to shape the oscillator waveform such that the ISF is negligible at certain oscillation periods and thus improves the oscillator phase noise due to the reduction in the $1/f$ noise up-conversion. The schematic and output waveform with expected ISF for a class-F oscillator are shown in Fig. 6.22 [53]. In a class-F oscillator, the oscillator waveform is shaped by injecting the second or third harmonics of the fundamental frequency realizing a class-F_2 or class-F_3 oscillator, respectively.

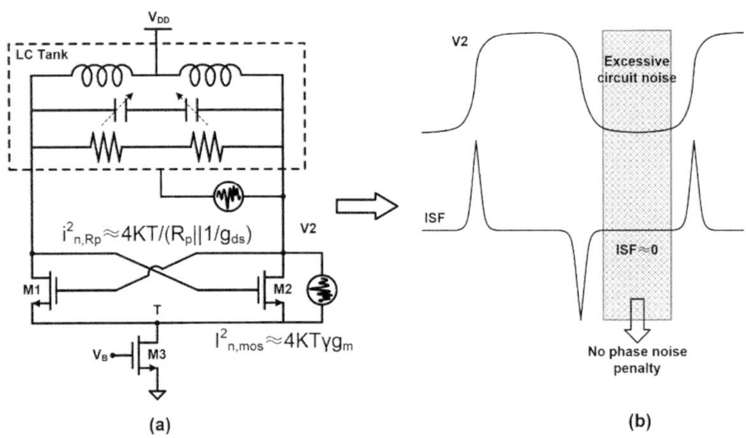

Figure 6.22 Class-F oscillator: (a) schematic and (b) output waveform and expected ISF.

The operation principle of a class-F oscillator is derived with an understanding of the special ISF properties of a square wave oscillation voltage as shown in Fig. 6.22(b) (see above). When the oscillator waveform is a square rather than sinusoidal, the special ISF properties of the square wave can be exploited [38] to achieve better phase noise and power efficiency. Though the g_m devices are in the triode region and inject more noise along with the loaded resonator into the tank, the ISF is expected to be negligible due to the zero derivative of the oscillation voltage [38]. In a typical oscillator, though the drain current is a square wave with fundamental and odd harmonics, the tank impedance has its peak at the fundamental frequency and filters out the odd harmonics. However, if the tank offers, additional impedance magnitude peak at the third harmonic as shown in Fig. 6.23, the third harmonic component of the drain current is not filtered out by the tank and a square wave across the LC tank can be realized.

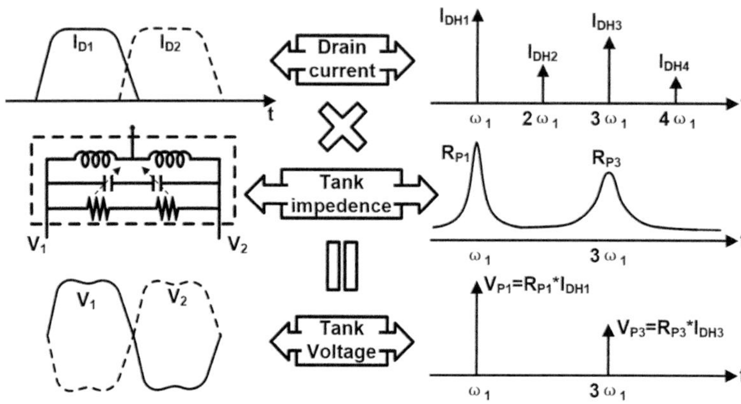

Figure 6.23 Class-F oscillator waveform in time and frequency domain [53].

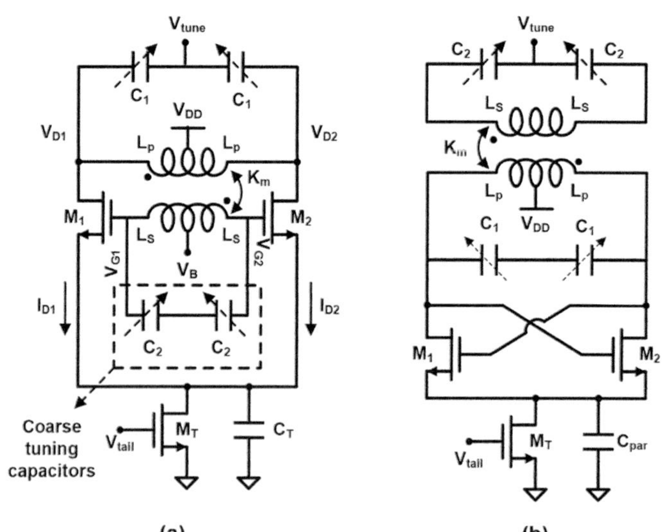

Figure 6.24 Class-F oscillator topology: (a) transformer coupled and (b) cross-coupled [53].

The typical implementation of the class-F oscillator is shown in Fig. 6.24 [53]. The class-F oscillator can either be in cross-coupled topology or transformer-coupled topology. In the transformer-based topology, the secondary is connected to the gate of the active transistors and the transistors sustain a larger voltage swing in comparison to the cross-coupled topology [53]. Also, the active devices generate a higher amount of third harmonic resulting in a sharper pseudo-square oscillation voltage with lower ISF rms value. In addition, the transformer-based topology guarantees oscillation only in the desired resonant frequency. However, due to the additional gate parasitic capacitance at the drain, the tuning range is lower in comparison to the cross-coupled topology.

6.4 Oscillator Design Methodology

The design of an oscillator starts with understanding the important performance parameters such as phase noise, tuning range, tuning sensitivity, and power consumption. Since the basic oscillator used in the RF/mm-Wave frequencies is the voltage-controlled oscillator requiring low phase noise and low power consumption, LC VCO design is outlined.

The VCOs can either be narrow-band or wide-band VCOs. Due to the current trends towards multi-band multi-standard VCO designs, the tuning range of the VCO needs to be ascertained to choose the values of the resonant circuit. A high quality factor of the resonant tank is required for a low phase noise in the VCO. However, in practical design, VCO is an integral component of the frequency synthesizer system, and the overall phase noise is also dependent on the tuning sensitivity of the VCO. When the VCO has a very high tuning sensitivity, the noise injected into the VCO tuning injection port from the preceding circuits can degrade the overall noise and generate undesired spurs in the VCO output. Therefore, VCO design should not only consider the tuning range coverage but also the tuning sensitivity.

(a) VCO tank circuit

In the LC VCO, the tank circuit is formed by the inductor and capacitor. For the on-chip inductors, the loss in metal traces is represented by the series resistance R_S. In low frequencies (<10 GHz), the quality factor of the inductor is worse than the quality factor of capacitors and dominates the overall tank quality factor. Therefore, assuming the loss from the capacitor is negligible, the quality factor of the inductor is given as

$$QL = \frac{\omega_{oL}}{R_s}, \tag{6.19}$$

where ω_o is the resonant frequency. The equivalent parallel tank impedance at resonance is given as

$$R_T = Q_T \omega_o L = \frac{(\omega_o L)^2}{R_s}. \tag{6.20}$$

From Eqs. (6.19) and (6.20), larger inductance results in higher quality factor and lower power consumption. Therefore, it is imperative to choose the largest inductor and small capacitor for the chosen resonant frequency satisfying the tuning range requirements. However, the large inductor has a high parasitic coupling capacitance to the substrate and limits the self-resonant frequency of the inductor. The typical choice of self-resonant frequency in practical VCO design is at least 3 times the resonant frequency. Also, the large inductor results in a larger frequency step and smaller tuning range. Therefore, finding the optimum value of the inductance satisfying the tuning range and self-resonant frequency is non-trivial and a few iterations are required after initial simulations.

In the current scenario covering multiple frequency bands in a single chip, tank inductors have been replaced by transformer or coupled inductor in the VCOs.

Several transformer-based oscillators have been reported in [54–71] to improve the oscillator tuning range without degrading its phase noise performance and FOM. The transformer-based oscillators are employed for multi-mode resonance [54] to improve the common-mode resonance without explicit inductors, dual/multi-band transformers [55–63] and multi-resonant mode and multi-tank oscillators [64–70] for a wide tuning range and optimized trade-off between oscillator tuning range, phase noise and power efficiency.

(b) Start-up condition

For any oscillator, the loss in the tank circuit needs to be compensated to ensure start-up and sustain oscillation. In LC VCOs, due to the non-linearity in the $-g_m$ transistors, oscillator start-up is a function of frequency [72]. To ensure start-up, the minimum transconductance required is given as

$$G_m \geq \frac{1}{R_T} = \frac{R_s}{(\omega_o L)^2}. \tag{6.21}$$

In practical implementations, the G_m is chosen at least 3 times the required minimum to ensure proper oscillator start-up. Arbitrarily increasing the size of $-g_m$ transistors to maximize G_m, leads to large parasitic capacitance limiting the tuning range and increasing the noise contribution. Also, based on the required G_m, the bias current of the VCO can be determined.

(c) Oscillator tuning

The frequency tuning range of an oscillator is the range of frequencies that an oscillator can track and in VCOs, it is linearly proportional to the tuning voltage given as

$$\omega_{\text{out}} = \omega_f + K_{\text{VCO}} V_{\text{ctrl}}, \tag{6.22}$$

where ω_f is the free running frequency, K_{VCO} is the tuning sensitivity in Hz/V and V_{ctrl} is the frequency control voltage. In RF/mm-Wave frequencies, the VCO needs to tune over a range of frequencies and discrete tuning methods with fine frequency step is typically employed for coarse tuning. Once the desired coarse-tuning frequency band is set, varactors are used for fine-tuning.

The maximum and minimum capacitance of the discrete tuning bank determines the tuning range of the VCO. Therefore, the discrete tuning banks need to have a high C_{max}/C_{min} with low loss. The typical implementation of the switched unit capacitance cell with an MIM / MOM capacitor with a switch in series is shown in Fig. 6.25. When the switch is closed, the series combination of the switch and the capacitor appears across the terminals. When the switch is open, the capacitors are disconnected and the switched capactor is in OFF state.

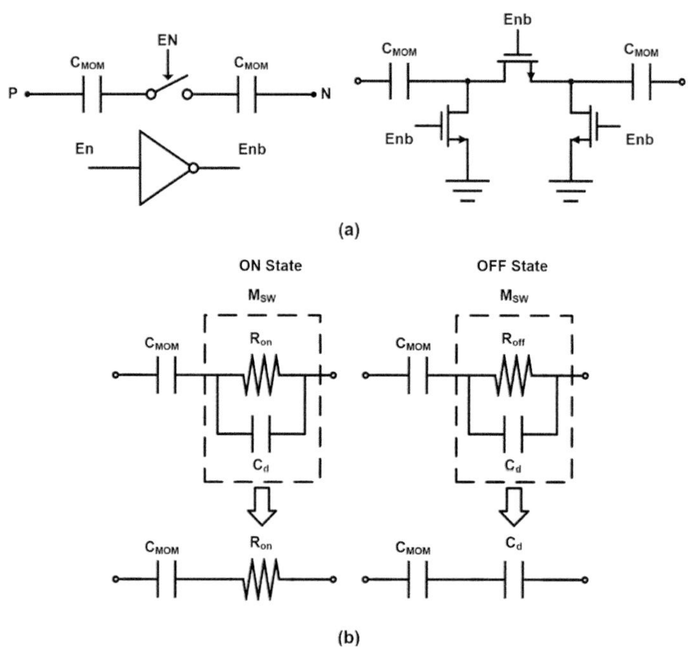

Figure 6.25 Switched capacitor: (a) typical configuration (left) and implementation (right) and (b) equivalent circuit in ON state and OFF state.

During the ON state, the switch can be replaced with the on-resistance (R_{on}) of the transistor and the quality factor is given as

$$Q_{ON} = \frac{1}{\omega R_{on} C_{on}}, \tag{6.23}$$

where $C_{on} = C_{MOM}/2$. Therefore, from Eq. (6.23), to maximize the quality factor, the switch R_{on} must be large. The switch on-resistance is given as

$$R_{on} = \frac{1}{\mu_n C_{ox}\left(\frac{W}{L}\right)(V_{gs} - V_{th})}. \tag{6.24}$$

From Eq. (6.24), it can be observed that (W/L) of switch transistor needs to be large to maximize R_{on}. However, in the OFF state, the parasitic capacitance C_d due to large (W/L) dominates the capacitance and the overall OFF state capacitance is given as

$$C_{OFF} = \frac{C_{off} C_d}{C_{off} + 2C_d}, \tag{6.25}$$

where $C_{off} = C_{MOM}$. Therefore from Eq. (6.25), an increase in (W/L) increases the switch parasitic capacitance in the OFF state and reduces the capacitance range. Therefore, the switch transistor (W/L) must be chosen based on an optimum trade-off between the quality factor in the ON state and OFF state capacitance.

(d) VCO simulation

Based on the proposed design methodology, an NMOS Class-B oscillator is designed as shown in Fig. 6.26. The VCO is composed of negative resistance transistor $M_{1,2}$, tank circuit comprising L_{tank}, C_{coarse} and continuous tuning varactors C_v and a tail-biased resistor. The varactors C_v are used for fine-tuning. The 3-bit switch cap is used for coarse tuning and the C_{max}/C_{min} ratio is 2.1. The negative resistance plot for oscillator start-up is shown in Fig. 6.27. As shown in Fig. 6.27, the VCO negative resistance is maximum at lower frequencies and starts to decrease at higher frequencies. If the negative resistance is not enough at the frequency of interest, then it needs to be maximized with additional tuning circuits [64]. As the loss in the VCO is compensated by the negative resistance, the

VCO starts to oscillate, and the output amplitude starts to increase before reaching a steady-state condition as shown in Fig. 6.28.

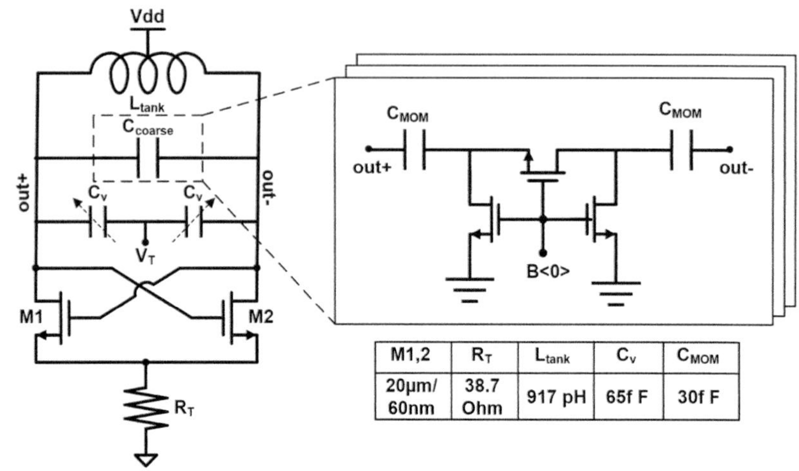

M1,2	R_T	L_{tank}	C_v	C_{MOM}
20μm/ 60nm	38.7 Ohm	917 pH	65f F	30f F

Figure 6.26 VCO schematic.

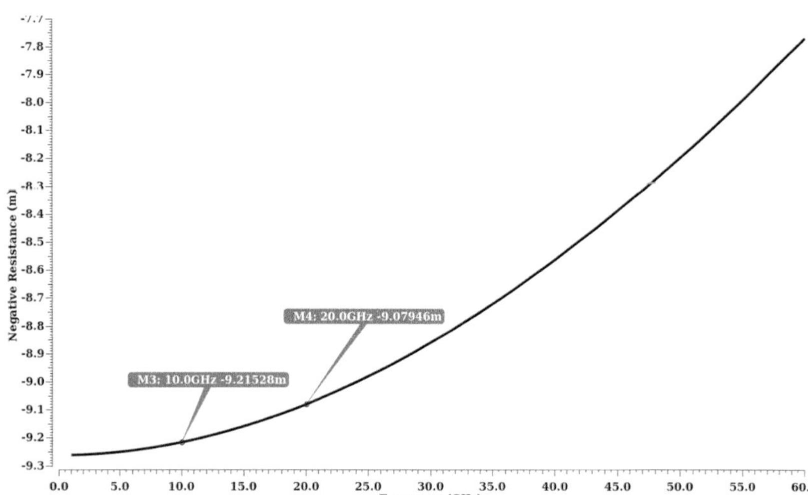

Figure 6.27 Simulated VCO negative resistance.

The tuning range of the VCO is shown in Fig. 6.29. The VCO achieves an overall tuning range from 13.23 GHz to 16.17 GHz with an overlap > 50% between the frequency bands. The overlap between the tuning bands is necessary to compensate for the frequency variation across the process, voltage, temperature (PVT) variations. Also, the overlap helps to compensate the irregular step variations due to the switched capacitance tuning as can be observed in the tuning step in Fig. 6.29. The frequency tuning range (FTR) of the VCO is 20%.

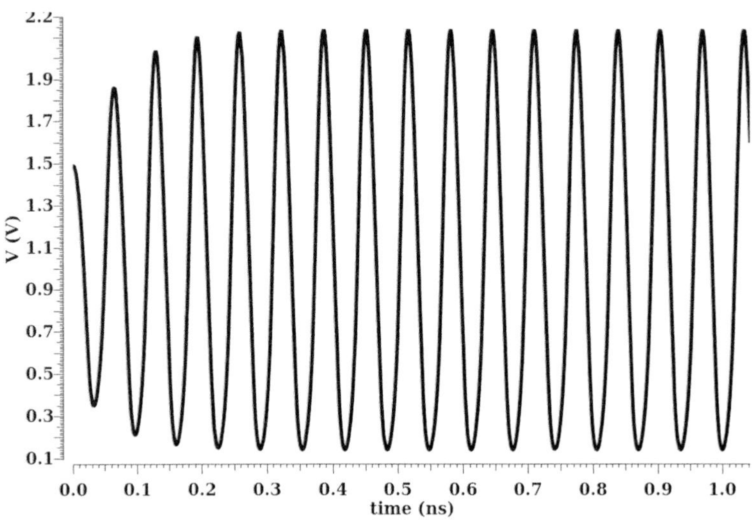

Figure 6.28 Simulated VCO start-up.

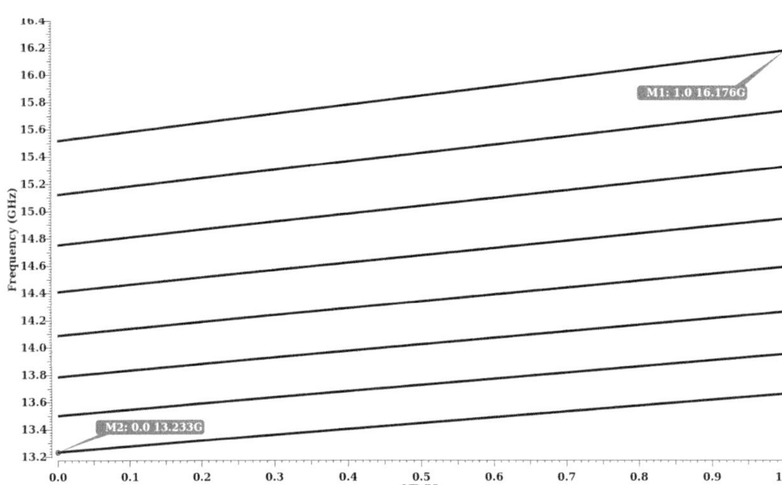

Figure 6.29 Simulated VCO tuning range.

The simulated phase noise of the VCO is shown in Fig. 6.30. The VCO achieves a phase noise of −113.6 dBc/ Hz and −110.7 dBc/Hz at 1 MHz offset from the carrier for the lower and higher frequency range, respectively. Based on the designed VCO, the VCO achieves an FOM and FOM_{FTR} of −185.1 dBc/Hz and −191.14 dBc/Hz, respectively, at 1 MHz offset for the highest frequency of 16.17 GHz.

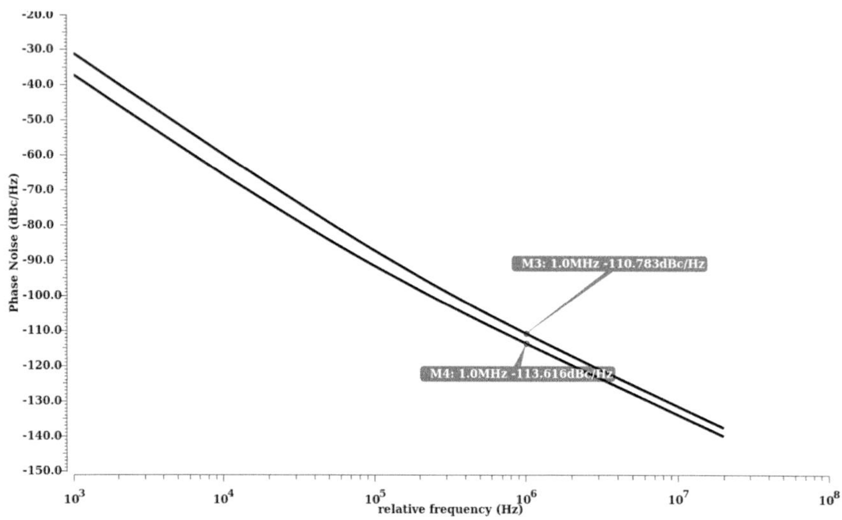

Figure 6.30 Simulated VCO phase noise.

6.5 Advanced Topics in Oscillator Design

6.5.1 Quadrature VCO

Quadrature VCO (QVCO) are class of coupled oscillator topology which has widespread usage in the RF/mm-Wave signal generation where complex modulation and demodulation schemes are employed to maximize the signal bandwidth with more data. Several QVCOs have been reported in the literature [73–82]. Apart from QVCOs, techniques such as frequency divider followed by differential VCO and polyphase filter have been proposed in the literature for quadrature signal generation [83–85]. However, they are lossy and power consumption is large.

The earliest QVCO was proposed in [73] by coupling two differential oscillators with coupling transistors to generate quadrature signals. The model of QVCO [74] and the typical QVCO structure is presented in Fig. 6.31. As shown in Fig. 6.31(a), G_1 and G_2 are fixed frequency differential oscillators, and they are coupled by a coupling network with coupling coefficients M_1 and M_2. In steady-state conditions, if the two oscillators are synchronous and generate a single output frequency, the output of the individual oscillators must satisfy the following condition [74]:

$$(X + M_2 Y)G_2(j\omega) = X \tag{6.26}$$

$$(Y + M_1 X)G_1(j\omega) = Y \tag{6.27}$$

Since identical oscillators are coupled in a QVCO, $G_1 = G_2 = G$ and $M_1 = M_2 = M$. Therefore, from Eqs. (6.26) and (6.27), it can be shown that $X^2+Y^2 = 0$ and $X = \pm jY$. Also, the oscillation frequency can be tuned by varying the coupling coefficients M_1 and M_2 [74].

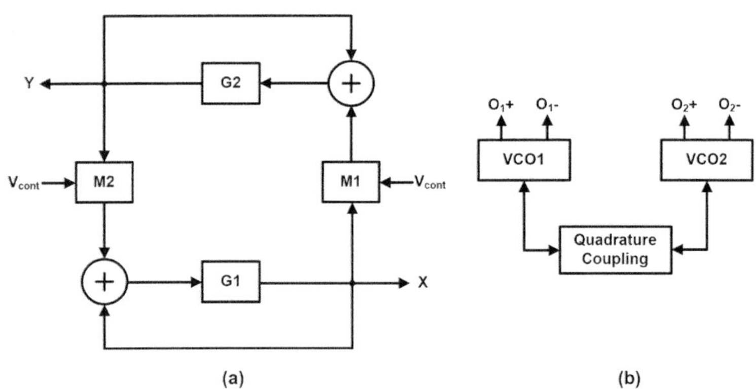

Figure 6.31 QVCO: (a) block diagram representation and (b) typical structure.

(a) Parallel QVCO

The parallel QVCO was proposed in [73] and is shown in Fig. 6.32. The parallel QVCO consists of two LC VCOs outputs coupled to each other using a coupling transistor network formed by M_5–M_8. The coupling transistors are connected in parallel to the switching transistors and can either be biased with the same current source or a separate current source [86] providing a degree of freedom to tune the coupling coefficient between the VCO cores and thus achieving a wider tuning range and additional tuning of oscillation frequency. To obtain quadrature outputs, the outputs of the first core are directly connected to the second core and the output of the second core is cross-coupled to the first core. One of the drawbacks of the parallel QVCO is increased phase noise performance due to the addition of coupling transistors. Also, since coupling transistors are biased separately for additional tuning of coupling or frequency, the power consumption is large. Also, the choice of coupling transistor size is important in parallel QVCO. The bigger size of coupling transistors adds additional parasitic capacitances to the QVCO reducing its frequency tuning range.

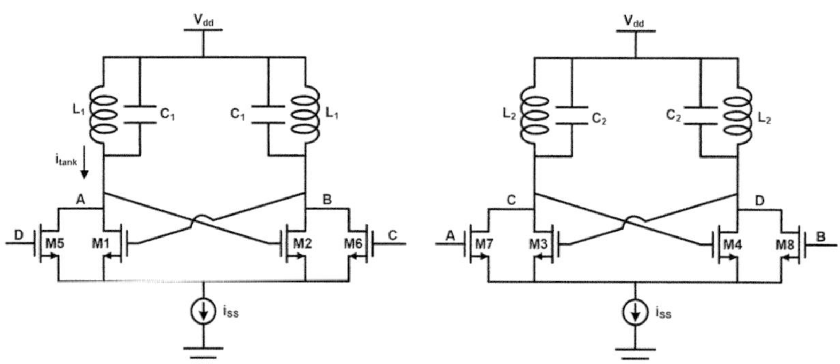

Figure 6.32 Parallel QVCO (NMOS only).

(b) Series QVCO

An alternative to parallel QVCO, series QVCO was proposed in [87] wherein the coupling transistors are placed in series to the switching transistors in cascode configuration as shown in Fig. 6.33. In comparison to the parallel QVCO, in the series QVCO, the switching transistor current is reused for the coupling transistors. The aspect ratio of coupling transistors is of interest in the series QVCO. The coupling coefficient of the series QVCO is given as

$$M = \frac{V_{DS3}}{V_{GS,3} - V_{th}} = \frac{V_{DS8}}{V_{GS,8} - V_{th}}, \tag{6.28}$$

where $V_{DS3,8}$ and $V_{GS3,8}$ are drain-source and gate-source voltage of coupling transistors. In the QVCO, it is important to maximize the coupling coefficient for matched IQ in the presence of transistor mismatches and non-linearity [81]. However, for low-power applications, it is difficult to control the coupling coefficient as V_{GS} of coupling transistor is coupled to oscillator output nodes and is dependent on the V_{GS} of cross-coupled PMOS switching transistors. The PMOS switching transistor can only be made small if the (W/L) is made small. A smaller PMOS (W/L) will in turn make NMOS (W/L) small if the PMOS size is chosen to be 2–3 times NMOS and also can reduce the negative conductance for oscillator start-up. Therefore, there is a trade-off and compromise between the size of switching transistors and the coupling efficient M in the series QVCO.

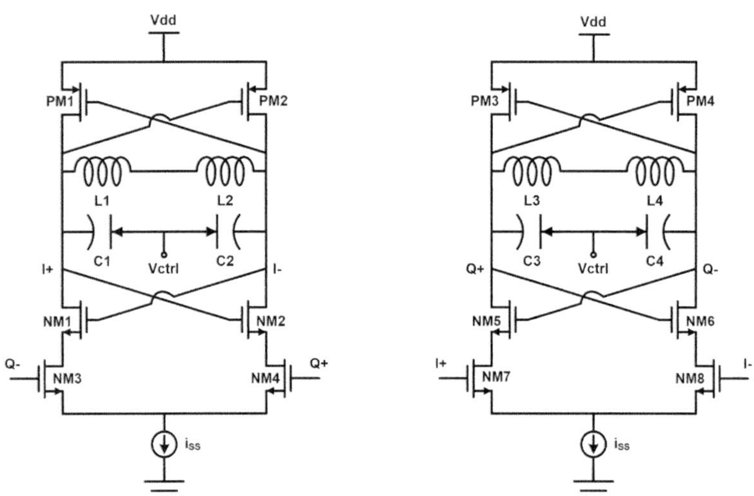

Figure 6.33 Series QVCO (CMOS).

Though ideally coupling transistor (W/L) should be made large to reduce the phase noise contribution [87, 88] the transistor parasitic capacitance will be large and load the

LC tank reducing its frequency and tuning range. Therefore, as a compromise, the coupling transistors' size is typically chosen to be same size as NMOS switching transistors. Also, it is difficult to design series QVCO in low supply voltages due to the stacking of transistors.

(c) Super-harmonic coupled QVCO

Though largely popular, one of the drawbacks of the series and parallel QVCO is bimodal oscillation, and the oscillation frequency is not the resonant tank frequency [74, 86]. Super-harmonic coupling for quadrature signal generation is proposed in [76] and is shown in Fig. 6.34. As shown in Fig. 6.34, a transformer is used to couple the VCO cores and for IQ signal generation, the direction of coupling needs to be anti-phase to each other. The $2f_{\mathrm{osc}}$ at the tail of the switching transistor gets down-converted to f_{osc} by the switching transistors. The inductors $L_{5,6}$ are chosen to resonate the parasitic capacitance of the frequency when driven in anti-phase such that the oscillator operates in the current mode regime and the oscillation amplitude increases [76]. The transistors $M_{6,9}$ are added to give directivity for the IQ thus resolving the ambiguity on the

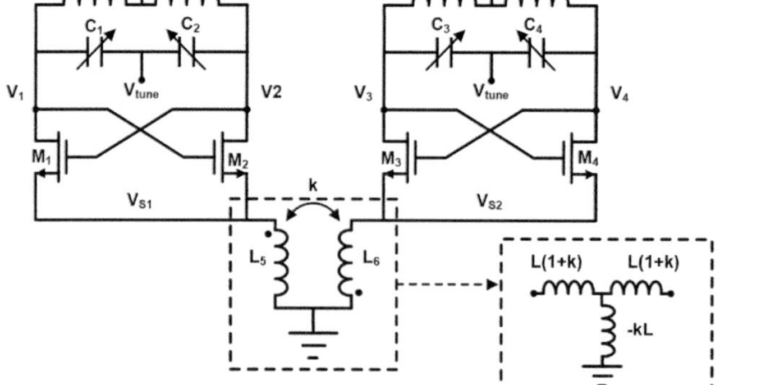

Figure 6.34 Super-harmonic QVCO [76].

I and Q phases. Since $M_{6,9}$ are small transistors, their current consumption is small. In comparison to series or parallel QVCO, the super-harmonic coupling VCO does not exhibit bimodal oscillation.

To overcome the die size limitation due to transformer-based super-harmonic coupling, capacitor-based coupling is proposed in [89] as shown in Fig. 6.35. When the VCO operates with a phase difference of $\pi/2$, the VCOs lock to each other to generate IQ signals. In this mode, the coupling is maximum and is equal to $4C_c\omega_0A_s$ [89] and the oscillation amplitude is maximum.

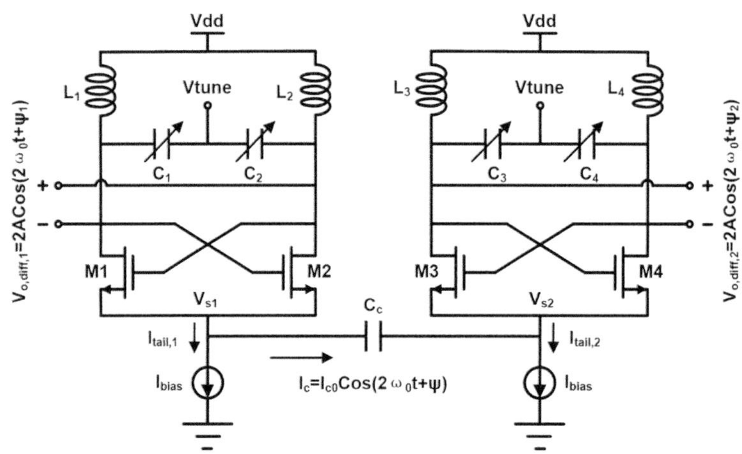

Figure 6.35 QVCO with coupling capacitor [89].

6.5.2 Digital Controlled Oscillators

To take advantage of the continuous scaling of semiconductor technology, digital signal generation techniques have been proposed [90]. In the digital signal generation circuits, a digital controlled oscillator (DCO) is implemented. Though there are quite a lot of implementation variations in the DCO, we will focus on the DCO implementation for high-frequency RF and mm-Wave applications.

In comparison to the voltage-controlled oscillators, the DCO does not have any analog tuning and the frequency is tuned with digital control word. The design of high-frequency DCO is analogous to the VCO described in previous sections for the inductor, active part and coarse tuning capacitor banks. The major difference in the DCO design is in the generation of oscillator output using the digital control bits and will be the focus of this section.

Since the output frequency of the DCO is typically generated with a single inductor and large number of digitally tuned capacitor cells, the output frequency is given as

$$f_{\mathrm{DCO}} = \frac{1}{2\pi\sqrt{L\sum_{i=0}^{N-1} C_k}}, \tag{6.29}$$

where C_k is the overall capacitance of the DCO to cover the required frequency tuning range. In addition to the frequency tuning range and phase noise, the frequency resolution is one of the important specifications of the DCO as it will determine the resolution of the DCO output frequency and also determine the overall phase noise of the digital phase-locked loops. The phase noise contribution of the DCO at an offset frequency Δf is given as [90]

$$L(\Delta f) = \frac{1}{12}\left(\frac{\Delta f_{\mathrm{res}}}{\Delta f}\right)^2 \frac{1}{f_s}\mathrm{sinc}\left(\frac{\Delta f}{f_s}\right)^2, \tag{6.30}$$

where f_{res} is the resolution of the DCO and f_s is the reference frequency. Therefore, for the quantization noise from DCO to be negligible, the resolution of the DCO needs to be small.

6.5.2.1 Design of DCO capacitor banks

The typical schematic of DCO is shown in Fig. 6.36. The DCO is composed of multiple digitally controlled capacitor banks split across the frequency tuning range often around 16–20 bits for a fine-resolution DCO. Since there are a large number of control bits in the DCO, it is impossible to design the capacitor bank with precision without error correction techniques due to component mismatch [91]. Therefore, the DCO operation mode is split into multiple capacitor banks as follows:

(a) PVT or coarse-tuning bank which has the largest resolution covering the required frequency tuning range

(b) Acquisition or medium-tuning bank which has a lower resolution than coarse-tuning bank to reduce the large PVT resolution to fine DCO resolution

(c) Tracking or fine-tuning bank which has the lowest resolution equal to the required DCO resolution

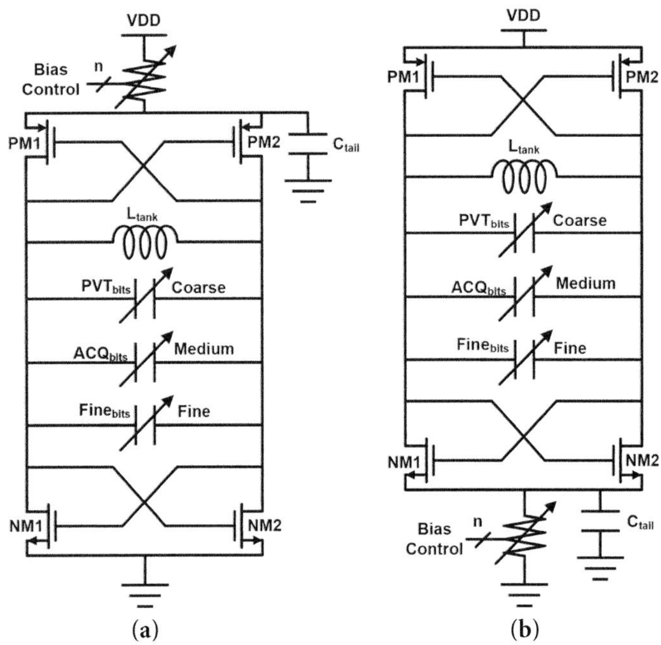

Figure 6.36 DCO schematic: (a) top bias and (b) bottom bias.

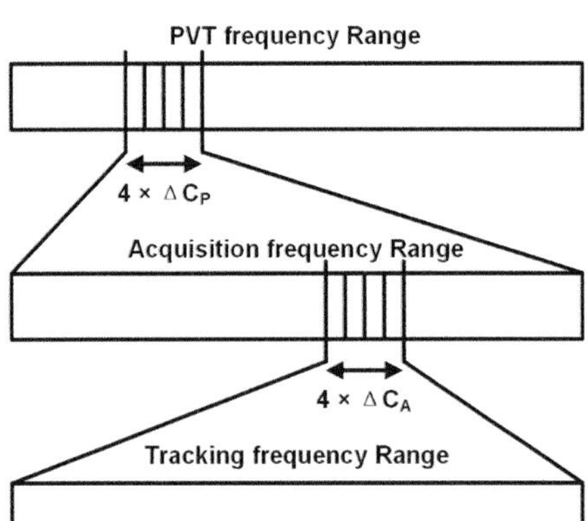

Figure 6.37 DCO frequency resolution and frequency range of capacitor banks.

The relationship between the frequency resolution of the individual banks and the required frequency range is given according to Eq. (6.31) [90] and is shown in Fig. 6.37. In practical DCOs, due to non-linearity and mismatch errors in the capacitor banks, a margin factor of around 4 is chosen. The lower margin factor results in discontinued frequency tuning, and the higher margin factor results in a large frequency overlap increasing the design complexity due to additional capacitor banks.

$$\Delta f = \left(2\pi^2 L f^3\right)\Delta C \qquad (6.31)$$

(a) Coarse-tuning bank

The design of capacitor banks for the DCO is dependent on the resolution requirement of the individual tuning banks and the coarse-tuning bank spans the whole DCO tuning range. The coarse-tuning bank is implemented as a switched capacitor as discussed earlier in Section 6.4 and the unit cell implementation is shown in Fig. 6.38 for completeness. During the ON state, the capacitors are connected across the output terminals and the equivalent capacitance is $C_{MOM}/2$. In the OFF state, the capacitance is determined by the parasitic of the MOS switch.

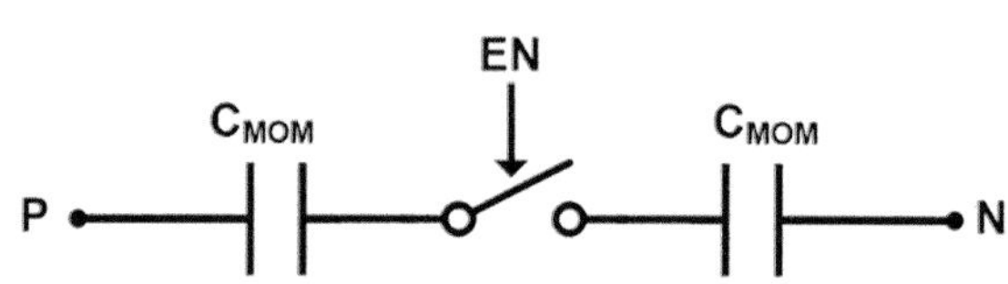

Figure 6.38 Switched MOM capacitor.

(b) Medium-tuning bank

The medium-tuning bank has a smaller resolution than the coarse-tuning bank and a higher resolution than the fine-tuning bank. In most cases, the implementation is similar to the coarse-tuning bank but with a smaller value of MOM capacitance. However, in a typical DCO, the medium-tuning bank has to be around 6–10 bits to cover the required frequency step and the unit cells are arranged in a matrix. In some cases, the medium-tuning banks are controlled by thermometer coding and the bits are used to drive the row and column decoders of the unit cell array [92]. An example of 6-bit medium-tuning bank controlling the 8×8 matrix of unit cell varactors is shown in Fig. 6.39.

(c) Fine-tuning bank

The fine-tuning bank is analogous to the continuous tuning varactor of the VCOs and locks the digital PLL to the target frequency with the required resolution. Therefore, to achieve a fine resolution, a very low capacitance in terms of attofarad (aF) is required. For example, for a frequency of 2.7 GHz with inductance of 7.7 nH and 50 kHz resolution, a capacitance of 16.7 aF is required for the fine-tuning bank. However, such low values of capacitance cannot be achieved either with simple switched capacitors or accumulation mode MOS varactors as it is difficult to accurately model such small values. The choice of different fine capacitance bank implementations is further presented.

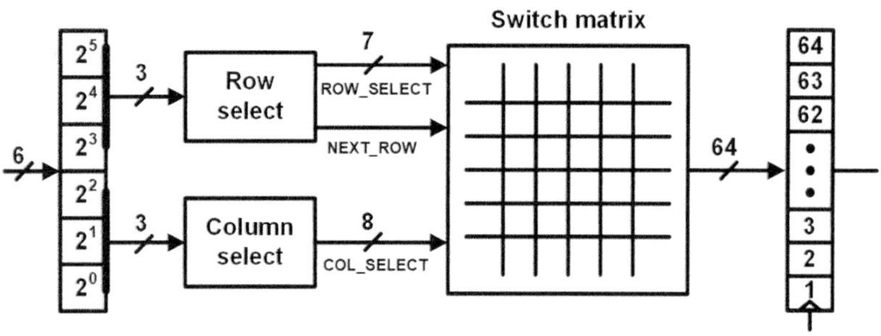

Figure 6.39 Block diagram of medium-tuning bank implementation.

i. Capacitor scaling

In practical DCOs, it is always better to choose a better modelled MOM capacitor and the capacitor scaling based on either the unit cell switching bank similar to coarse-tuning bank or unit varactor as shown in Fig. 6.40. [93, 94]. In the OFF state, the capacitors C_{mom} and C_b are in series and the effective capacitance is given as

$$C_{\text{off}} = \frac{C_b C_{\text{mom}}}{2(C_b + C_{\text{mom}})}, \tag{6.32}$$

When the switch M_1 is ON state, the step capacitance is obtained due to capacitance scaling and the effective capacitance is given as

$$C_{\mathrm{ON}} = \frac{C_{\mathrm{mom}}^2}{2(C_b + C_{\mathrm{mom}})}. \tag{6.33}$$

Therefore, for a smaller capacitance, a large value of C_b and a small value of C_{mom} are chosen. In the varactor-based implementation, the equivalent capacitance is given as [94]

$$C_{\mathrm{eq}} \approx \left(\frac{C_{\mathrm{mom}}}{C_v + C_b + C_{\mathrm{mom}}}\right)^2 \Delta C \approx \left(\frac{C_{\mathrm{mom}}}{C_b}\right)^2 \Delta C, \tag{6.34}$$

where C_v is the varactor capacitance. However, Eq. (6.34) is valid only if $C_b > C_{\mathrm{mom}} >> C_v$. Since the value of the feedback capacitor C_b is large and chosen based on the required resolution, choosing the fine-tuning bits to be > 6 bits makes the layout of the fine-tuning capacitor bank large. In such cases, to reduce the complexity, the fine-tuning bank also can be implemented as shown in Fig. 6.40 with additional decoder circuits.

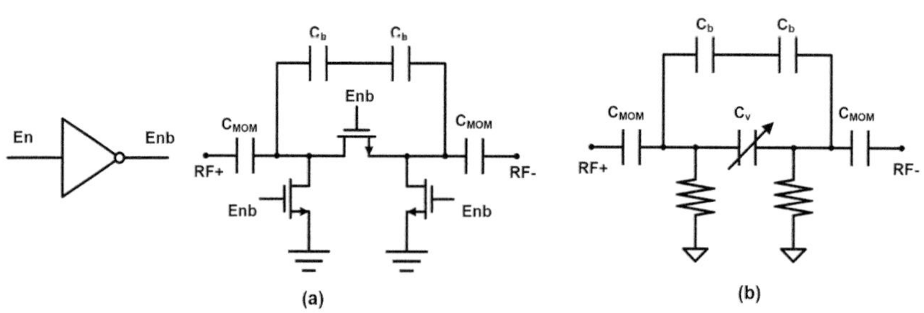

Figure 6.40 Fine-tuning bank capacitor divider (a) normal, and (b) with varactor.

ii. Sigma-Delta (ΣΔ) Dithering

In addition to the capacitor scaling, finer resolution in the fine-tuning capacitor bank is realized using ΣΔ dithering. In this technique, the least significant bit

(LSB) bit of the fine-tuning capacitor bank is switched ON and OFF using the high-speed clock in such a way that the effective capacitance is an average of the ON and OFF value resulting in a better resolution. With an n-bit fractional input, n ones and (n–1) zeros are applied to the control input of the LSB tuning capacitor. Since a high-speed clock is used, the dithering happens multiple times within one reference period achieving an average capacitor value and thus realizing fine resolution DCOs with ΣΔ dithering as presented in [95, 96]. However, one of the major drawbacks of ΣΔ dithering is the requirement of a high-speed clock increasing the complexity and power consumption of the DCO.

To overcome the drawbacks of ΣΔ dithering in DCOs, incremental MOS capacitor tuning is proposed in [97, 98] as shown in Fig. 6.41. In this tuning scheme, one set of MOS transistors with minimum size is chosen as one device and the incremental size is chosen for another device. In this way, the delta capacitance is realized. However, since is incremental size is process dependent, only fixed capacitance increment can be realized.

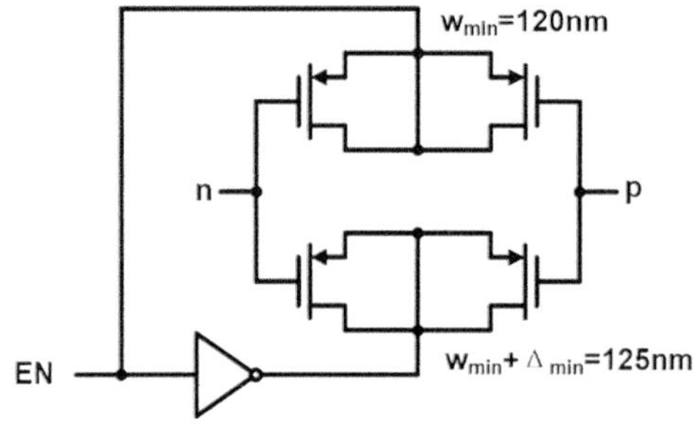

Figure 6.41 Fine-tuning with incremental capacitance.

6.6 Conclusion

This chapter begins with the introduction of oscillator theory and the basic concept of oscillator followed by the commonly used oscillator architectures. Then, the resonator-based oscillator is discussed in detail with focus on the oscillator start-up and phase noise as it is mostly used in the RF/mm-Wave frequencies.

Following the basic concepts, the chapter discusses the circuit design concepts and practical implementation of VCOs by first introducing the commonly used oscillator topologies including the more recently published class-F oscillators followed by the design methodology for the RF and mm-Wave VCO with a design example of 15 GHz VCO. In the last section, the design concepts for advanced oscillators such as quadrature VCO and digital oscillators are presented.

In conclusion, this chapter presented a comprehensive view of the oscillators from a practical standpoint with design insights for the design of high-performance RF and mm-Wave LC oscillators.

References

1. A. Hajimiri, S. Limotyrakis, and T. H. Lee, Jitter and phase noise in ring oscillators, *IEEE Journal of Solid-State Circuits*, vol. 34, no. 6, pp. 790–804, 1999.
2. R. Aparicio and A. Hajimiri, A noise shifting differential Colpitts VCO, *IEEE Journal of Solid-State Circuits*, vol. 37, no. 12, pp. 1728–1736, 2002.
3. A. Babcock, B. Loftin, P. Madhani, Xinfen Chen, A. Pinto, and D. K. Schroder, Comparative low frequency noise analysis of bipolar and MOS transistors using an advanced complementary BiCMOS technology, *IEEE Custom Integrated Circuits Conference*, pp. 385–388, May 2001.
4. L. Romanò, A. Bonfanti, S. Levantino, C. Samori, and A. L. Lacaita, 5 GHz oscillator array with reduced flicker up-conversion in 0.13 μm CMOS, *IEEE Journal of Solid-State Circuits*, vol. 41, no. 11, pp. 2457–2467, Nov. 2006.
5. A. Bonfanti, S. Levantino, C. Samori, and A. L. Lacaita, A varactor configuration minimizing the amplitude-to-phase noise conversion in VCOs, *IEEE Transactions on Circuits and Systems I: Regular Papers*, vol. 53, no. 3, pp. 481–488, Mar. 2006.
6. S. Levantino, C. Samori, A. Zanchi, and A. L. Lacaita, AM-to-PM conversion in varactor tuned oscillators, *IEEE Transactions on*

Circuits and Systems II: Analog and Digital Signal Processing, vol. 49, no. 7, pp. 509–513, Jul. 2002.

7. S. Levantino, C. Samori, A. Bonfanti, S. Gierkink, and A. L. Lacaita, Frequency dependence on bias current in 5 GHz CMOS VCO's: Impact of tuning range and flicker noise up-conversion, *IEEE Journal of Solid-State Circuits*, vol. 37, no. 8, pp. 1003–1011, Aug. 2002.
8. E. Hegazi and A. A. Abidi, Varactor characteristics, oscillator tuning curves, and AM-FM conversion, *IEEE Journal of Solid-State Circuits*, vol. 38, no. 6, pp. 1033–1039, Jun. 2003.
9. A. Ismail and A. A. Abidi, CMOS differential LC oscillator with suppression up-converted flicker noise, *IEEE International Solid-State Circuits Conference*, pp. 98–99, Feb. 2003.
10. N. N. Tchamov and N. T. Tchamov, Technique for flicker noise up-conversion suppression in differential LC oscillators, *IEEE Transactions on Circuits and Systems II: Express Briefs*, vol. 54, no. 11, pp. 481–488, Nov. 2007.
11. E. Hegazi, J. J. Rael, and A. A. Abidi, The Designer's Guide to High-Purity Oscillators, Springer, 2004.
12. J. J. Rael and A. A. Abidi, Physical processes of phase noise in differential LC oscillators, *IEEE Custom Integrated Circuits Conference*, pp. 569–572, May 2000.
13. B. Soltanian and P. Kinget, AM-FM conversion by the active devices in MOS LC-VCOs and its effect on the optimal amplitude, *IEEE Radio Frequency Integrated Circuits Symposium*, pp. 4–108, Jul. 2006.
14. A. Koukab, Reactive power imbalances in LC VCOs and their influence on phase-noise mechanisms, *IEEE Transactions on Microwave Theory and Techniques*, vol. 59, no. 12, pp. 3118–3128, Dec. 2011.
15. C. C. Boon, M. A. Do, K. S. Yeo, J. G. Ma, and X. L. Zhang, RF CMOS low-phase-noise LC oscillator through memory reduction tail transistor, *IEEE Transactions on Circuits and Systems II: Express Briefs*, vol. 51, no. 2, pp. 85–90, Feb. 2004.
16. A. Jerng and C. G. Sodini, The impact of device type and sizing on phase noise mechanisms, *IEEE Journal of Solid-State Circuits*, vol. 40, no. 2, pp. 360–369, Feb. 2005.
17. S-J. Yun, C-Y. Cha, H-C. Choi, and S-G. Lee, RF CMOS LC-oscillator with source damping resistors, *IEEE Microwave and Wireless Components Letters*, vol. 16, no. 9, 511–513, Sep. 2006.

18. M. A. Margarit, J. L. Tham, R. Meyer, and M. J. Deen, A low-noise, low-power VCO with automatic amplitude control for wireless applications, *IEEE Journal of Solid-State Circuits*, vol. 34, no. 6, pp. 761–771, Jun. 1999.
19. E. A. Vittoz, M. G. R. Degrawe, and S. Bitz, High-performance crystal oscillator circuits: Theory and application, *IEEE Journal of Solid-State Circuits*, vol. 23, no. 3, pp. 774–783, Jun. 1988.
20. A. Bevilacqua and P. Andreani, An analysis of $1/f$ noise to phase noise conversion in CMOS harmonic oscillators, *IEEE Transactions on Circuits and Systems I: Regular Papers*, vol. 59, no. 5, pp. 938–945, May 2012.
21. E. Hegazi, H. Sjoland, and A. A. Abidi, A filtering technique to lower LC oscillator phase noise, *IEEE Journal of Solid-State Circuits*, vol. 36, no. 12, pp. 1921–1930, Dec. 2001.
22. S-J. Yun, C-Y. Cha, H-C. Choi, and S-G. Lee, RF CMOS LC-oscillator with source damping resistors, *IEEE Microwave and Wireless Components Letters*, vol. 16, no. 9, pp. 511–513, Sep. 2006.
23. M. A. Margarit, J. L. Tham, R. Meyer, and M. J. Deen, A low-noise, low-power VCO with automatic amplitude control for wireless applications, *IEEE Journal of Solid-State Circuits*, vol. 34, no. 6, pp. 761–771, Jun. 1999.
24. A. Bevilacqua and P. Andreani, An analysis of $1/f$ noise to phase noise conversion in CMOS harmonic oscillators, *IEEE Transactions on Circuits and Systems I: Regular Papers*, vol. 59, no. 5, pp. 938–945, May 2012.
25. A. Rofougaran, G. Chang, J. J. Rael, J. Y.-C. Chang, M. Rofougaran, P. J. Djafari, M. K. Ku, E. W. Roth, A. A. Abidi, and H. Samueli, A single-chip 900 MHz spread-spectrum wireless transceiver in 1 μm CMOS - Part I: Architecture and transmitter design, *IEEE Journal of Solid-State Circuits*, vol. 33, no. 4, pp. 515–530, Apr. 1998.
26. B. De Muer, M. Borremans, M. Steyaert, and G. Li Puma, A 2 GHz low-phase noise integrated LC-VCO set with flicker-noise up-conversion minimizations, *IEEE Journal of Solid-State Circuits*, vol. 35, no. 8, pp. 1034–1038, Jul. 2000.
27. E. A. M. Klumperink, S. L. J. Gierkink, A. P. van der Wel, and B. Nauta, Reducing MOSFET $1/f$ noise and power consumption by switching biasing, *IEEE Journal of Solid-State Circuits*, vol. 35, no. 7, pp. 994–1001, Jul. 2000.
28. S. Levantino, M. Zanuso, C. Samori, and A. L. Lacaita, Suppression of flicker noise upconversion in a 65 nm CMOS VCO in the 3.0–3.6 GHz

band, *IEEE International Solid-State Circuits Conference*, pp. 50–51, Feb. 2010.

29. R. L. Bunch and S. Raman, Large-signal analysis of MOS varactors in CMOS-Gm LC VCOs, *IEEE Journal of Solid-State Circuits*, vol. 38, no. 8, pp. 1325–1332, Aug. 2003.
30. J. Groszkowski, The interdependence of frequency variation and harmonic content, and the problem of constant-frequency oscillators, *Proceedings of the Institute of Radio Engineers*, vol. 21, pp. 958–981, 1933.
31. A. Bevilacqua and P. Andreani, On the bias noise to phase noise conversion in harmonic oscillators using Groszkowski theory, *IEEE International Symposium on Circuits and Systems*, pp. 217–220, May 2011.
32. A. Bonfanti, F. Pepe, C. Samori, and et al., Flicker noise up-conversion due to harmonic distortion in Van der Pol CMOS oscillators, *IEEE Transactions on Circuits and Systems I: Regular Papers*, vol. 59, pp. 1418–1430, 2012.
33. A. Mazzanti and P. Andreani, Class-C harmonic CMOS VCOs with a general result on phase noise, *IEEE Journal of Solid-State Circuits*, vol. 43, no. 12, pp. 2716–2729, Dec. 2008.
34. K. Hoshino, E. Hegazi, J. J. Rael, and et. al, A 1.5 V, 1.7 mA 700 MHz CMOS LC oscillator with no upconverted flicker noise, *IEEE European Solid-State Circuits Conference*, pp. 337–340, 2001.
35. D. B. Leeson, A simple model of feedback oscillator noise spectrum, *Proceedings of the IEEE*, vol. 54, no. 2, pp. 329–330, 1966.
36. J. Craninckx and M. Steyaert, Low-noise voltage-controlled oscillators using enhanced LC-tanks, *IEEE Transactions on Circuits and Systems II: Analog and Digital Signal Processing*, vol. 42, no. 12, pp. 794–804, 1995.
37. J. J. Rael and A. A. Abidi, Physical processes of phase noise in differential LC oscillators, *IEEE Custom Integrated Circuits Conference*, pp. 569–572, 2000.
38. A. Hajimiri and T. H. Lee, A general theory of phase noise in electrical oscillators, *IEEE Journal of Solid-State Circuits*, vol. 33, no. 2, pp. 179–194, 1998.
39. T. H. Lee and A. Hajimiri, Oscillator phase noise: A tutorial, *IEEE Journal of Solid-State Circuits*, vol. 35, no. 3, pp. 326–336, 2000.
40. A. Hajimiri and T. H. Lee, Design issues in CMOS differential LC oscillators, *IEEE Journal of Solid-State Circuits*, vol. 34, no. 5, pp. 717–724, May 1999.

41. H. Darabi et al., A quad band GSM/GPRS/EDGE SoC in 65 nm CMOS, *IEEE Journal Solid-State Circuits*, vol. 46, no. 4, pp. 870–882, Apr. 2011.
42. D. Murphy, J. J. Rael, and A. A. Abidi, Phase noise in LC oscillators: A phasor-based analysis of a general result and of loaded Q, *IEEE Transactions on Circuits and Systems I: Regular Papers*, vol. 57, no. 6, pp. 1187–1203, Jun. 2010.
43. L. Fanori and P. Andreani, Highly efficient class-C CMOS VCOs, including a comparison with class-B VCOs, *IEEE Journal of Solid-State Circuits*, vol. 48, no. 7, pp. 1730–1740, Jul. 2013.
44. B. Soltanian and P. Kinget, Tail current-shaping to improve phase noise in LC voltage-controlled oscillators, *IEEE Journal of Solid-State Circuits*, vol. 41, no. 8, pp. 1792–1802, Aug. 2006.
45. W. Deng, K. Okada, and A. Matsuzawa, A feedback class-C VCO with robust startup condition over PVT variations and enhanced oscillation swing, *IEEE European Solid-State Circuits Conference*, pp. 499–502, Sep. 2011.
46. J. Chen, F. Jonsson, M. Carlsson, C. Hedenas, and L. R. Zheng, A low power, start-up ensured and constant amplitude class-C VCO in 0.18 μm CMOS, *IEEE Microwave and Wireless Components Letters*, vol. 21, no. 8, pp. 427–429, Aug. 2011.
47. K. Okada, Y. Nomiyana, R. Murakami, and A. Matsuzawa, A 0.114 mW dual-conduction class-C CMOS VCO with 0.2 V power supply, *IEEE Symposium on VLSI Circuits*, pp. 228–229, Jun. 2009.
48. L. Fanori, A. Liscidini, and P. Andreani, A 6.7-to-9.2 GHz 55 nm CMOS hybrid class-B/class-C cellular TX VCO, *IEEE International Solid-State Circuits Conference*, pp. 354–356, Feb. 2012.
49. L. Fanori, T. Mattsson, and P. Andreani, A class-D CMOS DCO with an on-chip LDO, *IEEE European Solid-State Circuits Conference*, pp. 335–338, 2014.
50. A. G. Roy, S. Dey, J. B. Goins, T. S. Fiez, and K. Mayaram, 350 mV, 5 GHz class-D enhanced swing differential and quadrature VCOs in 65 nm CMOS, *IEEE Journal of Solid-State Circuits*, vol. 50, no. 8, pp. 1833–1447, Aug. 2015.
51. Y. Yoshihara, H. Majima, and R. Fujimoto, A 0.171 mW, 2.4 GHz class-D VCO with dynamic supply voltage control, *IEEE European Solid-State Circuits Conference*, pp. 339–342, 2014.
52. L. Fanori and P. Andreani, class-D CMOS oscillators, *IEEE Journal of Solid-State Circuits*, vol. 48, no. 12, pp. 3105–3119, Dec. 2013.

53. M. Babaie and R. B. Staszewski, A class-F CMOS Oscillator, *IEEE Journal of Solid-State Circuits*, vol. 48, no. 12, pp. 3120–3133, Dec. 2013.

54. D. Murphy, H. Darabi, and H. Wu, 25.3 A VCO with implicit common-mode resonance, *IEEE International Solid-State Circuits Conference*, pp. 1–3, Feb. 2015.

55. L. Geynet, E. De Foucauld, P. Vincent, and G. Jacquemod, Fully integrated multistandard VCOs with switched LC tank and power controlled by body voltage in 130 nm CMOS/SOI, *IEEE Symposium on Radio Frequency Integrated Circuits*, pp. 129–132, Jun. 2006.

56. M. Demirkan, S. P. Bruss, and R. R. Spencer, 11.8 GHz CMOS VCO with 62% tuning range using switched coupled inductors, *IEEE Symposium on Radio Frequency Integrated Circuits*, pp. 401–404, Jun. 2007.

57. A. Bevilacqua, F. P. Pavan, C. Sandner, A. Gerosa, and A. Neviani, Transformer-based dual-mode voltage-controlled oscillators, *IEEE Transactions on Circuits and Systems II: Express Briefs*, vol. 54, no. 4, pp. 293–297, Apr. 2007.

58. B. Catli and M. Hella, A dual band, wide tuning range CMOS voltage-controlled oscillator for multiband radio, *IEEE Symposium on Radio Frequency Integrated Circuits*, pp. 595–598, Jun. 2007.

59. J. Borremans, S. Bronckers, P. Wambacq, M. Kuijk, and J. Craninckx, A single inductor dual-band VCO in a 0.06 mm^2 5.6 GHz multiband front-end in 90 nm digital CMOS, *IEEE International Solid-State Circuits Conference*, pp. 324–325, Feb. 2008.

60. A. Goel and H. Hashemi, Frequency switching in dual-resonance oscillators, *IEEE Journal of Solid-State Circuits*, vol. 42, no. 3, pp. 571–582, Mar. 2007.

61. A N. T. Tchamov, S. S. Broussev, I. S. Uzunovand, and K. K. Rantala, Dual-band LC VCO architecture with a fourth-order resonator, *IEEE Transactions on Circuits and Systems II: Express Briefs*, vol. 54, no. 3, pp. 277–281, Mar. 2007.

62. B. Sadhu, J. Kim, and R. Harjani, A CMOS 3.3–8.4 GHz wide tuning range, low phase noise LC VCO, *IEEE Custom Integrated Circuits Conference*, pp. 559–562, Sep. 2009.

63. Z. Safarian and H. Hashemi, Wideband multi-mode CMOS VCO design using coupled inductors, *IEEE Transactions on Circuits and Systems I: Regular Papers*, vol. 56, no. 8, pp. 1830–1843, Aug. 2009.

64. N. Mahalingam, K. Ma, K. S. Yeo, and W. M. Lim, A K-band high PAE wide tuning range VCO using triple-coupled LC tanks, *IEEE Transactions on Circuits and Systems II: Express Briefs*, vol. 60, no. 11, pp. 736–740, Nov. 2013.
65. Q. Zou, K. Ma, and K. S. Yeo, A low phase noise and wide tuning range millimeter-wave VCO using switchable coupled VCO-cores, *IEEE Transactions on Circuits and Systems I: Regular Papers*, vol. 62, no. 2, pp. 554–563, Feb. 2015.
66. C.-K. Hsieh, K.-Y. Kao, J. R. Tseng, and K.-Y. Lin, A K-band CMOS low power modified Colpitts VCO using transformer feedback, *IEEE International Microwave Symposium*, pp. 1293–1296, Jun. 2009.
67. H.-Y. Chang and Y.-T. Chiu, K-band CMOS differential and quadrature voltage-controlled oscillators for low phase-noise and low-power applications, *IEEE Transactions on Microwave Theory and Techniques*, vol. 60, no. 1, pp. 46–59, Jan. 2012.
68. Z.-D. Huang and C.-Y. Wu, The design of double-positive-feedback voltage-controlled oscillator, *IEEE Microwave and Wireless Components Letters*, vol. 21, no. 7, pp. 386–388, Jul. 2011.
69. J. Yang, C.-Y. Kim, D.-W. Kim, and S. Hong, Design of a 24 GHz CMOS VCO with an asymmetric-width transformer, *IEEE Transactions on Circuits and Systems II: Express Briefs*, vol. 57, no. 3, pp. 173–177, Mar. 2010.
70. C.-A. Lin, J.-L. Kuo, K.-Y. Lin, and H. Wang, A 24 GHz low power VCO with transformer feedback, *IEEE Radio Frequency Integrated Circuits Symposium*, pp. 75–78, Jun. 2009.
71. N. Mahalingam, et al., A fully integrated pulling mitigation synthesizer for NB-IoT transmitter, *IEEE Transactions on Microwave Theory and Techniques*, vol. 71, no. 12, pp. 5221–5232, Dec. 2023.
72. B. D. Muer, N. Itoh, M. Borremans, and M. Steyaert, A 1.8 GHz highly tunable low-phase-noise CMOS VCO, *IEEE Custom Integrated Circuits Conference*, pp. 585–588, May 2000.
73. A. Rofougaran, J. Rael, M. Rofougaran, and A. Abidi, A 900 MHz CMOS LC oscillator with quadrature outputs, *IEEE International Solid-State Circuits Conference*, pp. 392–393, Feb. 1996.
74. T. P. Liu, A 6.5 GHz monolithic CMOS voltage-controlled oscillator, *IEEE International Solid-State Circuits Conference*, pp. 404–405, Feb. 1999.
75. M. A. Do, R. Y. Zhao, K. S. Yeo and J. G. Ma, New wideband/dual band CMOS LC voltage-controlled oscillator, *IEEE Proceedings – Circuits, Devices and Systems*, vol. 150, no. 5, pp. 453–459, Oct. 2003.

76. S. L. J. Gierkink, S. Levantino, R. C. Frye, C. Samori, and V. Boccuzzi, A low phase noise 5 GHz CMOS quadrature VCO using superharmonic coupling, *IEEE Journal of Solid-State Circuits*, vol. 38, pp. 1148–1154, Jul. 2003.

77. A. Ravi, et al., An optimally transformer coupled 5 GHz quadrature VCO in a 0.18 μm digital CMOS process, *IEEE Symposium on VLSI Circuits*, pp. 141–144, 2003.

78. C. W. Yao and A. N. Willson, A phase-noise reduction technique for quadrature LC-VCO with phase-to-amplitude noise conversion, *IEEE International Solid-State Circuits Conference*, pp. 196–197, Feb. 2006.

79. X. Li, S. shekhar, and D. J. Allstot, Gm-boosted common-gate LNA and differential Colpitts VCO/QVCO in 0.18 μm CMOS, *IEEE Journal of Solid-State Circuits*, vol. 40, no. 12, pp. 2609–2619, Dec. 2005.

80. J. Crols and M. Steyaert, A fully integrated 900 MHz CMOS double quadrature downconverter, *IEEE International Solid-State Circuits Conference*, pp. 136–137, Feb. 1995.

81. P. Andreani, A 2 GHz, 17% tuning range quadrature CMOS VCO with high figureof-merit and 0.6º phase error, *IEEE European Solid-State Circuits Conference*, pp. 815–818, Aug. 2002.

82. S. Maheshwari and I. A. Khan, Current controlled third order uadrature oscillator, *IEEE Proceedings – Circuits, Devices and Systems*, vol. 152, no. 6, pp. 605–607, Dec. 2005.

83. J. P. Maligeorgos and J. R. Long, A low voltage 5.1–5.8 GHz image reject receiver with wide dynamic range, *IEEE Journal of Solid-State Circuits*, vol. 35, pp. 1917–1926, Dec. 2000.

84. D. Pache, J. M. Fournier, G. Billiot, and P. Senn, An improved 3 V 2 GHz BiCMOS image reject mixer IC, *IEEE Custom Integrated Circuits Conference*, Santa Clara, CA, pp. 95–98, 1995.

85. A. S. Porret, T. Melly, D. Python, C. C. Enz, and E. A. Vittoz, An ultra-low-power UHF transceiver integrated in a standard digital CMOS process: Architecture and receiver, *IEEE Journal of Solid-State Circuits*, vol. 36, no. 3, pp. 452–446, Mar. 2001.

86. S. Li, I. Kipnis, and M. Ismail, A 10 GHz CMOS quadrature LC-VCO for multi-rate optical applications, in *IEEE Journal of Solid-State Circuits*, vol. 38, no. 10, pp. 1626–1634, Oct. 2003.

87. P. Andreani, A. Bonfanti, L. Romano, and C. Samori, Analysis and design of a 1.8 GHz CMOS LC quadrature VCO, *IEEE Journal of Solid-State Circuits*, vol. 37, pp. 1737–1747, Dec. 2002.

88. A. W. L. Ng, and H. C. Luong, A 1 V 17 GHz 5 mW CMOS quadrature VCO based on transformer coupling, *IEEE Journal of Solid-State Circuits*, vol. 42, no. 9, Sep. 2007.

89. B. Soltanian and P. Kinget, A low phase noise quadrature LC VCO using capacitive common-source coupling, *IEEE European Solid-State Circuits Conference*, pp. 436–439, Sep. 2006.

90. R. B. Staszewski and P. T. Balsara, All-digital frequency synthesizer in deep-submicron CMOS, New York, NY, USA: Wiley Interscience, 2006.

91. S. H. Lee and B. S. Song, Digital-domain calibration of multistep analog-to-digital converters, *IEEE Journal of Solid-State Circuits*, vol. 27, pp. 1679–1688, Dec. 1992.

92. C. S. Salimath, Design of CMOS LC voltage controlled oscillators, Master's Thesis, Louisiana State University, Baton Rouge, LA, 2006.

93. W. Jiang, A. Tavakol, P. Effendrik, M. van de Gevel, F. Verwaal, and R. Staszewski, Design of ADPLL system for WiMAX applications in 40 nm CMOS, *IEEE International Conference on Electronics, Circuits and Systems*, pp. 73–76, 2012.

94. Y. Chen et al., A 9 GHz dual-mode digitally controlled oscillator for GSM/UMTS transceivers in 65 nm CMOS, *IEEE Asian Solid-State Circuits Conference*, pp. 432–435, Nov. 2007.

95. R. B. Staszewski, C. M. Hung, N. Barton, M. C. Lee, and D. Leipold, A digitally controlled oscillator in a 90 nm digital CMOS process for mobile phones, *IEEE Journal of Solid-State Circuits*, vol. 40, no. 5, pp. 2203–2211, Nov. 2005.

96. S. S. Yoo, Y. C. Choi, H. J. Song, S. C. Park, J. H. Park, and H. J. Yoo, A 5.8 GHz high frequency resolution digitally controlled oscillator using the difference between inversion and accumulation mode capacitance of PMOS varactors, *IEEE Transactions on Microwave Theory and Techniques*, vol. 59, no. 2, pp. 375–382, Feb. 2011.

97. J. Chen, L. Rong, F. Jonsson, and L. R. Zheng, All-digital transmitter based on ADPLL and phase synchronized delta sigma modulator, *IEEE Radio Frequency Integrated Circuits Symposium*, pp. 1–4, Jun. 2011.

98. J. Chen, L. Rong, F. Jonsson, G. Yang, and L. R. Zheng, The design of all-digital polar transmitter based on ADPLL and phase synchronized delta sigma modulator, *IEEE Journal of Solid-State Circuits*, vol. 47, no. 5, pp. 1154–1164, May 2012.

In unity, we find strength.
In collaboration, we discover our weaknesses.

—Kiat Seng Yeo

The rich invest and look for opportunities while
the poor save and look for work.

—Kiat Seng Yeo

A bright past does not guarantee a brilliant future.
Hence, we must plan and continue working hard.

—Kiat Seng Yeo

Chapter 7

CMOS Frequency Synthesizers

7.1 Introduction

The local oscillator in the RF transceiver generates signals for up-conversion and down-conversion by the transmitter and receiver, respectively. The important features of the local oscillator signal are its high stability, accuracy and programmability. The phase-locked loop (PLL) offers a universally acceptable solution to synthesize the local oscillator signal in the RF transceiver. The theory of PLL is well documented [1–5] and recent attention has been drawn on the CMOS/BiCMOS implementation towards fully integrated PLLs and system-on-chip (SoC) implementations [6–8]. The performance of PLLs is dependent on the various RF building blocks namely the voltage-controlled oscillators (VCO) as discussed in Chapter 3, frequency divider, phase frequency detector, charge-pump and loop filters to be discussed in this chapter. The PLL is also used in application areas such as demodulation of frequency-modulated (FM) signals [9–11], clock and data recovery (CDR) circuits [12–14]. This chapter focuses on the RF transceiver frequency synthesis as PLL-based local oscillator is one of the most challenging blocks to design in the RF transceiver.

CMOS RF and mm-Wave Transceivers and Synthesizers
Bharatha Kumar Thangarasu, Nagarajan Mahalingam, Kaixue Ma, and Kiat Seng Yeo

ISBN 978-981-5129-80-9 (Hardcover), 978-1-003-67356-9 (eBook)
www.jennystanford.com

7.2 Fundamentals of Frequency Synthesizers

In the RF transceiver shown in Fig. 7.1, the role of the frequency synthesizers is to generate local oscillator signal for frequency translation in the transmitter and receiver path. A frequency synthesizer is a device that can synthesize one or multiple output frequencies from a low-frequency reference. Frequency synthesizer can be implemented in either the direct [3, 15] or indirect frequency synthesis based on PLL topology. The ensuing discussions assume PLL-based frequency synthesis for the local oscillator signal as it is most suitable for the synthesis of high-frequency signals and henceforth PLL and frequency synthesizer are used to describe the PLL-based frequency synthesizer.

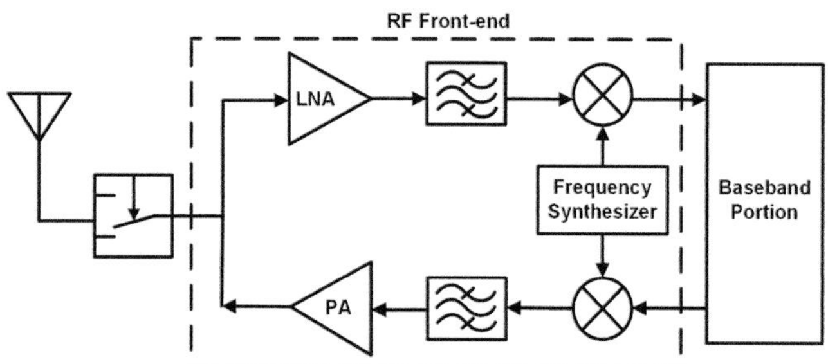

Figure 7.1 Block diagram of typical RF wireless transceiver.

The block diagram of the PLL is shown in Fig. 7.2. The PLL operates based on a negative feedback system and its main advantage is its simplicity. The output frequencies in the PLL can be synthesized with a single low-frequency reference which is typically an external crystal oscillator. The phase detector compares and aligns the reference phase with the VCO phase by controlling the VCO frequency through its tuning port. When the VCO and reference phase are aligned, the PLL achieves lock and in steady state condition, the following relationship holds:

$$f_{\text{out}} = N \times f_{\text{ref}} \tag{7.1}$$

Though the simple PLL-based frequency synthesis as shown in Fig. 7.2 is commonly used, slow switching time and frequency resolution equal to reference frequency, spectral purity and spurious signals are some of the limitations in practical implementations that have been actively researched over the years.

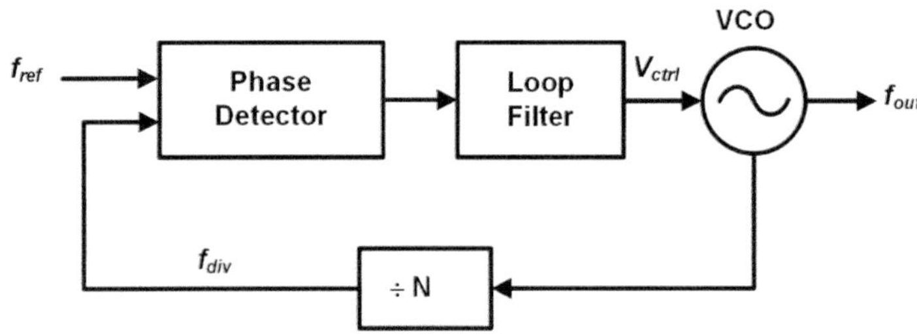

Figure 7.2 Block diagram of PLL-based indirect frequency synthesis.

The synthesizer shown in Fig. 7.2 is generally classified based on the implementation of the divider in the feedback path. The two main classifications of the synthesizer are as follows:

a. Integer-*N* PLL

The integer-*N* PLL is the most common and widely used architecture and the block diagram of the divider chain is shown in Fig. 7.3 [16]. In the integer-*N* PLL, the feedback division ratio is an integer and the output frequency which is an integer multiple of the input reference frequency given by

$$f_{\text{out}} = N_p \times P \times f_{\text{ref}}, \tag{7.2}$$

where N_p is the programmable division ratio and P is the division ratio of the fixed divider stages in the feedback divider. The major advantage of the integer-*N* PLL synthesizer is its simple and straightforward design approach. One of the main limitations of integer-*N* PLL is the achievable frequency resolution where the minimum frequency resolution and channel spacing is equal to the input reference frequency. Also, the settling time for the PLL is large when lower frequency resolution is desired. In addition, when the reference frequency is small, the division ratio is quite large. The large division ratio increases the power consumption of the divider and degrades the spectral purity in the lower frequency offsets commonly referred as in-band phase noise.

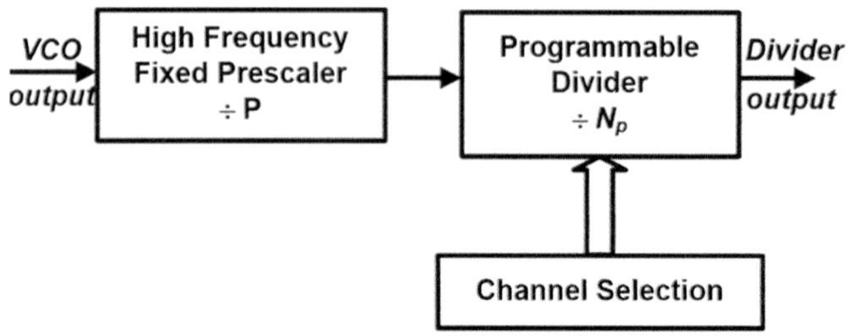

Figure 7.3 Integer-*N* frequency division.

b. Fractional-*N* PLL

In the fractional-*N* PLL, the feedback division ratio is a fractional number and the block diagram of the divider chain in the fractional-*N* PLL synthesizer is shown in Fig. 7.4 [16]. The programmable divider in the fractional-*N* PLL is similar to the integer-*N* topology but is controlled differently. To achieve a fractional division ratio, the $N/(N+1)$ divider is constantly varied based on the carry bit from the k-bit accumulator. The divider divides by N when the carry output is zero. When the carry output is 1 for every k cycles of reference frequency, the divider divides by $(N+1)$. The carry bit of the accumulator is dependent on the fractional input f, resolution of the accumulator and the reference frequency f_{ref}. The output frequency of the fractional-*N* PLL synthesizer is given by

$$f_{out} = \left((N \times P) + \frac{f}{2^k} \right) \times f_{ref}. \qquad (7.3)$$

The main advantage of fractional-*N* PLL over its integer-*N* counterpart is its ability to achieve a low-frequency resolution equal to $f_{ref}/2k$. Also, the fractional-*N* PLL synthesizer decouples the choice of reference frequency from the channel spacing and minimum frequency resolution. In the fractional-*N* PLL synthesizer, high input reference frequency can be chosen resulting in a fast settling of PLL loop with large loop bandwidth and small division ratio. The small division ratio results in a lower degradation of in-band phase noise. Although the fractional-*N* PLL synthesizers have a lot of advantages, the major downside is the appearance of fractional spurs in the output spectrum due to the fractional frequency resolution and increased circuit complexity.

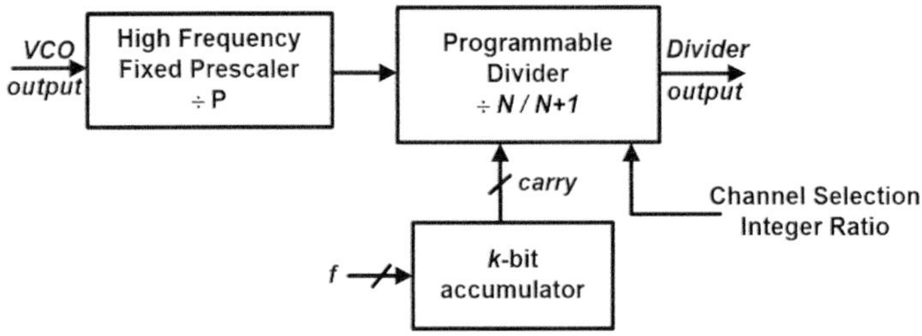

Figure 7.4 Fractional-*N* frequency division.

7.2.1 Frequency Synthesizer Performance Metrics

Irrespective of the synthesizer topology, the frequency synthesizer performance is generally characterized based on its frequency tuning range, frequency resolution, phase noise, output power and settling time. It is therefore important to understand the requirements of the synthesizer and optimize the specifications of the individual blocks in the design phase. The important PLL design parameters are discussed in brief in this section.

a. Frequency tuning range

The frequency tuning range is the basic requirement for the frequency synthesizer. The synthesizer must be able to generate all the LO frequencies required by the transceiver for frequency translation with sufficient frequency stability. The tuning range of the VCO is the important parameter that determines the frequency tuning range of the synthesizer. In order words, the VCO must have a wide tuning range. Also, the synthesizer frequency range sets the criteria for the locking range of the fixed high-frequency divider following the VCO. Therefore, the locking range of high-frequency divider is another important parameter that limits the frequency tuning range of the synthesizer.

b. Frequency resolution

The frequency resolution is an important parameter that determines the PLL architecture. In the integer-*N* PLL, the frequency resolution is same as the channel spacing and reference frequency. Therefore, the performance of integer-*N* PLL is greatly influenced by the required frequency resolution. However, in comparison, fractional-*N* PLL synthesizer can achieve a small frequency resolution with higher reference frequency. This can be advantageous for the PLL targeted for multi-standard wireless applications.

c. Phase noise

Phase noise is the most critical specification for the frequency synthesizer. Ideally, the local oscillator output spectrum from the synthesizer should be a single tone. But due to the presence of noise and non-ideal behavior of the individual blocks in the synthesizer, the pure single tone is degraded forming a skirt of noise power around the carrier frequency as shown in Fig. 7.5. Therefore, phase noise is typically defined as the difference between the carrier power and total noise power within a bandwidth of 1 Hz at a particular frequency offset from the carrier.

In the RF transceivers, the LO phase noise affects both the transmitter and receiver equally. In the receive path, the signal-to-noise ratio of the receiver is degraded due to the process of reciprocal mixing. As illustrated in Fig. 7.6(a), when the desired signal is demodulated using the LO signal with phase noise in the presence of a large unwanted signal, both the desired signal and unwanted interfering signal are down-converted to the same intermediate frequency. Similarly, when a signal with large phase noise is transmitted, the noise-skirt can fall within the bandwidth of the pass-band filter along with the desired signal in the receiver operating close to the transmitted frequency. This phenomenon is shown in Fig. 7.6(b).

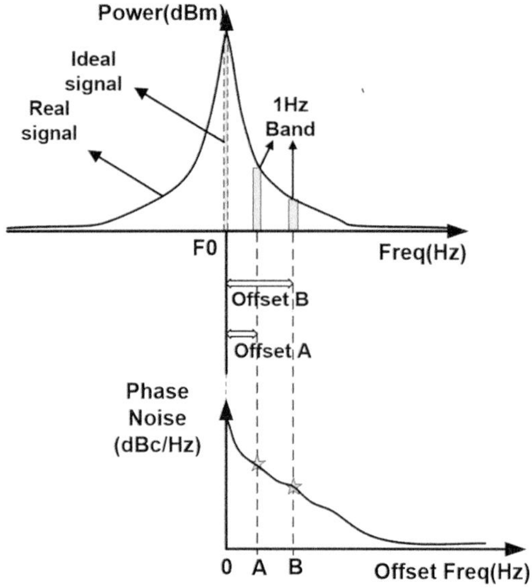

Figure 7.5 Typical PLL phase noise.

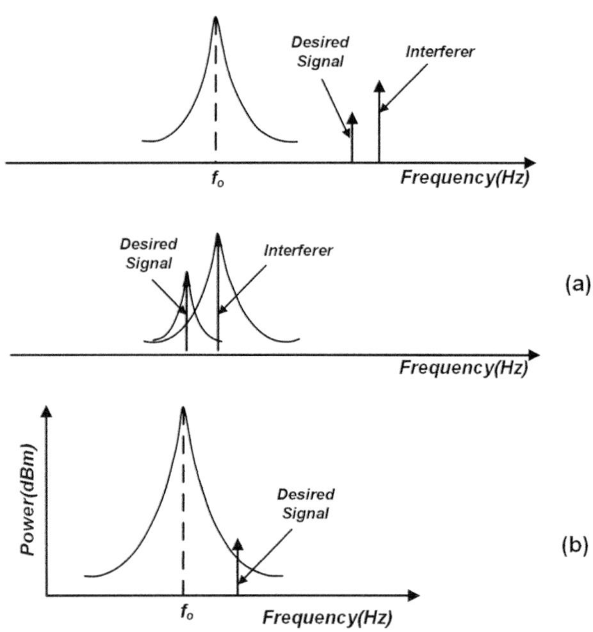

Figure 7.6 Effect of frequency synthesizer phase noise: (a) receive path and (b) transmit path.

In the PLL, the close-in phase noise within the channel bandwidth and far-away frequency offsets (> 1 MHz) are equally important and can have a significant impact on the transceiver system performance. Almost all of the PLL components influence the overall phase PLL noise. The in-band PLL noise is determined by the feedback divider ratio, quality of the reference signal, noise from the phase frequency detector (PFD), charge-pump and loop filter, whereas VCO typically dominates the far-away phase noise. As shown in Fig. 7.7, inside the loop bandwidth determined by the loop filter, the PLL phase noise is degraded by the factor of 20 log (N), where N is the overall feedback division ratio. In large frequency offset and outside the loop bandwidth, the PLL phase noise typically mimics the phase noise of the VCO.

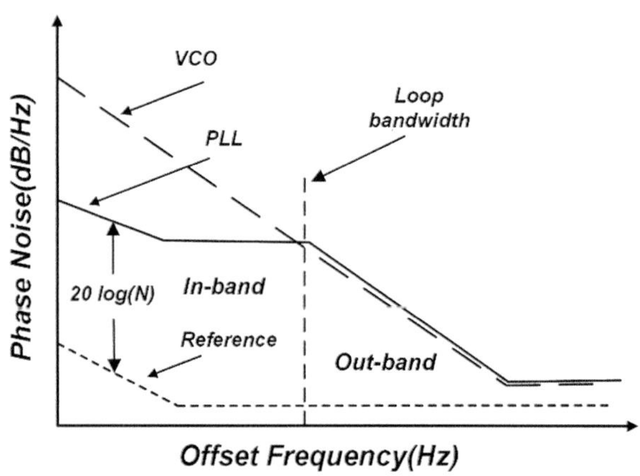

Figure 7.7 Typical frequency synthesizer phase noise.

d. Spurious signals

The spurious signals or spurs are unwanted signals that are generated within or outside the synthesizer. Unlike phase noise, spurs are generally considered interferers and have different amplitudes at different frequency offsets. The most common type of spur in the PLL is the reference spur arising mainly due to the switching noise and mismatch between the up and down currents in the charge-pump circuit. In the locked condition, ideally the output from the charge-pump is zero and a constant voltage is maintained at the VCO control line. But, due to the presence of parasitic resistance and capacitance in the charge-pump current switches and circuit non-linearity arising from finite switching time of voltage pulses from the PFD, a small amount of charge is leaked into the PLL loop. This leakage in turn modulates the VCO control line and up-converts to the carrier in the lower and upper sidebands at the intervals of reference frequency.

Like phase noise, the effect of spurs can be explained as in Fig. 7.8. If the desired signal in the presence of large interference is down-converted with LO signal that has spurs, the interfering signal can end up at the same intermediate frequency as the desired signal and degrade the signal-to-noise ratio [6]. Though spurs cannot be totally eliminated in the PLL, it can be reduced with careful design of charge-pump and choosing a lower bandwidth in the PLL synthesizer loop filter.

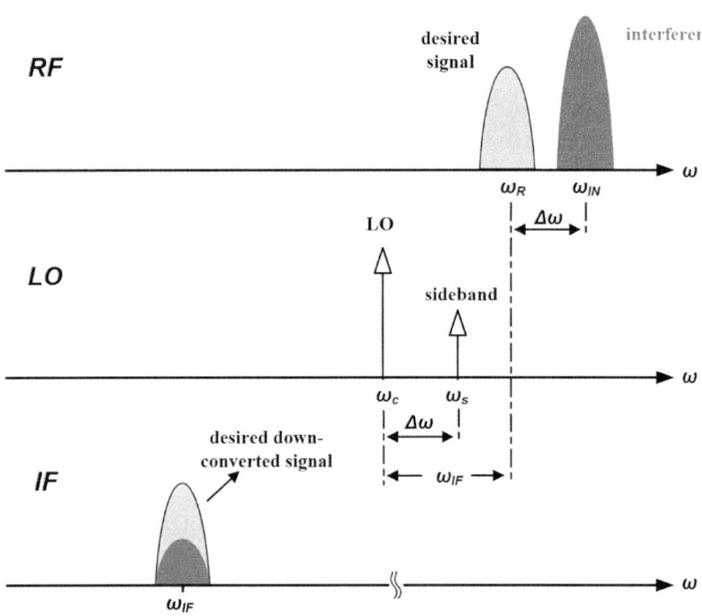

Figure 7.8 Effect of frequency synthesizer spurs.

e. **Settling time**

Settling time is the time it takes for the PLL to settle within a specified frequency tolerance after a change in the frequency control word. In the transmit and receive modes, the settling time is set by the system specification of a particular communication standard. For example, if a particular standard supports a data rate of 250 kbps considering each symbol is a word consisting of 4 bits, the supported symbol rate is 62.5 Ksymbols/sec. When the maximum RX-TX and TX-RX turnaround time is 12 symbol periods, and 16 channels, the worst-case settling time is 192 μs. The worst-case settling time is typically used to set PLL settling as only one PLL is used in the transceiver for frequency translation to save power and area in most cases. Since the PLL is a feedback system with the loop filter limiting its bandwidth, the settling time is directly related to the bandwidth of the PLL loop. The wider the loop bandwidth, the PLL settling time is less. Therefore, in a PLL, there is a strong relation and trade-off between settling time, phase noise and attenuation of reference spur.

f. **Output power and power consumption**

Often a fringe but important specification of the synthesizer is its output power. In the synthesizer, output power is related to the drive strength and power consumption of the transceiver. In the transceiver, the output of the synthesizer typically drives the up and down-conversion mixers in the transmitter and receiver path, respectively. Therefore, the synthesizer must deliver a high output power to eliminate or reduce the power consumption in the inter-stage amplifiers between the PLL and mixer stages in the transceiver. Also, with power management becoming essential in the mobile applications, reduction in PLL power consumption can increase the battery life due to overall reduction in power consumption of the transceiver. In the synthesizer, a large part of power dissipation occurs in the high-frequency blocks (VCO, fixed prescaler and divider chain) and the power efficiency of these blocks needs to be maximized.

g. PLL system parameters

Apart from the tuning range also known as the acquisition range, PLL has several other important system parameters. The parameters are described below and are shown in Fig. 7.9.

Lock-in range is defined as the range of frequencies which the PLL can lock within a single period of the beat frequency. The PLL loop will lock instantaneously if the frequency difference between the VCO and reference is less than the PLL loop bandwidth. The maximum range within which this frequency acquisition can occur is called the lock-in range.

Hold-in range is defined as the range of reference frequencies the PLL can maintain its phase tracking and lock. In the locked state, when the reference frequency is changed, the PLL should maintain its lock and not drift out of lock.

Pull-in range is defined as the maximum frequency difference between the VCO and the reference frequency for which the PLL loop can still achieve lock. The PLL pull-in range is practically limited by the VCO tuning range.

Pull-out range is defined as the range of frequency the PLL can secure lock again from unlocked state. When the phase tracking is lost, the range of frequencies the PLL can still achieve lock with the change in the frequency step is called the pull-out range.

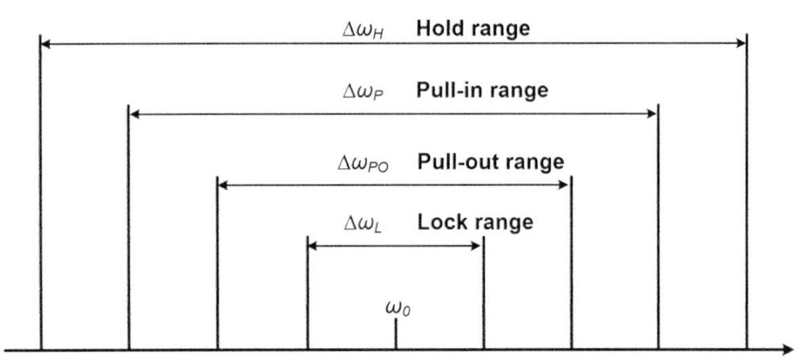

Figure 7.9 PLL system parameters range.

7.3 PLL Dynamics

The PLL-based frequency synthesizers are negative feedback loops which have both the frequency and phase-tracking capabilities. The PLL is analyzed based on control system theory and is linearly modelled though the PFD and the charge-pump are predominately discrete in nature. The linear approximation of the PLL is still valid as the PLL bandwidth is at least 1/10 of the frequency reference [1] and the linearized block diagram of the PLL is shown in Fig. 7.10. To analyze the PLL, the open- and closed-loop transfer functions are derived based on the transfer function of the constituent PLL blocks.

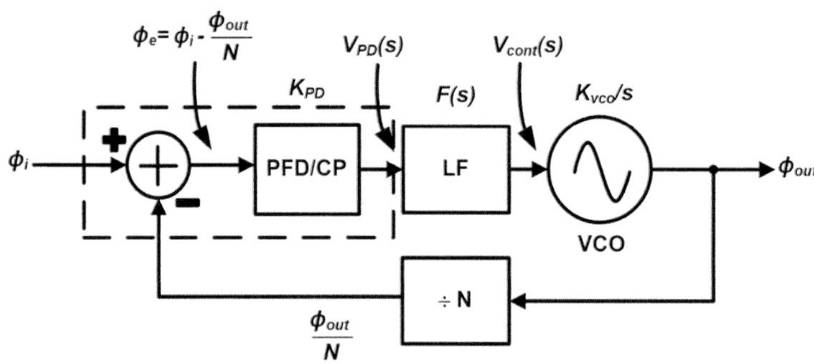

Figure 7.10 PLL linear block diagram.

The PLL predominantly operates based on phase and in the steady state or locked conditions, there is no excess phase or phase perturbation [17]. In the PLL, since the PFD and charge-pump work in tandem to convert the phase error to control voltage, the voltage at the output of the charge-pump can be expressed in the s-domain as

$$V_{\mathrm{PD}}(s) = K_{\mathrm{PD}}\varphi_e, \tag{7.4}$$

where K_{PD} is the PFD and charge-pump gain equal to $I_{\mathrm{CP}}/2\pi$. The loop filter following the charge-pump converts the phase error to the control voltage to control the VCO in addition to filtering the high-frequency components in the VCO control voltage. The voltage at the output of the loop filer can be expressed in s-domain as

$$V_{\text{cont}}(s) = F(s)V_{\text{PD}}(s). \tag{7.5}$$

As the integral of frequency is phase, the VCO output phase based on the control voltage V_{cont} is expressed as

$$\varphi_{\text{out}} = \int \omega_{\text{out}} dt = K_{\text{VCO}} \int V_{\text{cont}} dt = \frac{K_{\text{VCO}}}{s} V_{\text{cont}}(s), \tag{7.6}$$

where K_{VCO} is the VCO gain. The forward and feedback gain coefficients of PLL is given as

$$G(s) = \frac{\varphi_{\text{out}}}{\varphi_e} = K_{\text{PD}} \cdot F(s) \cdot \frac{K_{\text{VCO}}}{s} \tag{7.7}$$

$$H(s) = \frac{1}{N}. \tag{7.8}$$

From Eqs. (7.4) and (7.5), the open-loop and closed-loop PLL transfer functions are given as

$$G_{\text{oL}}(s) = G(s) \cdot H(s) = K_{\text{PD}} F(s) \frac{K_{\text{VCO}}}{s} \frac{1}{N} \tag{7.9}$$

$$G_{\text{CL}}(s) = \frac{G(s)}{1 + G(s) \cdot H(s)} = N \cdot \frac{G_{\text{oL}}(s)}{1 + G_{\text{oL}}(s)}. \tag{7.10}$$

The open-loop and closed-loop transfer functions of the PLL are important parameters that determine the dynamics of PLL. The frequency at which the gain of the open-loop transfer function is unity is defined as the PLL loop bandwidth. The PLLs are typically specified in order and type. The type of the PLL refers to the number of integrators and the order refers to the highest power of "s" in the PLL closed-loop transfer function. Since the VCO in the PLL is an integrator, all PLLs are at least type-I PLLs. The order of the PLL is one above the order of the loop filter in the PLL.

7.3.1 Loop Filter

In the PLL, the loop filter acts as low-pass filters which filter the unwanted high-frequency signals that generate spurious signals.

It can either be an active or passive loop filter. In the charge-pump PLL, passive loop filter is more commonly used as compared to an active loop filter that requires op-amps. Moreover, the noise contributed by the op-amps can be more than that of the passive loop filter. In the charge-pump PLL, either the second-order or third-order loop filter is typically used is shown in Fig. 7.11. The loop filter primarily consists of resistor R_2 in series with capacitor C_2. In the locked state, the charge-pump is at high impedance state and there is no discharge from the capacitor C_2. Therefore, the passive filter acts like an integrator when driven by a tri-state source. The resistor R_2 is used to generate the control voltage to tune the VCO frequency based on the output from the charge-pump. The resistor also adds a zero to improve the stability of the PLL loop. As shown in Fig. 7.11, an additional capacitor C_1 is added to the loop to smoothen the voltage drop across the resistor and reduces ripple in the VCO control voltage. However, the additional capacitor C_1 increases the order of the PLL and can create stability issues. Therefore, the capacitor C_1 is chosen 1/10 of C_2 to maintain the PLL stability [5, 18]. The transfer function of the loop filter and the poles and zeros is expressed as [19] assuming $C_2 >> C_1$, C_3 and $R_2 > R_3$ for the third-order loop filter.

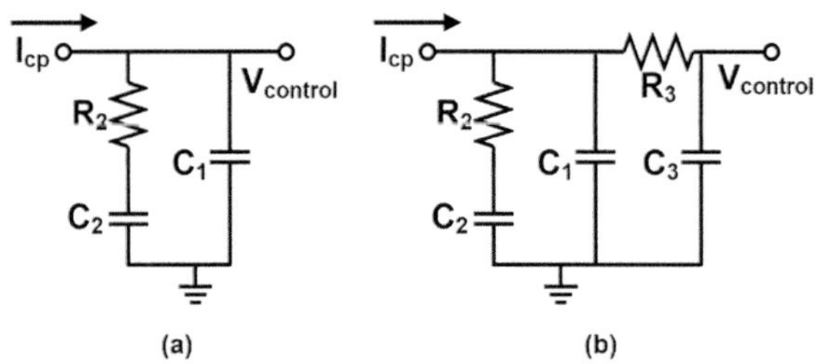

Figure 7.11 PLL passive loop filter: (a) second order and (b) third order.

Substituting the loop filter parameters in Table 7.1 in Eq. (7.9) and rearranging, the open-loop transfer function for the second-order loop filter can be expressed as

$$G_{\text{oL},2}(s) = G(s) \cdot H(s) = \frac{K_{\text{PD}} K_{\text{VCO}}}{N s^2 C_1} \left[\frac{(1 + s\tau_2)\,\tau_1}{(1 + s\tau_1)\,\tau_2} \right], \quad (7.11)$$

where the time constants $\tau_2 = R_2 C_2$ and $\tau_1 = R_2 C_1 C_2 / (C_1 + C_2)$. The poles and zeros of the PLL can be calculated from the loop filter parameters and is shown in Fig. 7.12. The PLL bandwidth and phase margin can be calculated by replacing $s = j\omega$ in the open-loop transfer function in Eq. (7.11) and observed from the bode-plot as shown in Fig. 7.13.

Table 7.1 Passive loop filter parameters and transfer function

Loop filter parameters	Second order	Third order
$F(s)$	$\left(R_2+\frac{1}{sC_2}\right)\|\|\frac{1}{sC_1}$	$\left(R_2+\frac{1}{sC_2}\right)\|\|\frac{1}{sC_1}+\left(R_3+\frac{1}{sC_3}\right)$
Zero frequency (ω_z)	$\frac{1}{R_2C_2}$	$\frac{1}{R_2C_2}$
Pole frequency ω_{p2}	$\frac{C_1+C_2}{R_2C_1C_2}$	$\approx\frac{1}{R_2(C_1+C_3)}$
Additional pole frequency ω_{p3}	—	$\approx\frac{(C_1+C_3)}{R_2C_2C_3}$

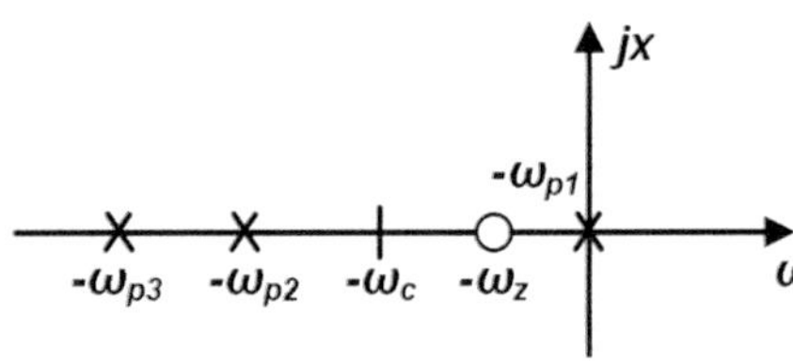

Figure 7.12 Pole zero location of third-order loop filter.

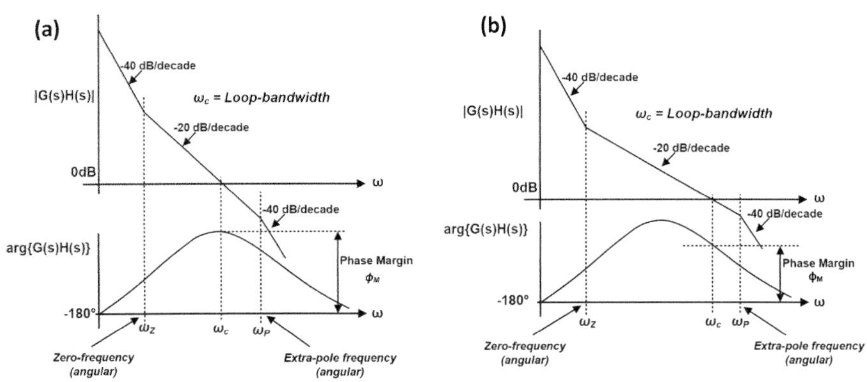

Figure 7.13 PLL magnitude and phase response: (a) best phase margin, and (b) sub-optimal phase margin.

The phase margin can be expressed as [20]

$$\varphi_m = \varphi_z - \varphi_{\omega p2} = \tan^{-1}\left(\frac{\omega_c}{\omega_z}\right) - \tan^{-1}\left(\frac{\omega_c}{\omega_{p2}}\right), \tag{7.12}$$

where ω_c is the PLL loop bandwidth and $\omega_z < \omega_c$. The PLL bandwidth can be approximated as

$$\omega_c \approx \frac{K_{\text{PD}} K_{\text{VCO}} R_2}{N}. \tag{7.13}$$

From Eq. (7.9), the optimal phase margin at PLL loop bandwidth can be given as [20]

$$\varphi_{m,\max} = \tan^{-1}\left(\sqrt{\frac{\omega_{p2}}{\omega_z}}\right) - \tan^{-1}\left(\sqrt{\frac{\omega_z}{\omega_{p2}}}\right). \tag{7.14}$$

Therefore, from Eq. (7.14), the maximum phase margin is determined by the ratio of the second pole (ω_{p2}) and zero (ω_z) and is shown in Fig. 7.14 [21].

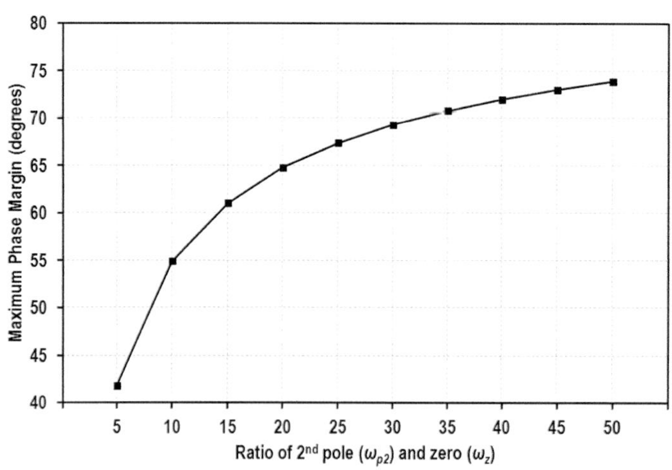

Figure 7.14 Maximum phase margin vs. ratio of ω_{p2}/ω_z.

For the third-order loop filter as shown in Fig. 7.11(b), the additional pole consisting of R_3 and C_3 provides extra attenuation on the reference spurious signal and its harmonics does not impact the phase margin as it is far away from the zero frequency [19]. Like the second-order loop filter, the phase margin of the third-order loop filter is given as

$$\varphi_m = \varphi_z - \varphi_{\omega p2} = \tan^{-1}\left(\frac{\omega_c}{\omega_z}\right) - \tan^{-1}\left(\frac{\omega_c}{\omega_{p2}}\right) - \tan^{-1}\left(\frac{\omega_c}{\omega_{p3}}\right), \quad (7.15)$$

where ω_c is the PLL loop bandwidth which is approximately equal to

$$\omega_c \approx \frac{K_{\text{PD}} K_{\text{VCO}} R_2}{N} \frac{C_2}{C_1 + C_2 + C_3}. \quad (7.16)$$

Analogous to the second-order loop filter, the maximum phase margin is a function of the capacitor ratio given by $b = C_2/(C_1 + C_3)$ and occurs at [19]

$$\omega_{c,\varphi m} \approx \sqrt{\omega_z \omega_{p2}}. \quad (7.17)$$

7.3.2 PLL Phase Noise Analysis

In the PLL, all the building blocks contribute to overall phase noise and PLL is analyzed in the phase domain using linear techniques [22–24]. The closed-loop transfer function is used to define the PLL output phase change due to a change in the input phase. The equivalent phase domain model of the PLL with noise sources and the respective noise transfer function is shown in Fig. 7.15 and Table 7.2.

The output PLL noise transfer in Table 7.2 is shown in Fig. 7.16. The transfer function for the reference, PFD and charge-pump and the feedback divider are low-pass in nature and the PLL output noise within the bandwidth is low-pass filtered and dominates the in-band phase noise. As can be seen from the transfer function, the noise from the PFD and charge-pump can be reduced by increasing the gain of the phase detector. In other words, increasing the charge-pump current at the expense of increased power consumption. The noise injected by the VCO has high pass filter characteristics and dominates the PLL output phase noise outside the PLL bandwidth. The noise contribution of the low-pass filter exhibits a band-pass characteristic. The loop filter noise can be improved with a reduction in VCO gain K_{VCO}.

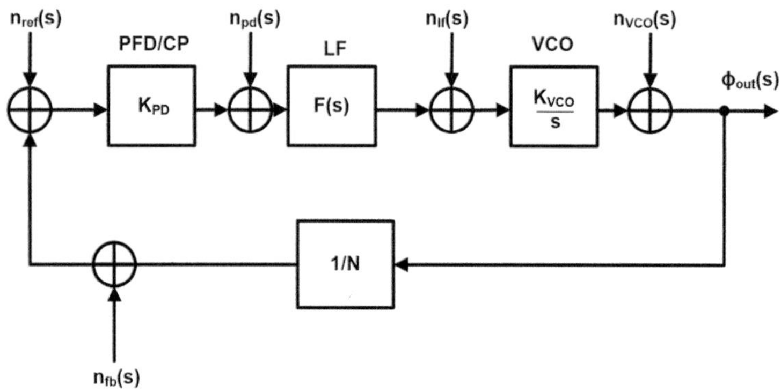

Figure 7.15 Frequency domain model of PLL with noise sources.

Table 7.2 PLL noise transfer function

PLL block	Noise transfer function	
Reference	$\frac{\varphi_{out(s)}}{n_{ref}(s)}$	$N\cdot\frac{G_{OL}(s)}{1+G_{OL}(s)}$
PFD and charge-pump	$\frac{\varphi_{out(s)}}{n_{pd}(s)}$	$\frac{N}{K_{PD}}\frac{G_{OL}(s)}{1+G_{OL}(s)}$
Loop filter	$\frac{\varphi_{out(s)}}{n_{pd}(s)}$	$\frac{K_{VCO}}{s}\frac{1}{1+G_{OL}(s)}$
Feedback divider	$\frac{\varphi_{out(s)}}{n_{pd}(s)}$	$N.\frac{G_{OL}(s)}{1+G_{OL}(s)}$
VCO	$\frac{\varphi_{out(s)}}{n_{vco}(s)}$	$\frac{1}{1+G_{OL}(s)}$

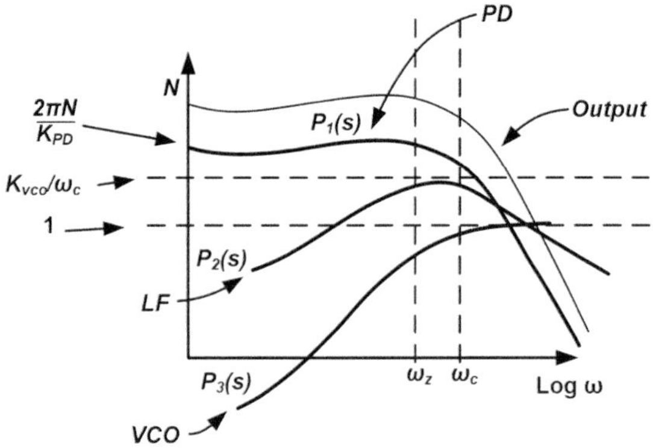

Figure 7.16 PLL noise transfer function.

7.4 Architecture of PLL Components

The phase-locked loop frequency synthesizer is a feedback system with multiple low-frequency and high-frequency blocks. Though the above discussions fairly cover the basic aspects of the PLL design, significant performance improvements in the PLL can be achieved by optimizing the individual circuit blocks in the frequency synthesizer. This section provides the description and design details of the individual blocks of the frequency synthesizer.

7.4.1 Prescaler and Frequency Divider

The frequency divider is one of the most versatile and important blocks in the frequency synthesizer as it down-converts the high-frequency VCO signals to low-frequency signals for phase comparison in the phase detector. Since it operates on high-frequency VCO signals much attention and efforts are made to reduce the power consumption and area of the frequency dividers. Also, the frequency divider in the PLL feedback loop, dominates the overall phase noise performance of the frequency synthesizer. The frequency divider can be classified based on the frequency of operation and is shown in Fig. 7.17. The high-frequency dividers mostly have fixed frequency division ratio, and the programmability is implemented in the low-frequency dividers to synthesize multiple channel frequencies. The description of different frequency dividers is discussed in this section.

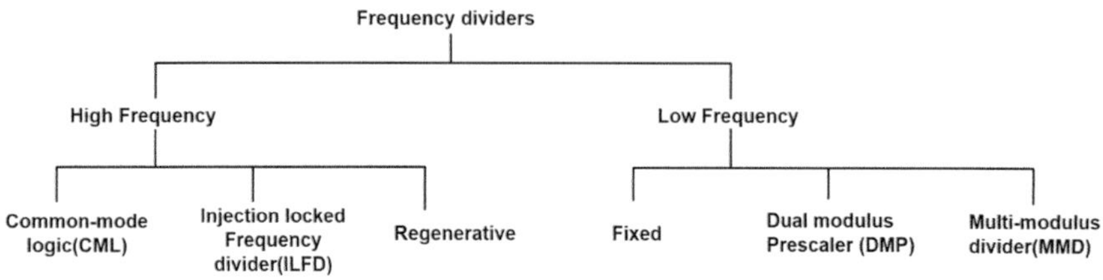

Figure 7.17 Frequency divider classification.

7.4.1.1 High-frequency dividers

a. Common-mode logic (CML) frequency dividers

CML frequency dividers are the most commonly used fixed high-frequency dividers in the divider chain. CML dividers are a class of static frequency dividers based on ring oscillator topology [25, 26]. A frequency divide-by-*N* is realized by cascading *N* latches in the positive feedback configuration with the cross-coupling of final stage differential outputs to the input of the first stage. The schematic of the CML latch in the frequency divider is shown in Fig. 7.18.

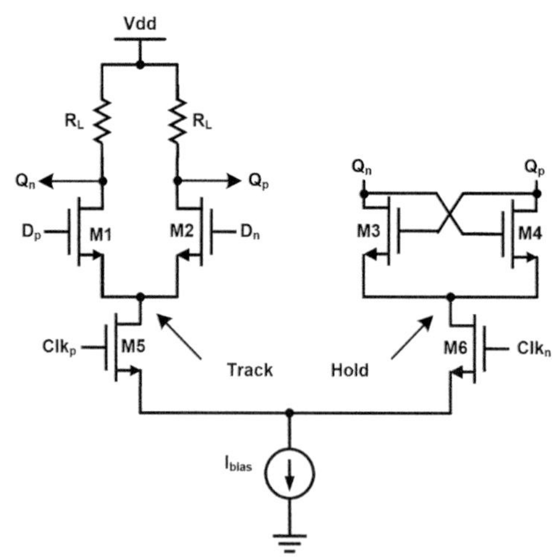

Figure 7.18 CML latch schematic.

The CML divider is a differential circuit using complementary input signals and consists of two branches—the differential pair for tracking and the latch transistors for holding the output. When the input amplitude is zero, the CML divider self oscillates and the oscillation frequency (f_{SO}) decreases with the increase in the size of the latch transistor $M_{3,4}$. In the positive half cycle, when the clock is high, the input is tracked by M_1 and M_2 and passed to the output directly. In the negative half cycle, when the clock is low, the latch M_3 and M_4 holds the current output. The maximum frequency of operation of the CML dividers is based on the total capacitance at the output and the load resistance R_L given as

$$F_{\max} = \frac{1}{2\pi R_L C_L}. \tag{7.18}$$

Therefore, the maximum frequency of operation for CML dividers can be maximized by reducing the output capacitance and load resistance. However, there is a design trade-off with the choice of R_L and C_L. As the load resistance R_L is reduced, the overall gain of the differential pair decreases and needs a higher current. Also, the output swing reduces and the CML divider fails to drive the following divider stages.

The advantage of CML dividers is their high divider sensitivity, which increases with increased input amplitude. One of the major disadvantages of the CML dividers is its requirement for higher supply voltage (current) as the frequency of operation increases. Therefore, in practical designs, inductive peaking can be used to increase the operation frequency [27–29]. However, this also increases the die area.

In advanced process technology nodes, the current source transistor stages are removed to maximize the frequency and voltage drop for low supply voltage operation. Also, the choice of transistor size for the differential pair and latch transistors is important to increase the frequency of operation. The minimum size for the latch transistor can increase the operation frequency. Also, the condition $g_{\mathrm{mL}}R_L > 1$, needs to be satisfied for proper division operation. Though several CML frequency dividers with operating frequencies up to 40 GHz have been reported in the literature [28–33], it is still a challenge to design CML dividers at frequencies beyond 50 GHz using bulk-CMOS technologies.

b. Regenerative frequency dividers

The regenerative frequency divider or the miller divider is a high-frequency divider based on a mixer with feedback [34–36] and the general block diagram is shown in Fig. 7.19(a). The input signal f_i is mixed with the output signal f_{out} by the mixer generating two frequency components at $f_i + f_{\text{out}}$ and $f_i - f_{\text{out}}$. The low-pass filter following the mixer, filters out the high-frequency component at $f_i + f_{\text{out}}$ leaving the difference frequency component $f_i - f_{\text{out}}$, where $f_{\text{out}} = f_i/2$.

The CMOS implementation of the regenerative frequency divider is shown in Figs. 7.19(b) and 7.19(c). The divider can either be formed by only the Gilbert cell mixer [37] or Gilbert cell mixer with an amplifier [25]. In comparison with the static frequency divider, when no input is present, the regenerative frequency divider does not oscillate, and the output remains switched off. When the clock input is zero, there is no loop gain to start oscillations as the outputs are fed back to the mixer inputs with opposite phases. Though the regenerative frequency dividers are often used when the input frequency is very high [37, 38] the locking range is less, and the power consumption is high. Also, the input power required to achieve a wider locking range for the regenerative divider is considerably higher [39].

c. Injection-locked frequency divider (ILFD)

The most promising and attractive type of analog frequency divider suitable for high frequency is the injection-locked frequency divider (ILFD). The ILFD is similar to a regenerative frequency divider but the ILFD locks the injection signal to the oscillator core signal forcing the oscillator to work at a different frequency from its core [40]. In RF and mmWave implementations, the ILFD is typically employed in frequency division ratios of 2 to 6 and the general block diagram of ILFD divide-by-2 and divide-by-6 is shown in Fig. 7.20.

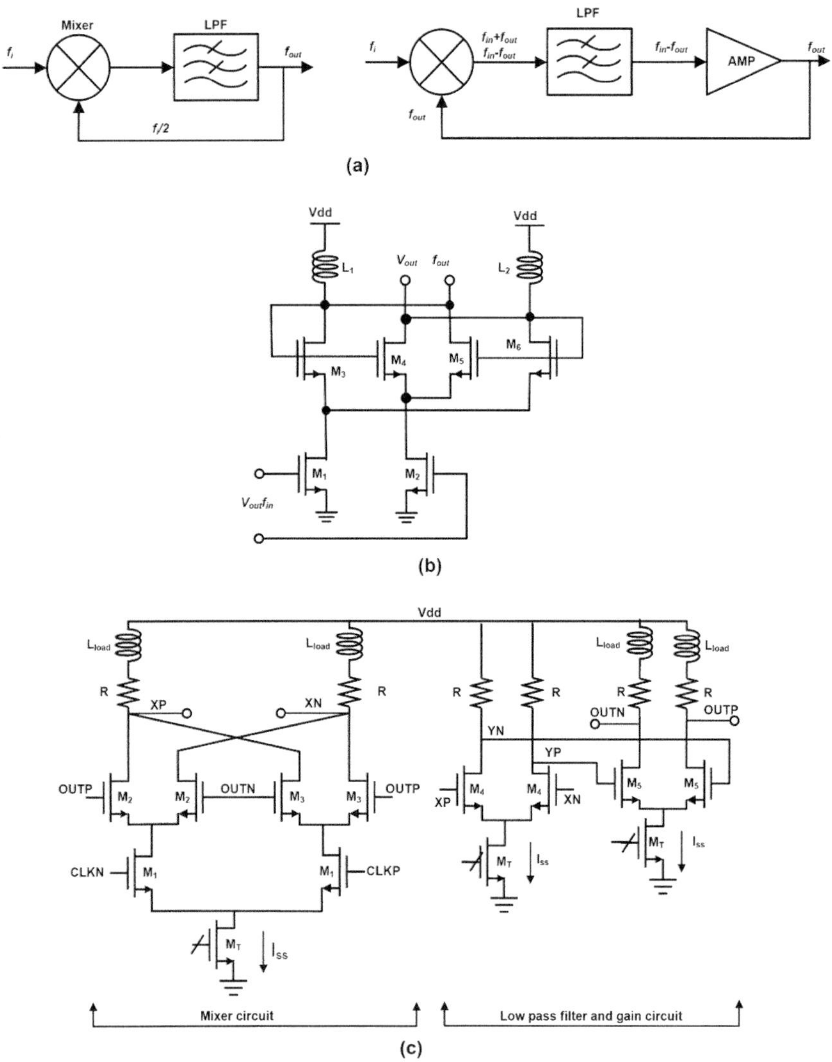

Figure 7.19 Regenerative frequency divider: (a) block diagram, (b) schematic without amplifier, and (c) schematic with amplifier.

In the ILFD, the injection transistor acts as a mixer, mixing the injected signal (f_i) with oscillation signal. This mixing operation by the injection transistor results in mixing products at oscillation frequencies (for example, at frequencies of $f_i/2$, $3f_i/2$ for ILFD divide-by-2). The frequency selective tank in the ILFD acts as a band-pass filter and selects the frequency of interest at the output of the ILFD.

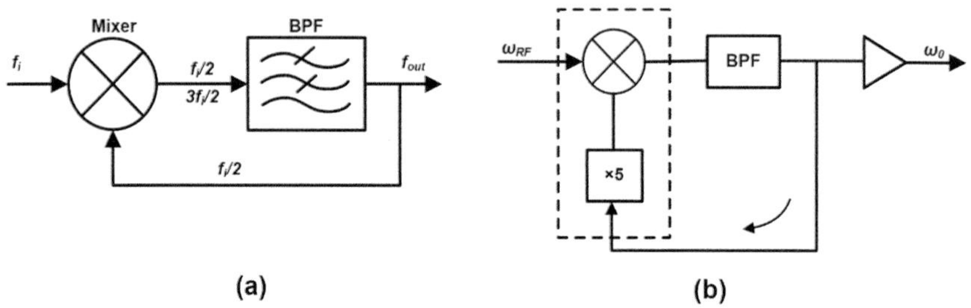

Figure 7.20 Conceptual block diagram of ILFD: (a) divide-by-2, and (b) divide-by-6.

The operation of ILFD is shown in Fig. 7.21 [21]. With no injection, the ILFD is equivalent to the free-running oscillator as shown in Fig. 7.21(a). When an input signal with appropriate amplitude and frequency is injected into the free-running oscillator, the beats of both the injected frequency f_{inj} and output frequency f_{out} are observed at the output. When f_{inj} is close to f_{out}, the beats disappear, and the oscillator starts to oscillate at the injected frequency rather than the output frequency. The injection locking happens when is f_{inj} close to f_{out}. Therefore, the range of frequencies at which the injection locking happens is defined as the locking range of ILFD as shown in Fig. 7.21(c) and is given as [40]

$$\frac{\Delta\omega}{\omega_o} = \frac{I_{\text{inj}}}{2QI_{\text{osc}}}, \tag{7.19}$$

where I_{inj} is the injection current, I_{osc} is the oscillation current and Q is the quality factor of the tank circuit.

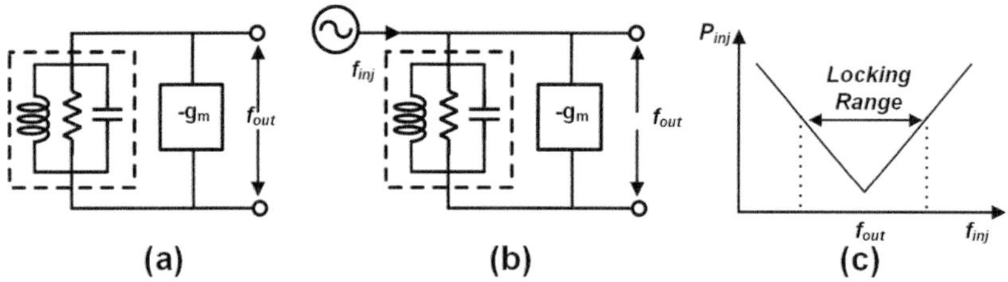

Figure 7.21 ILFD model (a) free-running oscillator, (b) under injection, and (c) locking range.

The smallest division ratio that can be realized in the ILFD is frequency divide-by-2 and the CMOS ILFD implementation is shown in Fig. 7.22. The ILFD looks similar to a cross-coupled oscillator and as shown in Fig. 7.22(a), the injection port is at the gate of the tail current transistor M_3 which acts as both the tail current transistor and injection transistor. The major drawback of this circuit is the low locking range due to the low injection efficiency. As shown in Eq. (7.19), for a given tank quality factor and oscillator current for low power consumption, to increase the locking range with higher injection currents, the size of M_3 is usually quite large to provide a higher transconductance and tail current. However, this leads to a high parasitic capacitance at the drain of M_3 and creates a ground path for higher injection frequencies resulting in low injection efficiency. Though, this issue can be addressed by adding an inductor in parallel to resonate out the parasitic capacitance, the silicon area is increased. Also, another drawback of the circuit in Fig. 7.22(a) is single-ended injection. Typically, the input of the divider is driven from a differential VCO and half the output power from the VCO is not utilized.

An alternative and most commonly used direct injection ILFD topology is shown in Fig. 7.22 (b). In the direct injection ILFD, the signal injection is achieved by placing the injection transistor across the ILFD tank circuit. Differential injection can be achieved by employing differential injectors across the ILFD tank as reported in [19, 21, 41]. In the direct injection ILFD, transistor M_{inj} acts as a mixer and optimization is focussed on improving the transconductance of the mixer usually achieved with increasing the width of the injection transistor. Typically, the general aim of improving the locking range of ILFD is to decrease the quality factor of the tank circuit or improve the injection efficiency by which more injected signal is available for the mixing operation. However, since decreasing the ILFD tank quality factor can increase power consumption, the focus is on increasing the ILFD injection efficiency. Several works are reported in the literature to increase the ILFD locking range with improvement in injection efficiency with a low power consumption [41–46] and higher division ratios [47–54].

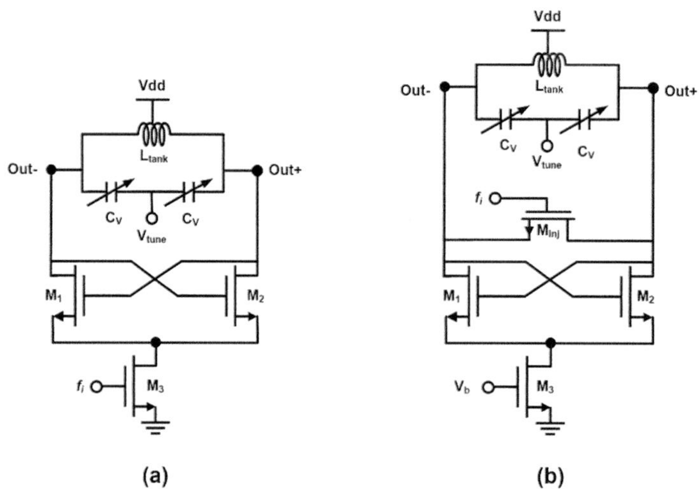

Figure 7.22 ILFD schematic (a) tail injection, and (b) direct injection.

7.4.1.2 Low-frequency dividers

Following the high-frequency dividers, the frequency synthesizer divider chain consists of a series of low-frequency dividers to reduce the VCO frequency down to the reference frequency for comparison in the phase frequency detector. The different low-frequency divider is discussed in brief in this section.

a. **Fixed frequency divider**

In the frequency synthesizers, the back-end dividers operate at a few hundred MHz. Therefore, digital dividers are more popular due to their small size and low power consumption.

The true single-phase clock (TSPC) divider is one of the common digital dividers used in the back end of the divider chain and its schematic is shown in Fig. 7.23. The divider uses only one clock signal and therefore, no clock-skew problem exists [55]. Also, with proper design, the TSPC divider can operate until a few GHz range.

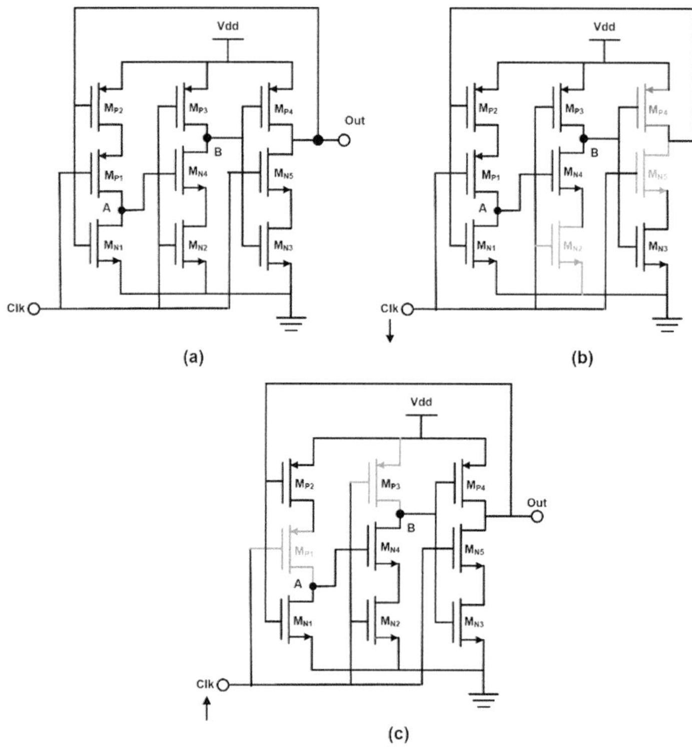

Figure 7.23 TSPC divider: (a) schematic, (b) operation when input *clk* is low, and (c) operation when input *clk* is high.

The operation principle of the TSPC divider when the input clock is low and high is shown in Figs. 7.23(b) and 7.23(c), respectively. When the *clk* input is low (hold mode), node *B* is pre-charged to a high and output is maintained at the previous state. However, when *clk* input is high (evaluation mode), the output is dependent on value at node *A*. When node *A* is high, node *B* is low and output is high.

While node A is low, node B is high, and output is low. Therefore, the divide-by-2 operation is realized when node A switches the state of node B for every two rising clock edges. To further improve the operation speed of the TSPC DFF, extended TSPC. (E-TSPC) was proposed in [56]. In comparison to the TSPC DFF, E-TSPC DFF uses only two transistors in each stage. However, E-TSPC DFF has higher static current and lower noise immunity in comparison to TSPC DFF.

b. Dual-modulus prescaler (DMP)

In an RF frequency synthesizer, the frequency synthesis function is realized by inserting a programmable frequency divider in the low-frequency divider chain of the PLL such that the PLL output frequency is varied by varying the feedback division ratio. A more straightforward solution is to insert a programmable down counter to preset the division ratio. However, this method poses practical implementation issues when the feedback division ratio is large. For example, when the operating frequency is 2.4 GHz and the required comparison frequency is 1 MHz, the required division ratio is 2.4 GHz/1 MHz = 2400. To synthesize a frequency of 2.4 GHz with a division ratio of 2400, a 12-bit programmable counter is required. In practice, the complexity of the counter increases due to the large operation frequency and also the power consumption increases drastically.

In practice, the DMP with pulse and swallow counter is used as programmable frequency divider [57] and the block diagram is shown in Fig. 7.24. The main components of the DMP are the divide-by-$N/(N+1)$ prescaler, programmable pulse (P) and swallow (S) counters. With the DMP implementation, the larger counter bits can be split to smaller value P- and S-counter. In the DMP, the programmable counters are loaded with the desired values, and the divider divides by $(N + 1)$ for S pulses and for $(P - S)$ pulses, the division is by N. The total division ratio of the divider is given by

$$N_{\text{total}} = (N+1)S + (P-S)N = PN + S. \tag{7.20}$$

The main limitation of the dual-modulus prescaler is in the choice of values for the P and S counters. From Eq. (7.20), the value of the P-counter must be always greater than the value of the S-counter. The required division ratio can be achieved using different combinations of N, P and S and the choice of N is the initial step in the divider design. For ease of fixed divide-by-N implementation in the DMP, values based on cascaded divide-by-2 stages are chosen. The values of P and S is given by

$$P = \text{Integer}\left[\frac{N_{\text{total}}}{N}\right] \tag{7.21}$$

$$S = N_{\text{total}} - (P \times N). \tag{7.22}$$

The basic block in the implementation of the DMP is the dual-modulus divider and the simplest divide-by-2/3 is shown in Fig. 7.25. The mode-control "*MC*" controls the divide-by-2/3 and the timing diagram is shown in Fig. 7.25. When MC = 1, the output of the OR gate is always high and FF1 is isolated from the circuit. When the AND gate shorts the output of FF2 to its input, the divide-by-2 operation is achieved. When MC = 0, the output of the OR gate is the same as the output of FF1. The input of the AND gate is the output of FF1 and the inverted output of FF2 forming a divide-by-3 circuit operation.

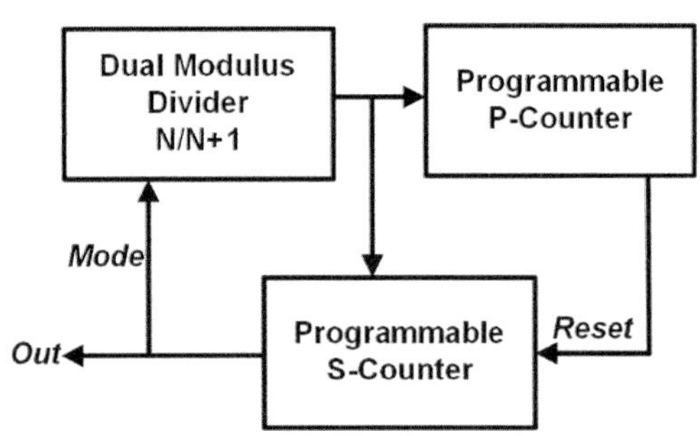

Figure 7.24 Dual-modulus prescaler block diagram.

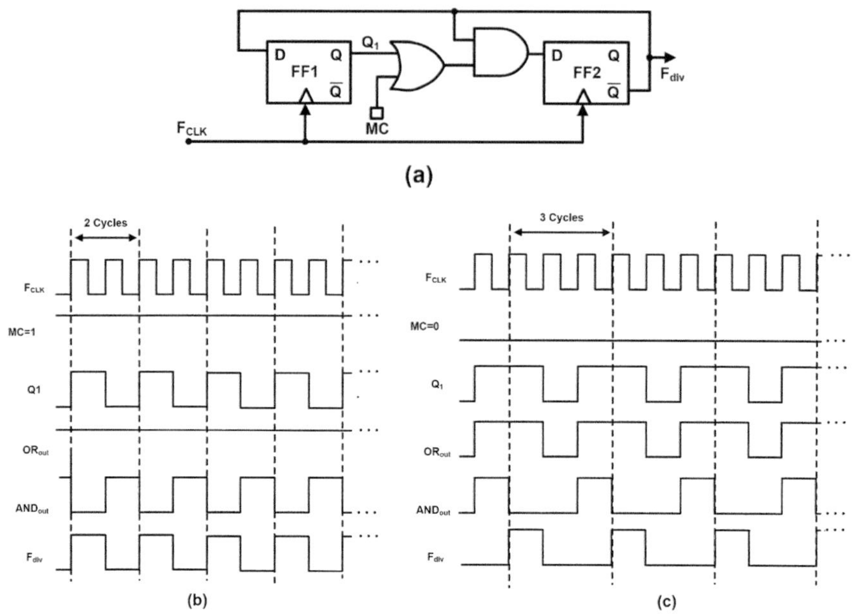

Figure 7.25 Dual-modulus prescaler: (a) 2/3 prescaler, (b) /2 timing diagram, and (c) /3 timing diagram.

The divide-by-2/3 divider circuit forms the basic divider cell in the DMP and the following cells are cascading stages of divide-by-2. The basic block diagrams divide-by-16/17 and divide-by-32/33 prescaler are shown in Figs. 7.26 and Fig. 7.27, respectively.

As shown in Fig. 7.27, the modulus control for divide-by-32/33 is generated by NAND and NOR gates rather than multi-input AND gates due to the large voltage headroom required for AND gates in low voltage applications [58]. As shown in Figs. 7.27(b) and 7.27(c), when *MOD* = 1, *MC* will always be high and the divide-by-2/3 functions as chain of divide-by-2 and divider-by-32 operation is achieved. When *MOD* = 0, *MC* will be low for three output cycles and high for the remaining 30 cycles realizing a divide-by-33 operation. Due to the scaling of semiconductor technology, the input frequency of the DMP is consistently increased and several high-speed DMP is reported in the literature [59–62] for further reading.

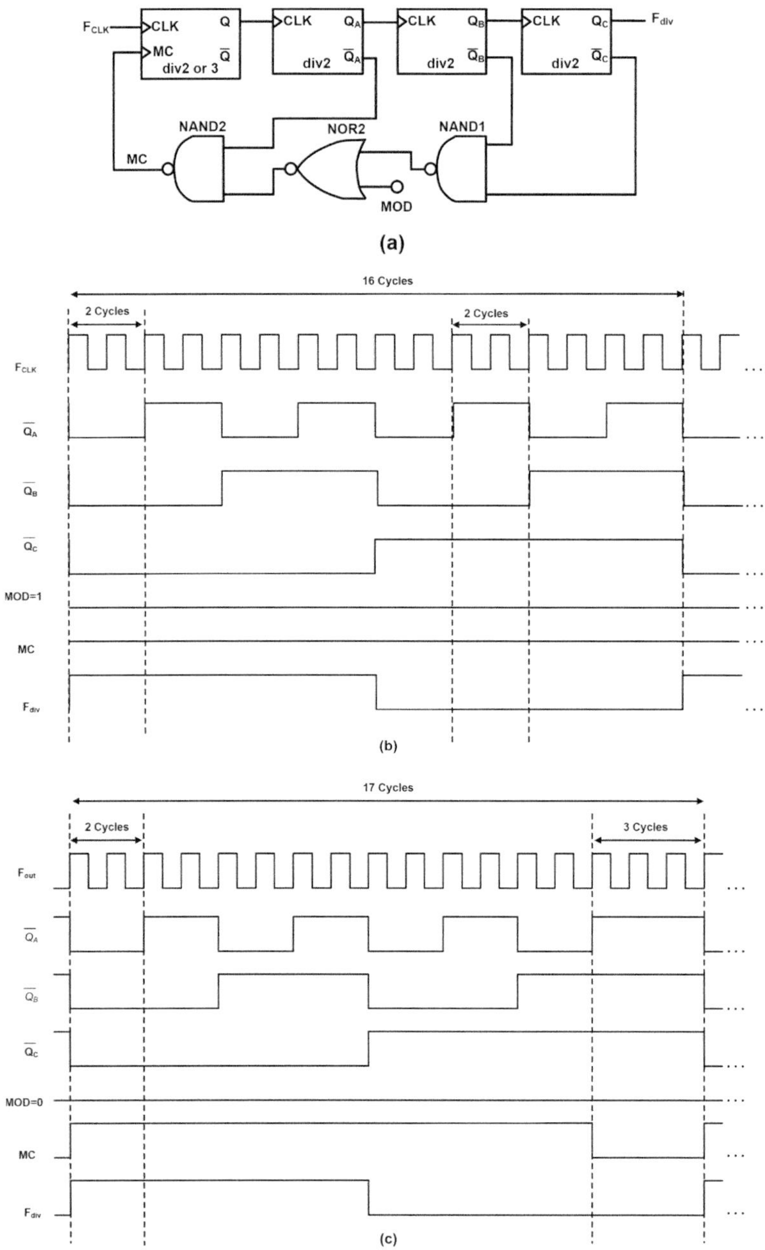

Figure 7.26 Dual-modulus divide-by-16/17 prescaler: (a) block diagram, (b) /16 timing diagram, and (c) /17 timing diagram.

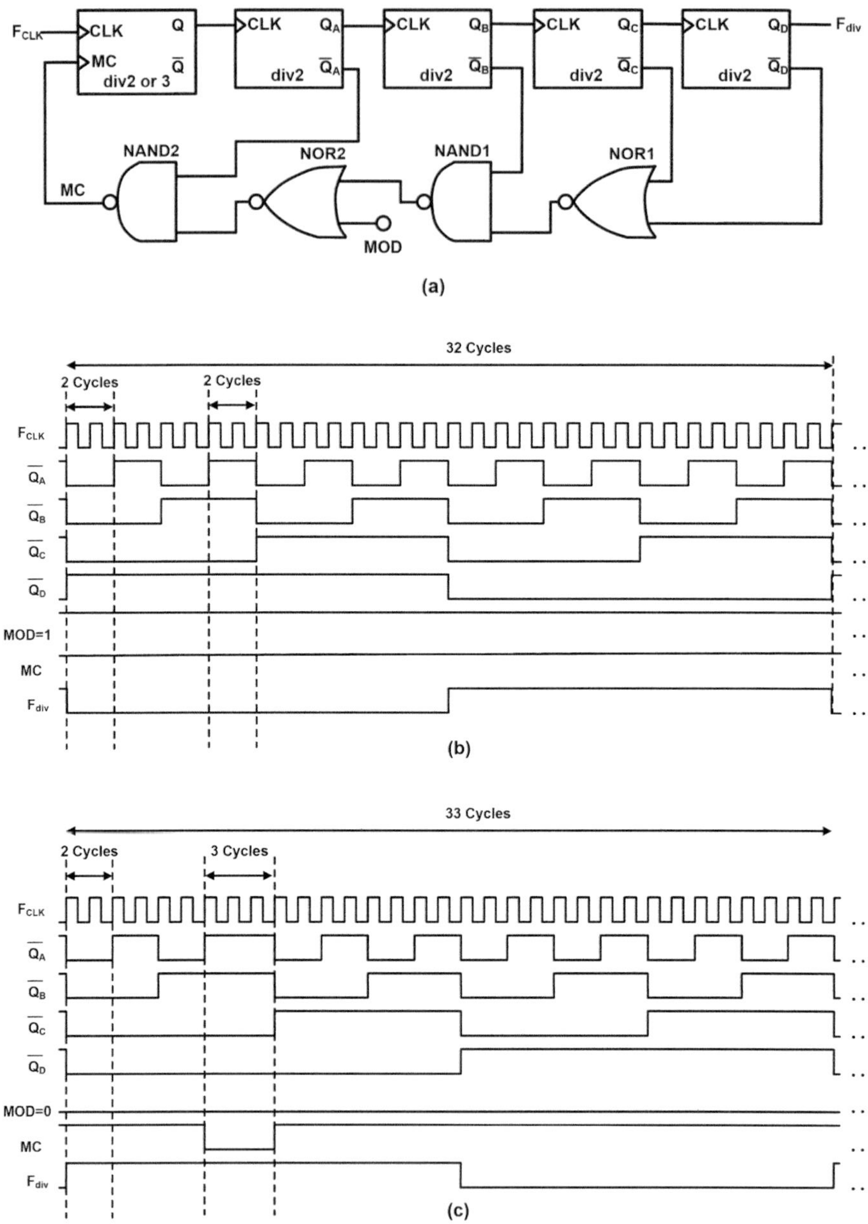

Figure 7.27 Dual-modulus divide-by-32/33 prescaler: (a) block diagram, (b) /32 timing diagram, and (c) /33 timing diagram.

The typical implementation of *P*-counter (4-bit) and *S*-counter (3-bit) in the feedback divider is shown in Fig. 7.28. The counters are loadable down counters which are used to synthesize multiple frequencies in the frequency synthesizer. The *P*-counter is designed as an asynchronous down counter with reloadable DFF and a NOR DFF and additional logic as an end-of-count detector. In the initial state, the corresponding values are loaded in the *P*-counter. Based on the clock signal from the DMP, the counter starts to count from the loaded value. As it reaches 0000, the load signal goes high as detected by end-of-count and resets the counter inputs.

The *S*-counter in the feedback divider is also implemented an asynchronous down counter and the operation of the S-counter is similar to the *P*-counter and starts operation only after the $(P - S) \times N$ clock cycles. When the output of the S-counter is low, then $SP = 1$ and remains in the state until the *P*-counter $LD = 1$.

The schematic of the DFF in the *P*-counter and *S*-counter is shown in Fig. 7.29. As the input is loaded into the *P*-counter and when $LD = 0$, the counter begins to count down. As $LD = 0$, the transistor M_5 is off and only the transistors M_{1-14} are functional. The DFF functions as a divide-by-2 until the counter reaches 0010. When $LD = 1$, M_5 is turned on, node *A* is always at logic "0" and the transistors $M_{15\text{-}24}$ are functional. Under this condition, when $PI = 0$, node *D* is at logic "1" and the value in node *B* is latched to the output at the next rising edge of the clock. When $PI = 1$, the output switches to logic "1" on the clock rising edge. The DFF of the *S*-counter is similar to the *P*-counter except the SP logic is added to transistor M_{14}. In the *S*-counter all the states are used and the *S*-counter stops counting when $SP = 0$. During this time, when $LD = 0$, the mode-control signal MC goes high causing the odd-division ratio by swallowing one clock pulse. The output timing of the feedback divider for the division ratio of 199 in combination with the DMP, *P*-counter and *S*-counter is shown in Fig. 7.30.

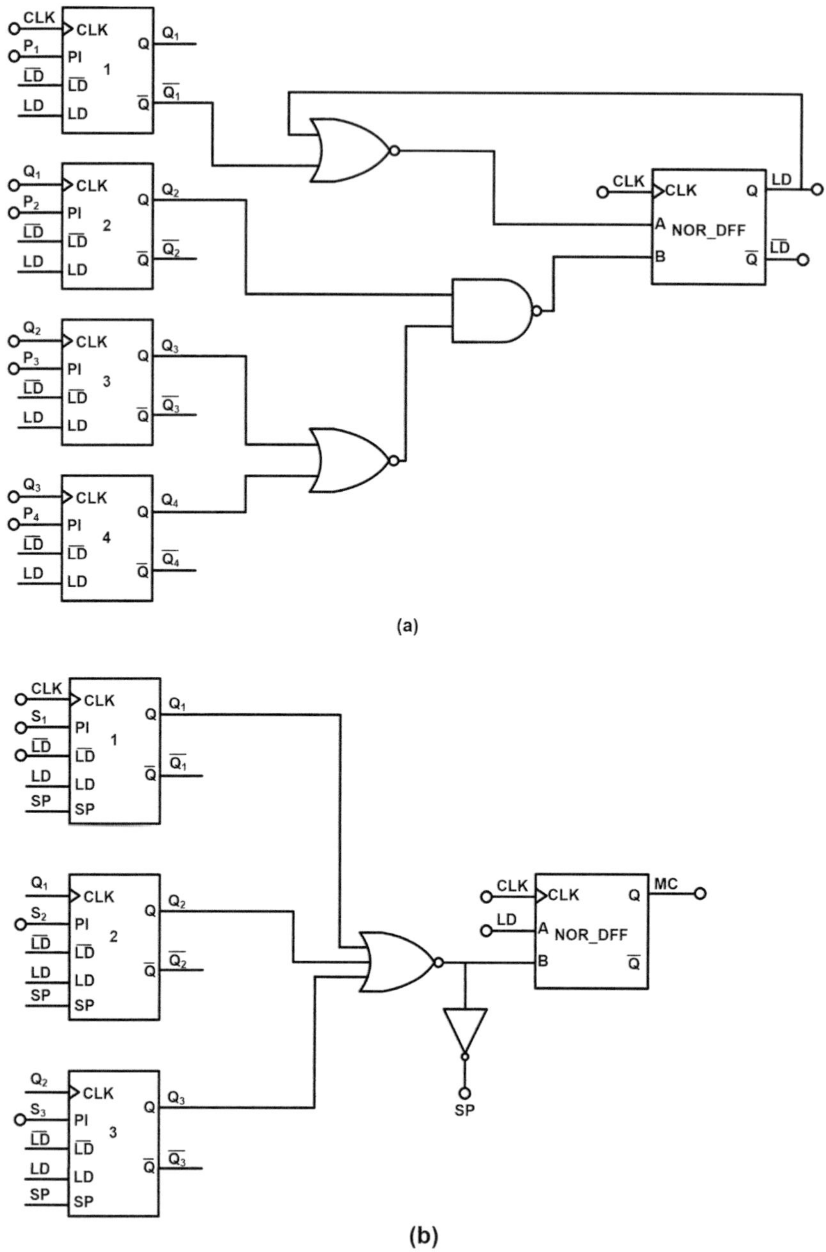

Figure 7.28 Asynchronous down counter: (a) *P*-counter (4-bit) and (b) *S*-counter (3-bit).

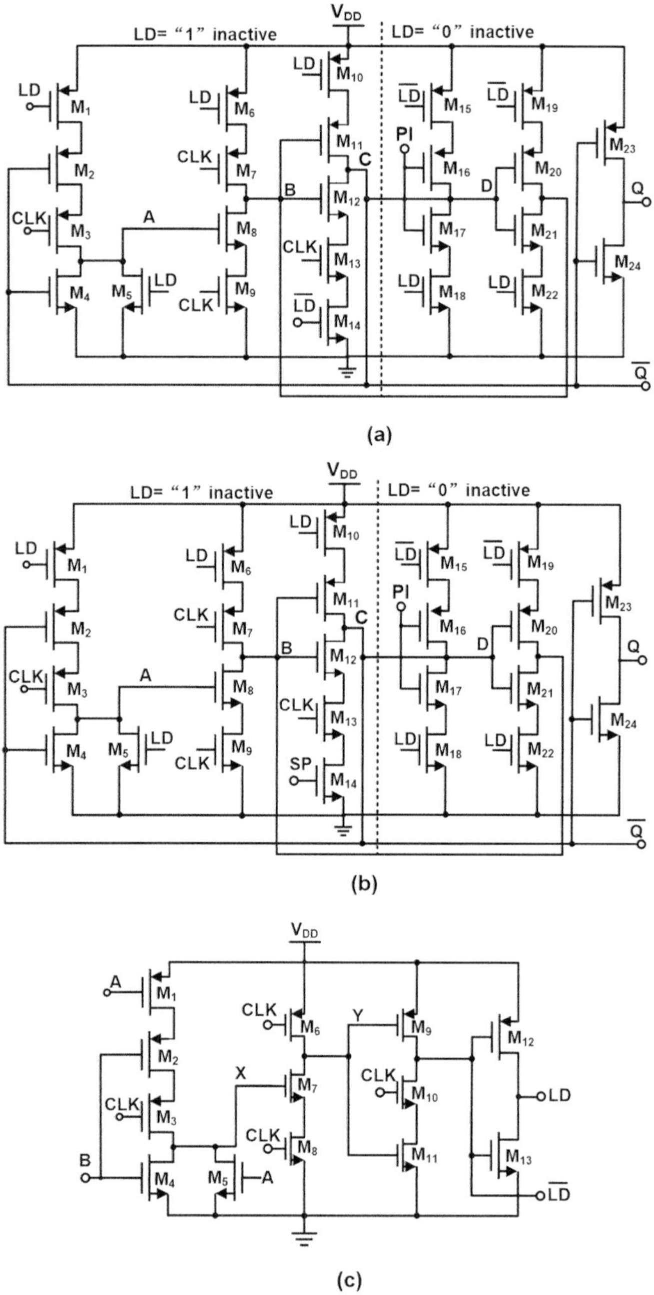

Figure 7.29 Bitcell: (a) *P*-counter DFF, (b) *S*-counter DFF, and (c) NOR DFF.

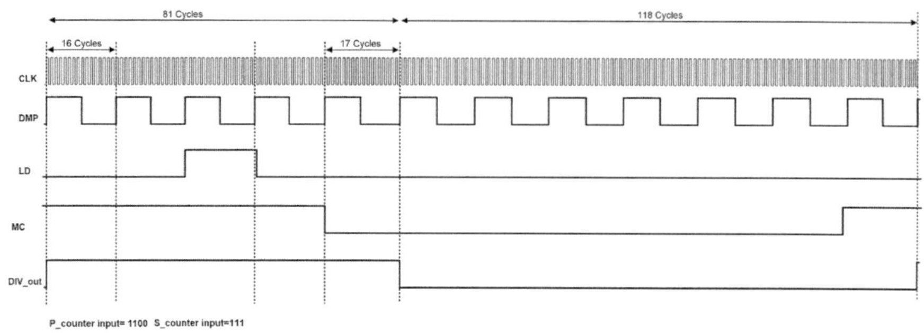

Figure 7.30 Feedback divider timing diagram.

c. Multi-modulus divider (MMD)

In addition to the DMP, the asynchronous PLL divider commonly used in the synthesizer divider chain is the MMD [63]. The basic cell of the MMD is the divide-by-2/3 as used in the DMP and the MMD selectively swallows the pulses from the previous stages realizing the required division ratio. For example, the schematic of 6-bit MMD is shown in Fig. 7.31. The division ratio of the corresponding 6-bit MMD shown in Fig. 7.31 is given as

$$N_{\text{div}} = 64 + 2^5 d_5 + 2^4 d_4 + 2^3 d_3 + 2^2 d_2 + 2^1 d_1 + 2^0 d_0. \tag{7.23}$$

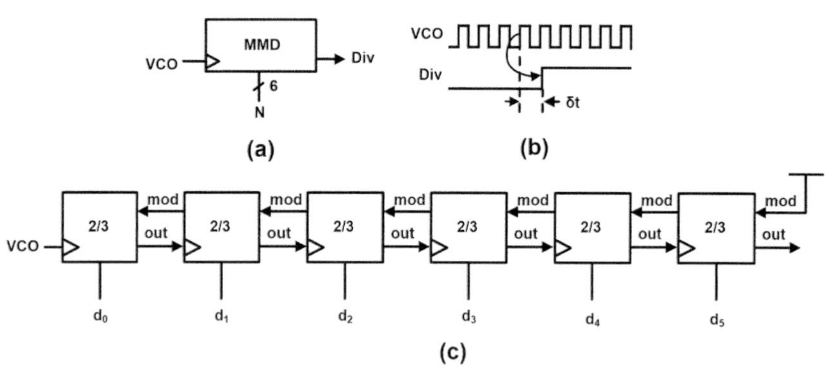

Figure 7.31 Multi-modulus divider: (a) top-level, (b) input/output timing, and (c) block diagram of 6-bit MMD.

Since MMD is an asynchronous divider, there is a delay between the VCO output and the divider output. Therefore, the MMD must be resynchronized if multiple divider phase outputs are derived from the divider output.

7.4.2 Sigma-Delta Modulator (SDM)

In the frequency synthesizer, the programmable frequency divider generates the multiple channel frequencies and determines the frequency resolution or the frequency step of the synthesizer. For example, in the fractional-N frequency synthesis, for a very low-frequency step, the fractional part is realized with an accumulator of size F clocked at the reference frequency. The overflow or the carry output is used to change the programmable division between N and $(N + 1)$ and occurs only K times over F cycles. However, this switching of division ratio between N and $(N + 1)$ to obtain an average division fractional division ratio generates undesired noise and this needs to be moved to higher frequencies that can be filtered out by the PLL loop. This can be obtained by noise shaping based on the sigma-delta modulator [64, 65].

The sigma-delta modulator for frequency synthesis has been quite consistent over the years. For the frequency synthesis, when the lower-order modulators are used, the power of fractional spurs is high. Therefore, higher-order modulators have been preferred. However, the stability is an issue with higher-order modulators. In the PLL frequency synthesis, typically the order of the modulator chosen is less than or equal to the PLL loop filter order. Since most PLLs employ a third-order loop filter, the third-order modulator is generally implemented in multi-stage noise shaping (MASH) structure with either cascade of single-stage modulators as MASH 1-1-1 or single and 2-stage modulators as MASH 2-1 or MASH 1-2 structure. Though the MASH 1-1-1 or MASH 2-1 modulator architecture have similar circuit hardware complexity, MASH 1-1-1 is always preferred for on-chip implementation of fractional synthesis due to its ease of implementation, stability over the whole input range in comparison to second-order modulator [66] and third-order quantization noise shaping.

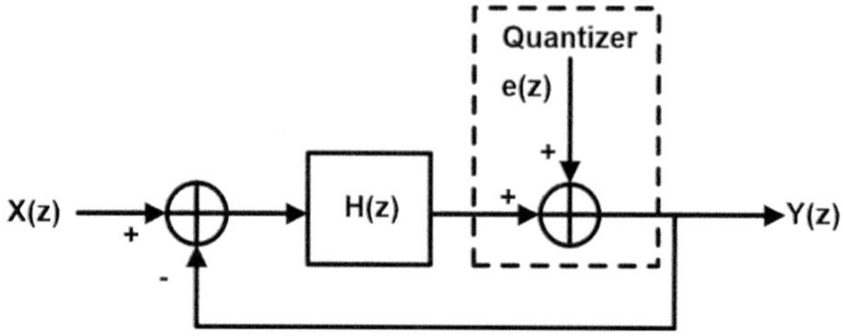

Figure 7.32 Linear model of sigma-delta modulator.

The sigma-delta modulator is a feedback system with a loop filter and a quantizer. The general linear model of the modulator is shown in Fig. 7.32. The output of the modulator in z-domain is given by

$$Y(z) = \frac{H(z)}{1+H(z)} X(z) + \frac{1}{1+H(z)} e(z). \tag{7.24}$$

The signal and noise transfer function can be derived when $e(z) = 0$ and $X(z) = 0$, respectively, and is given as

$$H_{\mathrm{STF}}(z) = \frac{H(z)}{1+H(z)} \text{ and } H_{\mathrm{NTF}}(z) = \frac{1}{1+H(z)}. \tag{7.25}$$

The output $Y(z)$ can be represented in terms and signal and noise transfer functions as

$$Y(z) = H_{\mathrm{STF}}(z)X(z) + H_{\mathrm{NTF}}(z)e(z). \tag{7.26}$$

From Eq. (7.26), the output signal is dependent on $H(z)$ and quantization noise. Therefore, to reduce the in-band quantization noise with high pass characteristics, $H(z)$ should have a unity gain across the frequency of interest.

7.4.2.1 Hardware implementation

The hardware implementation of the first-order modulator was presented in [9]. The linear model and the corresponding hardware implementation of the first-order modulator are shown in Fig. 7.33.

The linear model is an error feedback structure, and the output $Y(z)$ is given as

$$Y(z)=A(z)+(1+Z^{-1})e(z). \tag{7.27}$$

The digital accumulator implementation is best suited for the hardware implementation of the linear model in Fig. 7.33(a) [67]. A m-bit adder adds the input signal and the feedback signal. The accumulator carry then generates the quantized signal. The quantization noise is represented by the sum of the accumulator with a negative sign.

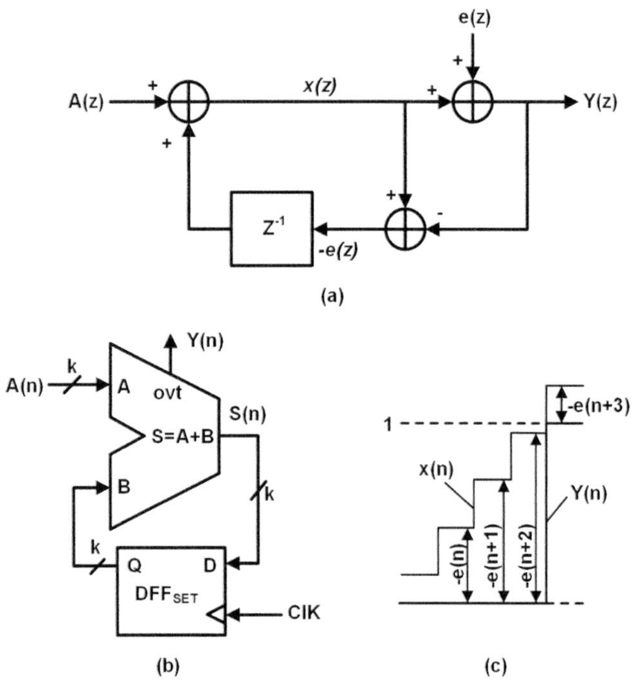

Figure 7.33 First-order sigma-delta modulator (a) error feedback model, (b) hardware implementation, and (c) quantization noise behavior.

7.4.2.2 MASH 1-1-1 sigma-delta modulator

The MASH sigma-delta modulator proposed in [67] is a popular architecture for the implementation of the fractional-N frequency synthesis [68]. The MASH 1-1-1 modulator consists of cascaded stages of lower-order 1-stage modulator as discussed previously and is shown in Fig. 7.34. In the MASH 1-1-1 architecture, the quantization errors of the previous stages are canceled and only the quantization error of the last stage remains.

Based on the linear model outputs of the first-order modulator, the MASH 1-1-1 outputs can be derived as

$$Y_1(z) = A(z) + (1 - z^{-1})e_1(z) \tag{7.28a}$$

$$Y_2(z) = -e_1(z) + (1 - z^{-1})e_2(z) \tag{7.28b}$$

$$Y_3(z) = -e_2(z) + (1 - z^{-1})e_3(z) \tag{7.28c}$$

$$Y(z) = y_1(z) + y_2(z)(1 - z^{-1}) + y_2(z)(1 - z^{-1})^2. \tag{7.29}$$

From Eqs. (7.25) and (7.26), the output $Y(z)$ of the MASH 1-1-1- can be given as

$$Y(z) = A(z) + (1 - z^{-1})^3 e_3(z), \tag{7.30}$$

where $A(z)$ is the input and $(1 - z^{-1})^3$ is the noise transfer function of the modulator. For controlling the feedback division ratio in frequency synthesis, the integer division ratio is added to the fractional part such that the overall output frequency is governed by the relation

$$F_{\text{VCO}} = (N + \Delta N)F_{\text{ref}}, \tag{7.31}$$

where ΔN is the fractional division ratio, which depends on the input of the MASH 1-1-1 sigma-delta modulator. Since in the frequency synthesizer, the MASH 1-1-1 is in the feedback path, in practice, the quantization noise introduced by the sigma-delta modulator needs to be mapped to the PLL overall phase. The general expression for the sigma-delta quantization noise to PLL noise is given by [69]

$$L(f)=\frac{(2\pi)^2}{12f_{\mathrm{ref}}}\left|2\sin\left(\frac{\pi f}{f_{\mathrm{ref}}}\right)\right|^{2(n-1)}\ \mathrm{rad}^2/\mathrm{Hz}, \tag{7.32}$$

where n is the order of the modulator. When the offset frequency $f << f_{\mathrm{ref}}$, Eq. (7.29) can be written as

$$L(f)=\frac{(2\pi)^2}{12f_{\mathrm{ref}}}\left|2\pi\left(\frac{f}{f_{\mathrm{ref}}}\right)\right|^{2(n-1)}\ \mathrm{rad}^2/\mathrm{Hz}. \tag{7.33}$$

Therefore, from Eqs. (7.32) and (7.33), the quantization noise is shaped and is low-pass filtered by the PLL loop filter. Therefore, the sigma-delta modulator noise adds to the trade-off in the PLL loop bandwidth and for better noise filtering of sigma-delta modulator noise, the order of the modulator is chosen to be equal to or less than the order of PLL loop filter.

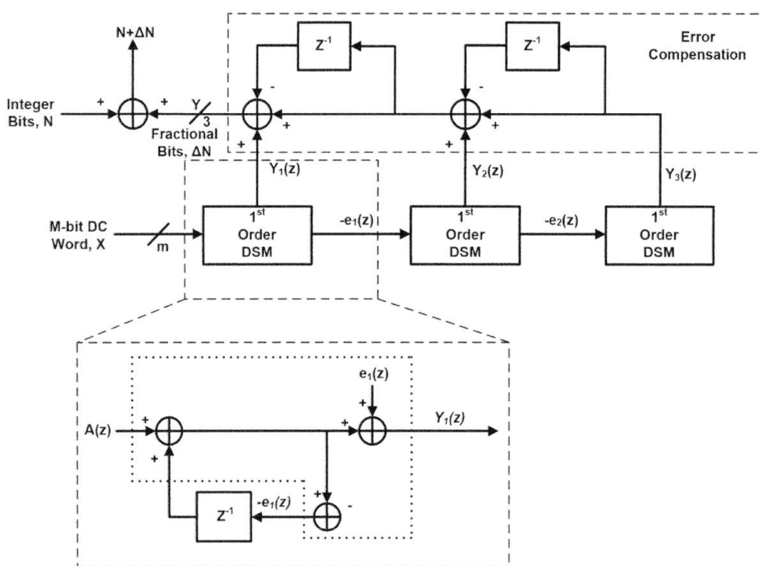

Figure 7.34 Linear model of MASH 1-1-1 SDM.

In the design of MASH 1-1-1 for frequency synthesis, the choice of input bits is important. Due to the periodic nature of the MASH 1-1-1 output, the fractional spurs appear at fixed offsets in the PLL output. Therefore, to improve the PLL output spurs and reduce the modulator periodicity, a high-capacity input bit in the range of 20 bits to 24 bits with an active LSB is chosen in practical implementation [69]. Also, the fractional frequency resolution is better with a higher number of input bits.

7.4.2.3 Hardware implementation of MASH 1-1-1

The hardware implementation of the MASH 1-1-1 follows the first-order modulator design. The typical implementation with the same number of bits in all three stages and the quantization error cancellation network based on differentiator and adders is shown in Fig. 7.35. In the implementation as shown in Fig. 7.35, the accumulator is the basic cell in the MASH 1-1-1 and the accumulator consists of repetitions of a 1-bit cell consisting of a D-flipflop and full adder. The carry input of the first stage is zero and the carry output of the first stage is connected to the input of the second stage. The MASH 1-1-1 generates 3 carry bits and the output sequence needs to be mapped accordingly as the output levels vary from –3 to +4 [70] and the typical mapping is shown in Table 7.3.

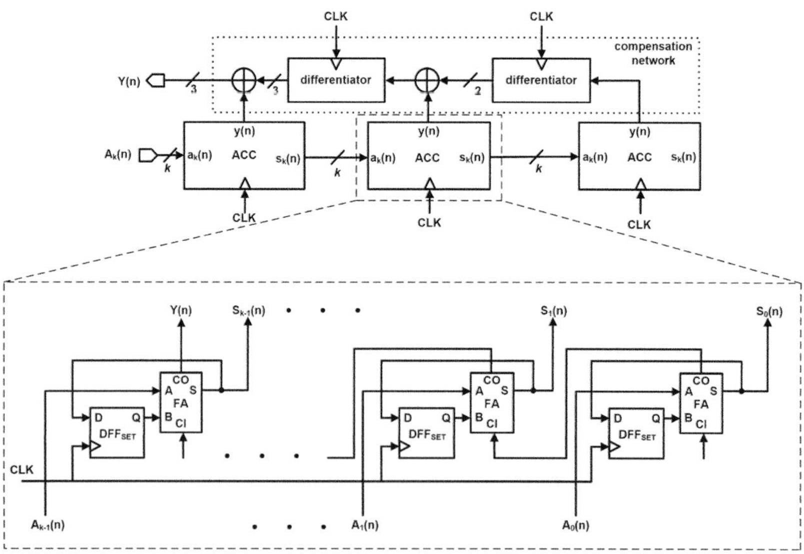

Figure 7.35 Hardware implementation of MASH 1-1-1 SDM.

Table 7.3 MASH 1-1-1 output mapping

Output level	Y2	Y1	Y0
–3	1	0	1
–2	1	1	0
–1	1	1	1
0	0	0	0
1	0	0	1
2	0	1	0
3	0	1	1
4	1	0	0

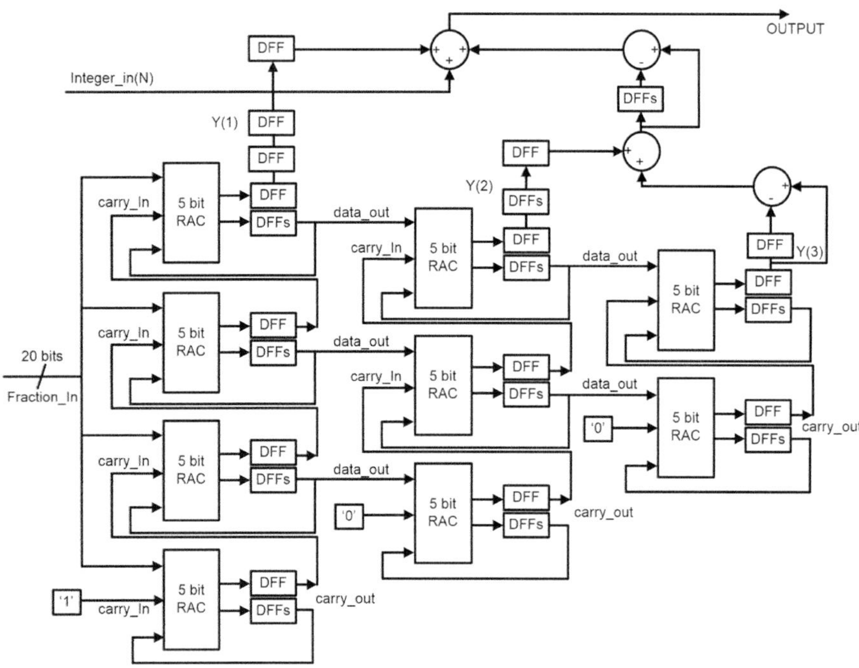

Figure 7.36 Logic structure of MASH 1-1-1 SDM [71].

There are different ways in the implementation of the MASH 1-1-1 in literature optimizing the number of bits and power consumption. The optimized implementation of MASH 1-1-1 sigma-delta modulator is shown in Fig. 7.36. Instead of the same *k*-bit inputs in every stage of MASH 1-1-1, variable input word-length with pipeline architecture and ripple-carry adder accumulator [71, 72]. The variable word-length in the MASH 1-1-1 stages achieves the same noise spectral density in comparison to equal word-length in all stages as shown in Fig. 7.37. Also, the complexity and power consumption of the sigma-delta modulator is reduced up to 25% [21].

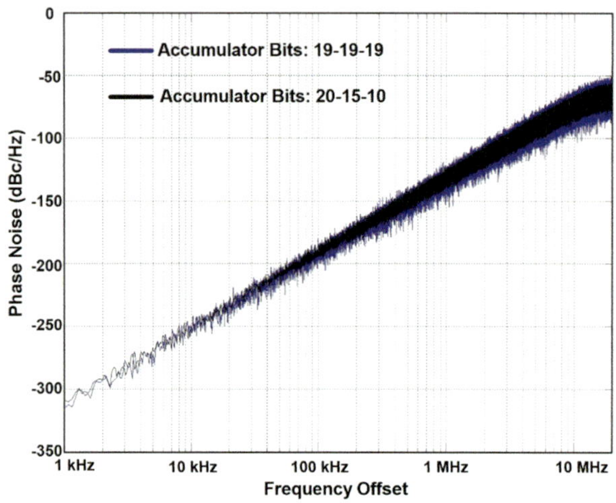

Figure 7.37 MASH 1-1-1 SDM with variable input bits [21].

7.4.3 Phase Frequency Detector

The low-frequency blocks in the frequency synthesizer are largely reusable blocks and therefore, a large number of architectures are available in literature based on the application criterion such as frequency synthesis, CDR, frequency locked loops (FLL), etc. The phase detector in the PLL compares the VCO phase with the reference signal phase and can be broadly classified as analog and digital phase detectors based on their implementation scheme. In this section, a brief description of different phase detectors for frequency synthesis is presented.

7.4.3.1 Analog phase detector

The simple analog multiplier such as Gilbert multiplier [1], which generates sum and difference frequencies, is commonly used as a multiplier phase detector in the PLL. Assuming $v_1(t)$ and $v_2(t)$ are the reference and VCO signals given by

$$v_1(t) = V_1 \sin(\omega_1 t + \varphi_1)$$
$$v_2(t) = V_2 \cos(\omega_1 t + \varphi_2). \tag{7.34}$$

Based on the trigonometric identity, the output of the multiplier can be given as

$$v_{out}(t) = \frac{V_1 V_2}{2}\left[\sin(2\omega_1 t + \varphi_1 + \varphi_2) + \sin(\varphi_1 - \varphi_2)\right], \tag{7.35}$$

where the phase detector gain $K_{PD} = V_1V_2/2$. When the inputs of the multiplier have the same frequency, their frequency difference is zero and the output frequency is proportional to phase (phase lock). Since the multiplier also generates the sum component of the input signals, this high-frequency component needs to be filtered out by the low-pass filter. Therefore, Eq. (7.32) can be approximated as

$$v_{out}(t) = \frac{V_1 V_2}{2}\left[\sin(\varphi_1 - \varphi_2)\right]. \tag{7.36}$$

The characteristic of the multiplier phase detector is shown in Fig. 7.38 [73]. For the multiplier phase detector, it can be noted that K_{PD} is related to the amplitude of the input signals and is constant only in the small range of $\pm\pi/2$.

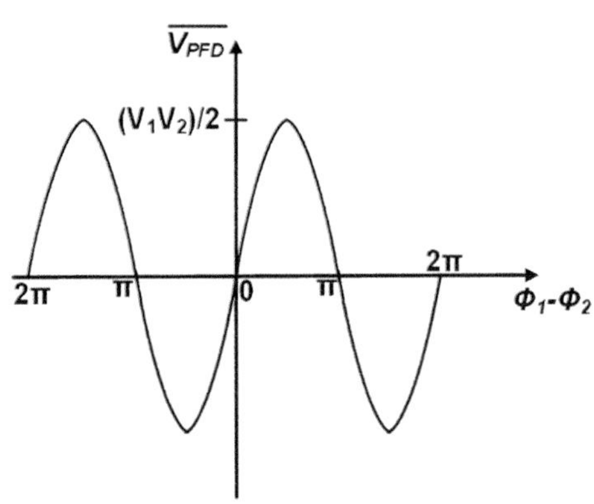

Figure 7.38 Multiplier PD characteristics.

7.4.3.2 XOR phase detector

In a frequency synthesizer, the VCO is followed by a chain of digital low-frequency dividers and the corresponding output is a square wave rather than a sine wave. Therefore, an XOR which is a digital equivalent of the multiplier is another commonly used phase detector for frequency synthesis. The characteristics of the XOR phase detector is shown in Fig. 7.39 [73]. As shown in Fig. 7.39, K_{PD} for XOR phase detector is no longer a function of the signal amplitude as in the multiplier phase detector as the digital signals are typically rail-rail and the amplitude of the square wave is between 0 and V_{dd}.

One of the important advantages of the XOR PD is its simplicity and high operational speeds which is particularly useful in clock and data recovery PLL. However, when the XOR phase detector is used in the frequency synthesizer for wireless applications, the XOR PD does not detect the frequency difference between the input signals as the phase difference is obtained based on the zero crossing of the input signals. For any frequency difference, the phase difference is accumulated either in the positive or in the negative direction.

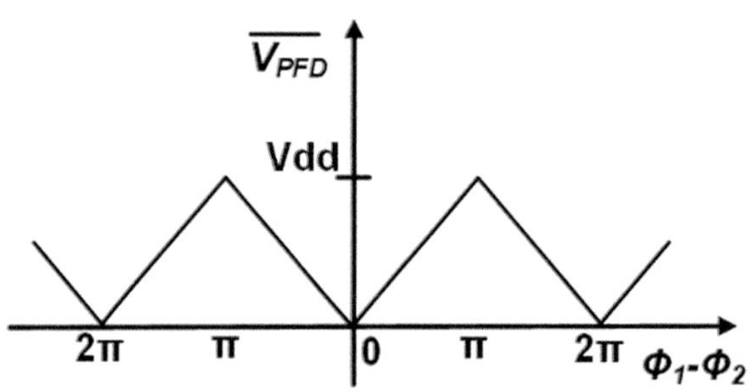

Figure 7.39 XOR PD characteristics.

7.4.3.3 Phase frequency detector

In contrast to multiplier and XOR phase detectors, more often frequency synthesizers employ PFD, which includes the frequency information along with the phase difference to speed up the PLL frequency acquisition and phase locking process. The PFD employs a D-flipflop as its main component and the main disadvantage is its low frequency of operation due to CMOS implementation. Since PLLs are used for multiple applications such as high-speed clock generators and RF frequency synthesis, the design of PFD also varies. The block diagram of the PFD used for frequency synthesis and the corresponding timing diagram is shown in Fig. 7.40.

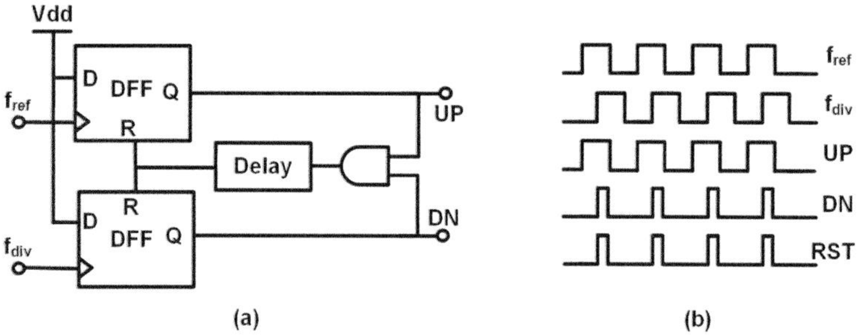

Figure 7.40 Phase frequency detector: (a) block diagram and (b) operation when $f_{ref} > f_{div}$.

The operation of the PFD can be explained as follows. During the initial phase acquisition process, the reference (f_{ref}) and feedback divider (f_{div}) signals are usually at different frequencies. When f_{ref} leads f_{div} as shown in Fig. 7.40(b), the DFF in the f_{ref} path is triggered and UP signal is switched while DN remains in low state. The UP signal maintains its value until it is reset by the rising edge of the feedback divider signal. When f_{div} leads f_{ref}, the DN signal is switched and UP is pulled to the lower state. Again, the DN is pulled to the lower state with the next rising edge of the *UP* signal. To convert the PFD outputs to usage voltage information for the VCO, a charge-pump is used in the frequency synthesizer. Based on the activation of UP or DN signal, the VCO frequency is steered in the corresponding direction either to increase or decrease the VCO frequency. When the PLL is in locked condition, the PFD outputs equal and identical UP and DN signals. In comparison to multiplier and XOR phase detectors, the PFD has a larger linear range of $\pm 2\pi$ as shown in Fig. 7.41. Though advantageous, several important non-idealities need to be considered in the PFD design for frequency synthesis.

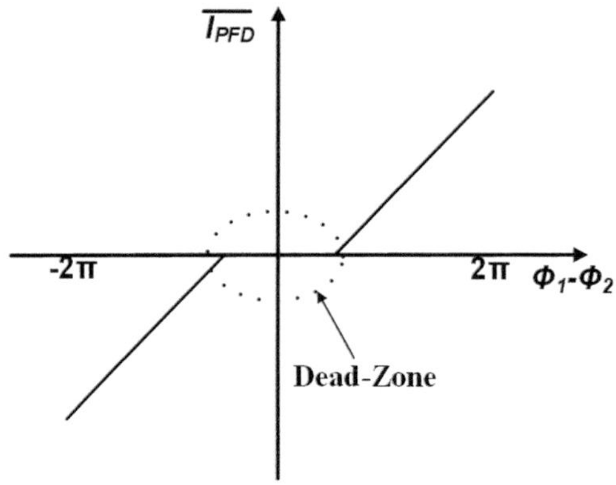

Figure 7.41 PFD transfer characteristics.

First, the PFD suffers from the issue of "dead-zone" where the PLL fails to respond to the phase difference between the reference and the feedback divider. When the PLL synthesizer is in the locked or close to the locked state, the phase difference between f_{ref} and f_{div} is either zero or close to zero. At this instant, the pulse width of UP and DN are at its lowest determined by the delay in the reset path as shown in Fig. 7.40(a). However, the charge-pump following the PFD does not react to the short pulse and no current is injected into the loop filter to stabilize the VCO output. As a result, the gain of the PFD, K_{PD} is zero and the PLL loop fails to converge. The transfer characteristic of PFD with dead-zone is shown in Fig. 7.41. The simplest solution to avoid the dead-zone in the PFD is to add more delays in the reset path. However, the downside is an increase in the reference spur. Therefore, the reset path must be chosen with careful trade-off with the reference spur and the maximum reset delay for a PFD should be less than half of the period of the reference clock signal [74].

Secondly, the PFD suffers from the clock-skew issue. The charge-pump following the PFD generally requires complementary signals for UP and DN signals. However, when the propagation delay in the PFD output path is different, the output of PFD can be switched at different times causing unwanted glitches in the output. These glitches interfere with the operation of charge-pump and manifest themselves as disturbances in the VCO control line generating spurious signals in the VCO output. The straightforward solution to reduce the clock-skew in the PFD is to add pass transistors in the UP or DN path to match the propagation delay.

7.4.4 Charge-Pump

The charge-pump follows the PFD in the PLL and either source or sinks current to the loop filter based on the PFD output. The ideal charge-pump consists of two current sources that either charge or discharge current switched by PFD UP and DN signals as shown in Fig. 7.42. In the locked state when UP and DN are the same, the charge-pump is in infinite impedance and there is no change in the output voltage as there is no charging or discharging current to the loop filter. However, a practical charge-pump suffer from several non-idealities.

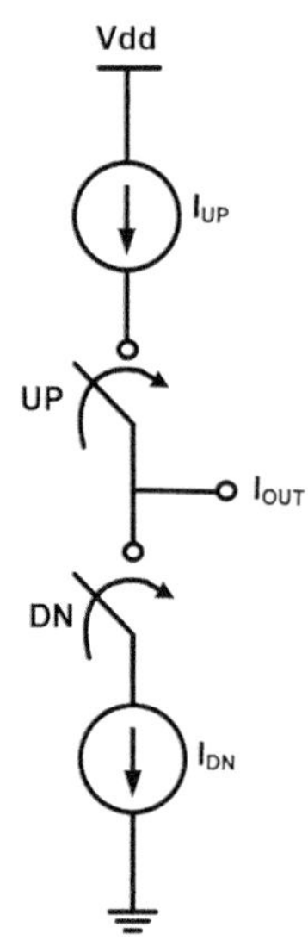

Figure 7.42 Ideal charge-pump schematic.

7.4.4.1 Current mismatch

As discussed in the previous section, the delay cells added in the PFD to eliminate the dead-zone causes the UP and DN outputs to have a constant pulse width in the locked state. Since, charge-pump is typically based on CMOS, the UP and DN current paths are implemented by PMOS and NMOS transistors, respectively. In the locked state, both the UP and DN current paths are turned ON for the period of the reset delay. In the ideal case, the net source and sink current is exactly the same and there is no current injection into the loop filter.

However, as shown in Fig. 7.43, some amount of charge is injected into the loop filter due to the current and switching time mismatch caused by the PMOS and NMOS transistors. Since the occurrence of the current mismatch is over a reference period, it manifests as a spurious tone in the PLL output. Therefore, to reduce the current mismatch in the charge-pump, the reset delay in the PFD must be minimized.

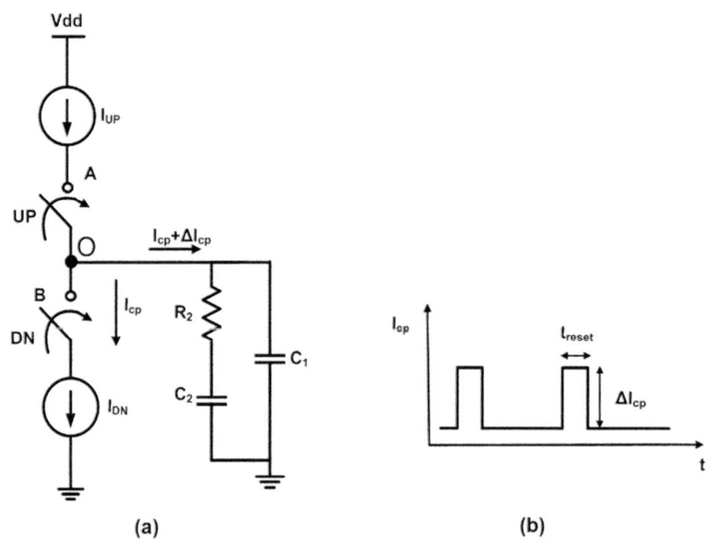

Figure 7.43 Charge-pump non-idealities: (a) conceptual schematic and (b) current mismatch.

7.4.4.2 Charge sharing

The issue of charge sharing in the charge-pump is due to the current switches being turned off after it was in the on-state. As shown in Fig. 7.43(a), when the UP and DN switches are turned off, node A is pulled to V_{dd}, node B is at GND and output node O is floating. However, due to mismatches in the switching of UP and DN, the current switches are ON for a small instance. This results in the change in voltages at nodes *A* and *B* and consequently charge sharing between capacitors at node *A*, node *B* and the output capacitance due to the difference in voltages. Charge sharing in the charge-pump causes glitches in the voltage at the loop filter output and thus degrades the PLL synthesizer spurious performance. The issue of charge sharing can be reduced by using unity gain amplifiers to maintain the voltages at nodes *A* and *B* to be the same as the output voltage.

7.4.4.3 Leakage current

The leakage current issue in the charge-pump is caused if there is a small amount of current when both the UP and DN switches are OFF. In the submicron CMOS process, the leakage current can be in the order of 1 nA [75] and the consequence is the generation of reference spurs. Under locked conditions, the average current over one reference period must be zero. However, in the presence of leakage current, there exists a constant phase error in the charge-pump error given by [75]

$$\varphi_e = 2\pi \cdot \frac{I_{\text{leak}}}{I_{\text{CP}}} (\text{rad}). \tag{7.37}$$

The current leakage in the charge-pump causes a ripple in the PLL control voltage line resulting in increased spurs in the synthesizer output. The charge-pump leakage can be reduced by cascode transistors and large transistors to reduce transistor mismatches. In addition, a large overdrive voltage may also reduce leakage [76].

7.4.4.4 Switching and transient characteristics

In the charge-pump, both PMOS and NMOS current switches are controlled by the UP and DN signals from the PFD. However, since the UP path controls the PMOS switches, an additional inverter logic is used in comparison to the DN path. Therefore, this additional delay introduces a mismatch in switching times resulting in a current mismatch in the charge-pump. This switching delay can be overcome by the use of a transmission gate to reduce the timing mismatch in the UP and DN signals [77].

In addition, due to the use of complementary current switches, the I_{UP} and I_{DN} experience different rise and fall times. Due to this difference in the rise and fall transient characteristics, spikes are observed in the VCO tuning voltage resulting in spur tones in the PLL output. The switching and transient characteristics of charge-pump is shown in Fig. 7.44.

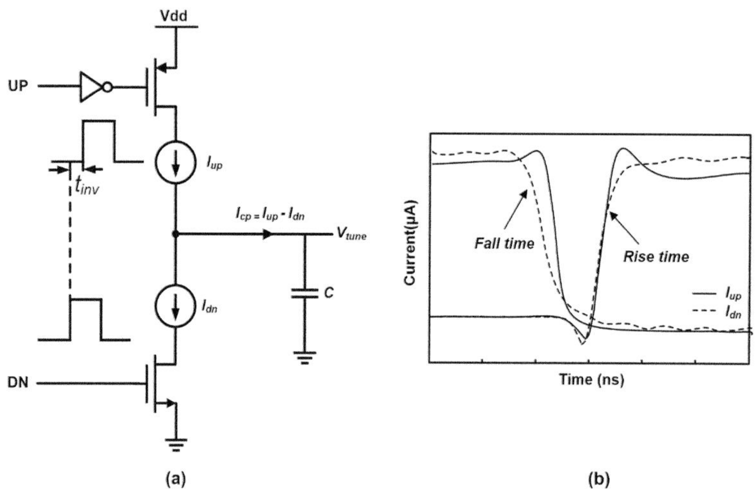

Figure 7.44 Charge-pump characteristics: (a) switching delay, and (b) transient.

7.4.4.5 Clock feedthrough

The issue of clock feedthrough is due to the coupling of high-frequency clock through the parasitic C_{gd} and C_{gs} of the UP and DN switch transistors in the charge-pump. The error occurs when the switches in the charge-pump are turned off due to periodic switching of UP and DN signals. When the current switches are turned off, the parasitic C_{gd} and C_{gs} form a capacitive divider with the capacitance in the output node resulting in a portion of the clock signal appearing across the capacitance in the output node. Therefore, the clock feedthrough disturbs the output voltage in the charge-pump. The issue of clock feedthrough can be reduced by using complementary switches or placing the current switches at the source and drain nodes farther away from the charge-pump output node.

7.4.4.6 Charge injection

In the charge-pump, when the current switches are ON, a finite amount of charge is held under the gate (in the channel). Since the current switches are always operated in the triode region, when the switches are turned off, some amount of charge is injected into the source and drain nodes of the switch transistor. As there exists a finite time for the switch control signal to make the low to high transition, this charge is injected into the charge-pump output causing disturbance to the VCO control voltage. Therefore, to reduce the effect of charge injection, a minimum transistor size is adopted for the switching transistors to increase the switching speed [78].

7.4.4.7 Output voltage range

In frequency synthesis, charge-pump output is used to control the VCO control voltage and the charge-pump behavior with varying output voltage is shown in Fig. 7.45. Typically, cascode current transistors are used in the charge-pump for higher linearity. Therefore, the UP and DN transistors are OFF for a particular time and the current mismatch is maximum at the minimum and maximum end of the VCO control voltages. The current mismatch

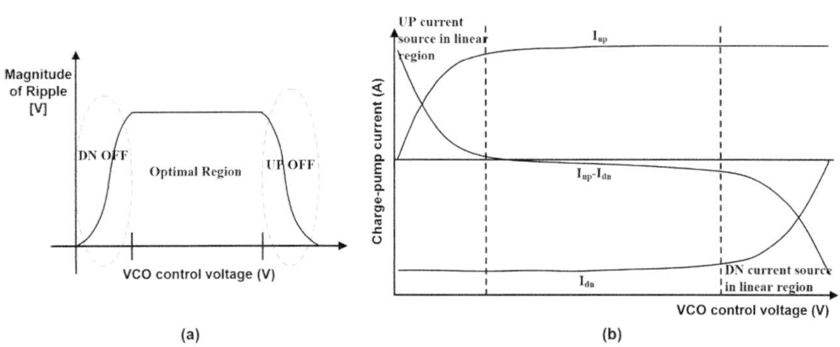

Figure 7.45 Charge-pump characteristics: (a) switching delay and (b) transient.

is ideally zero at the mid-point of the VCO control voltage and the mismatch increases as we move away from the mid-point of the control voltage. However, in practical charge-pumps, there is always a small amount of current mismatch between I_{UP} and I_{DN} due to transistor non-idealities as shown in Fig. 7.45(b).

7.4.4.8 Charge-pump architectures

a. Single-ended charge-pump

The commonly used charge-pump in the PLL is the single-ended charge-pump due to its simple structure, low power consumption, tri-state operation and lower chip area as it requires only one loop filter. Based on the location of the current switches, the single-ended charge-pumps are classified as source switching, drain switching or gate switching as shown in Fig. 7.46 [21]. However, due to non-idealities in the charge-pump, differential current switches are used in the PLL and the improved single-ended charge-pump with reduced non-idealities is shown in Fig. 7.47 [21].

The current-steering charge-pump is shown in Fig. 7.47(a) and is widely adopted in the design of frequency synthesizers [73]. The current-steering charge-pump is useful when the parasitic capacitance is comparable to the capacitor in the loop filter [75]. The main advantage of the current-steering charge-pump is its high-speed operation due to the use of current switches and reducing switching transients. Also, the charge-pump output does not stay in transition as the switches are turned off due to the use of complementary control signals, thus reducing the unwanted charging and discharging of the loop filter [79]. However, due to the use of PMOS and NMOS switches and the inherent difference in switching speeds, there are still current mismatches and leakage in the current-steering charge-pump. Since the current-steering charge-pump is based on drain switching, one of the major drawbacks is charge sharing between the drain of the switching transistor and loop filter capacitance when both the UP and DN switches are turned on. The one possible solution to reduce the charge sharing effect is to keep the output node and node n_1 at the same voltage by adding a unity gain buffer as shown in Fig. 7.47(b).

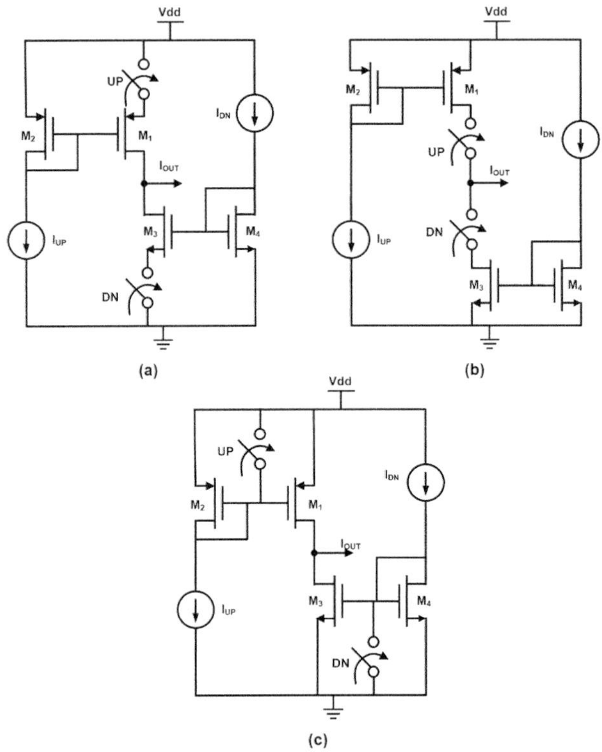

Figure 7.46 Single-ended charge-pump schematic: (a) source switching, (b) drain switching, and (c) gate switching.

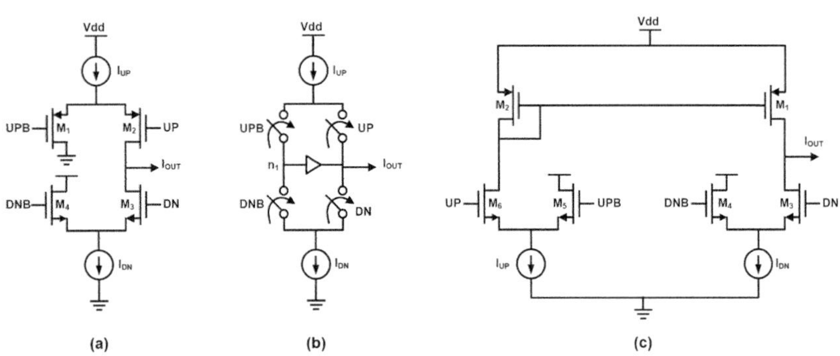

Figure 7.47 Improved single-ended charge-pump schematic: (a) current-steering, (b) with unity buffer, and (c) NMOS only.

As discussed earlier, one of the non-idealities in the charge-pump is the current mismatch due to the use of NMOS and PMOS switches. This can be overcome by the use of NMOS-only switches in the charge-pump as shown in Fig. 7.47(c). In the NMOS-only charge-pump, as the NMOS UP switch is turned on, the current is mirrored to the output using the PMOS current mirror and the current is discharged when the DN switches are turned on. Due to the use of the same type of transistor for the UP and DN switches, the switching transients and timing characteristics are the same in comparison to current-steering charge-pumps. However, the performance is limited by the PMOS current mirror transistors M_1 and M_2 when the switches are OFF and cause current mismatches due to low output impedance. The use of high cascode current mirrors can improve the output impedance of the current mirror, thus reducing the current mismatch non-idealities [79].

b. Differential charge-pump

The differential charge-pump has a lot of advantages over the single-ended charge-pump and a fully differential charge-pump as presented in [75]. Firstly, the current mismatch can be easily reduced with the use of both PMOS and NMOS switches. Secondly, the clock-skew and timing errors in the UP and DN paths can be eliminated as complementary transistors can be employed for both the normal and inverted outputs. Thirdly, with the differential charge-pump, the VCO control voltage can be doubled in comparison to the single-ended charge-pump. This can be useful to achieve a wide tuning range in the VCO with low VCO sensitivity. However, the major trade-off in the design of differential charge-pump is the increase in the overall chip area and circuit complexity due to the use of two loop filters and common-mode feedback circuitry also increasing the overall power consumption.

7.4.5 Practical Considerations for PFD and Charge-Pump

In the frequency synthesizer, PFD and charge-pump work in tandem and the design parameters of these blocks directly determine the synthesizer spur performance. Therefore, several important practical design aspects of PFD and charge-pump to improve the synthesizer performance are discussed in brief.

7.4.5.1 Clock-skew

As discussed in the previous section, the charge-pump uses differential switch control to reduce the non-idealities. However, to generate differential signals from PFD outputs—UP and DN—inverter and buffers are used. Due to the different timing characteristics and switching transients, the problem of clock-skew affects the PFD complementary signals controlling the charge-pump. Therefore, to avoid timing mismatch and reduce clock-skew, XOR gate-based differential signal generation in the PFD is shown in Fig. 7.48 [73].

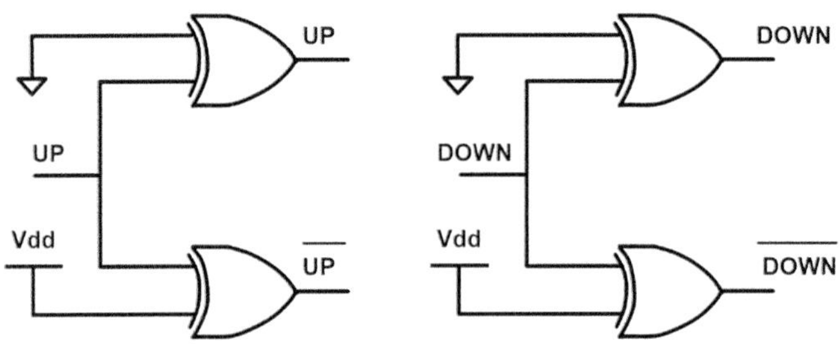

Figure 7.48 Charge-pump control signal generation from PFD outputs: (a) UP signal, and (b) DN signal.

7.4.5.2 Blind zone

The blind zone in the PFD occurs when the phase difference between the input reference and feedback signal is near 2π and arises due to the finite delay and response timing of digital circuit components such as D-flipflop and NAND gates used in the PFD. The effect of the blind zone in the dead-zone free PFD is shown in Fig. 7.49. In the PFD, the UP signal is triggered when the reference signal leads the feedback signal. As the feedback signal becomes high, the DN signal is triggered and the UP signal becomes low. As both the signals are active, UP and DN are reset. However, due to finite delay in the reset signal and the D-flipflops, when another reference signal is received within the settling time, the error in phase detection occurs leading to a blind zone in the dead-zone free PFD. The blind zone can be expressed as

$$T_{\mathrm{Bz}} = T_{\min} + T_{\mathrm{DFF}} + t_d, \tag{7.38}$$

where $T_{\min}$ is the minimum pulse width of UP/DN signal, T_{DFF} is the DFF response time and t_d is the delay of the reset pulse generation circuits, typically NAND gates and buffer chain. Therefore, the maximum phase detection range of the dead-zone free PFD with the blind zone is given as

$$\varphi_{\max} = 4\pi\left(1 - \frac{T_{\mathrm{Bz}}}{T_{\mathrm{ref}}}\right). \tag{7.39}$$

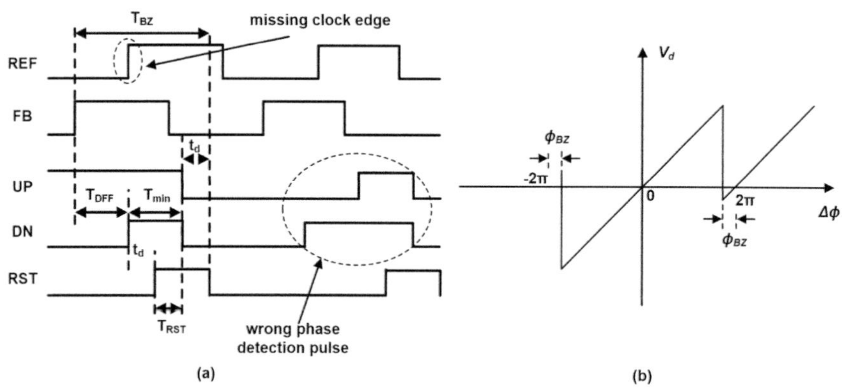

Figure 7.49 Blind zone in PFD: (a) timing diagram, and (b) transfer characteristics.

7.4.5.3 Reference spur

Unlike phase noise, spurious signals or spurs in the frequency synthesizer occurs at regular frequency intervals and inter-modulate with other signals to produce higher-order harmonic products that fall in the desired frequency channel in the communication system. Reference spur is a common spurious signal in the synthesizer output that appears at the intervals of the reference frequency. The main source of reference spurs in the PLL is the ripple in the VCO control voltage which is caused by the non-idealities in the PFD and charge-pump.

In the conventional PLL with PFD and charge-pump, there is a finite delay pulse in the PFD reset path to avoid the dead-zone. When the PLL is in the locked stage, due to this finite reset pulse, the pulse width of the UP and DN signals in the PFD is equal to the pulse width of this reset pulse. As shown in Fig. 7.50, assuming the charge-pump current pulses are perfectly matched, i.e. $I_{UP} = I_{DN} = I$, the net current at the charge-pump output is zero and the voltage at the output of the loop filter is fixed voltage. Therefore, to maintain the locking condition,

$$I_{UP}T_{UP} = I_{DN}T_{DN}. \tag{7.40}$$

However, since in practical charge-pumps, there is a mismatch in the charge-pump current sources due to the difference in the turn-on times of the PMOS and NMOS current source transistors and finite output impedances, a finite amount of current is dumped to the loop filter every reference cycle. Therefore, to maintain charge equilibrium and satisfy Eq. (7.40), the pulse width of either the UP or DN pulse has to increase leading to static phase offset (SPO) between the PFD UP and DN pulses. In the presence of SPO, the VCO control voltage exhibits disturbances causing reference spur in the synthesizer output as shown in Fig. 7.51. The magnitude of the control voltage ripple to the PFD reset delay is given as

$$\Delta_v = \frac{\Delta I}{C} t_d, \tag{7.41}$$

where ΔI denotes the difference in I_{UP} and I_{DN}, C is the loop filter capacitance and t_d is the PFD reset delay. To improve the current mismatch and reduce the reference spur, the discharge path is designed with slightly lesser current than the charging path. Apart from the charge-pump current mismatches, the control voltage ripple also leads to reference spur. The contributors for the control voltage ripple are (1) charge-pump charge injection when UP and DN pulses are OFF, and (2) the finite parasitic capacitance

(C_{gd}) of the switch transistors causing clock feedthrough. The charge injection can be reduced with the addition of dummy switches in series with the current sources. While the clock feedthrough can be reduced by using a smaller size for the switching transistors.

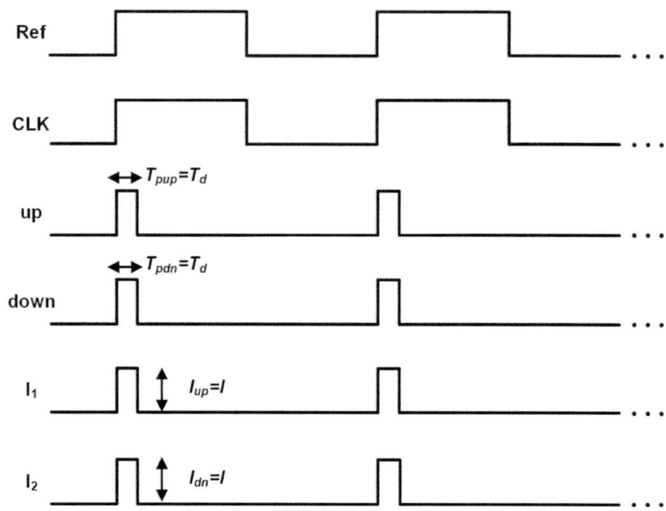

Figure 7.50 Ideal charge-pump output in locked condition.

The modelling of the reference spur has been a hot research topic for a long time, as spurs are detrimental to the performance of the communication system. The reference spurs are modelled based on narrow-band frequency modulation [80]. The VCO control voltage ripple is the main cause of the reference spur in the PLL. Due to the difficulty in modeling the VCO control voltage ripple, reference spur comparison with simulation and measurements cannot be exactly verified though the model can provide a quantitative prediction. In the conventional charge-pump PLL, the ratio of spur power to the carrier power is given by [81]

$$\frac{P_{\text{spur}}}{P_{\text{carrier}}} = \left(\frac{K_{\text{VCO}} \left| I(\omega_{\text{ref}}) \right| \left| Z(j\omega_{\text{ref}}) \right|}{2\omega_{\text{ref}}} \right)^2, \tag{7.42}$$

where K_{VCO} is the VCO tuning sensitivity, $I(\omega_{\text{ref}})$ is the charge-pump current and $Z(j\omega_{\text{ref}})$ is the loop filter bandwidth. Therefore, the reference spur can be improved with a lower loop bandwidth in the PLL. Indirectly, a smaller charge-pump current and low K_{VCO} improve the reference spur performance as they directly reduce the PLL loop bandwidth.

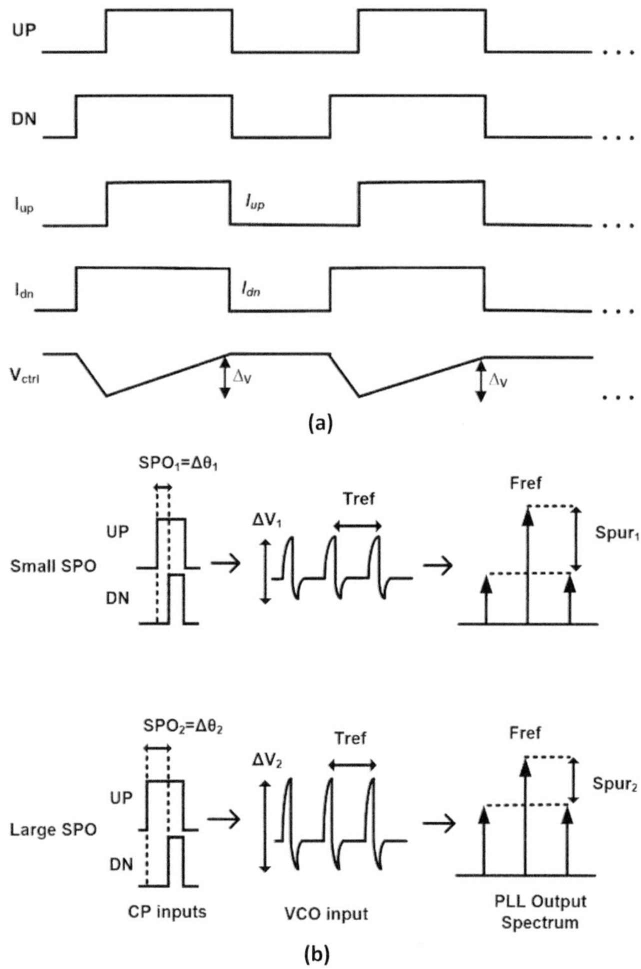

Figure 7.51 Static phase offset: (a) control voltage ripple, and (b) reference spur due to control voltage ripple.

Besides reducing the reference spur with low PLL loop bandwidth and reduced VCO sensitivity, several techniques are proposed in the literature to minimize the effect of reference spur [82–86]. One of the popular techniques is either to add additional circuits to the PFD and the charge-pump [82, 83] or to add a switch capacitor network between the charge-pump and the loop filter [85]. Also, moving the spur frequency to much higher frequencies than the reference frequency reduces the reference spur [84, 87]. In addition, several analog and digital calibration techniques have been proposed in the literature to minimize the effect of reference spur [81, 88, 89]. The techniques to minimize the reference spurs have either used variable delay elements to control the pulse width of the delay element [88] or introduced compensating currents based on the SPO [81].

7.5 Advanced Topics in Frequency Synthesizer

Phase-locked loops are important building blocks in the wireless and wireline communication systems. With the increase in the data rates with each generation of advancement in communication technology, the performance requirements of the PLLs have also become more stringent, especially in terms of phase noise, jitter, spurious tone reduction and power consumption. Figure 7.52 shows the recent advancements in the PLL development to achieve a better performance in the PLLs [90].

Among the PLL architectures, the charge-pump-based PLLs have been widely employed due to their robust locking range performance in the presence of small disturbances, better PLL performance and simple implementation. The main advantage of the charge-pump PLL is its inherent frequency and phase acquisition ability. The acquisition range of the charge-pump PLL is equal to its loop bandwidth. After the PLL has achieved its lock, the type II loop may still be able to correct itself if there are any frequency drifts due to aging, temperature drifts, etc. However, the main limitation of the charge-pump PLL is its low PFD gain of $1/2\pi$, which makes it difficult to suppress the in-band phase noise required for next-generation communication systems [90]. Therefore, to achieve a low in-band noise and low jitter, a low-noise, high-current charge-pump is required to increase the PFD gain. This makes charge-pump PLLs not attractive for low-jitter application, typically required for next-generation wireless and wireline communications such as 5G, 6G and PAM4.

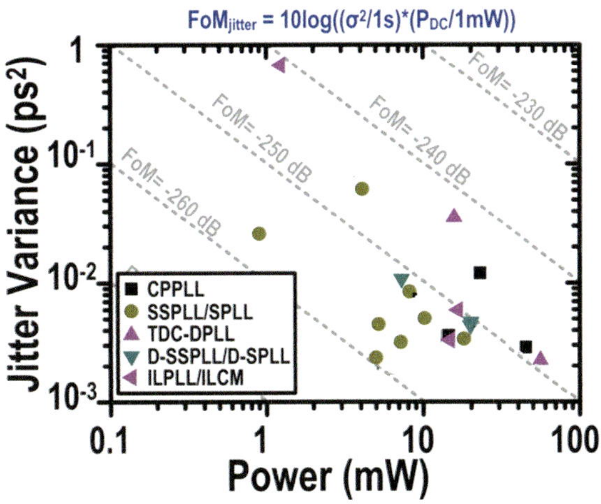

Figure 7.52 Frequency synthesizer jitter performance of recently published PLLs [90].

As shown in Fig. 7.52, more recently, sub-sampling or sampling PLLs have gained significant attention in the design of low-jitter PLLs due to their ability to achieve a low in-band phase noise because of the large gain of phase detectors. In the sub-sampling PLLs, a sub-sampling phase detector (SSPD) samples the VCO output and thus achieves a high phase detector gain equal to the slope of the VCO signal [91–97]. However, the phase detection range of SSPD is much smaller in comparison to the charge-pump PLLs. Sampling PLLs based on a sampling phase detector improve on the SSPD detection range by sampling the sharp edge of the divider output [98, 99].

7.5.1 Sub-Sampling PLL

The next-generation communication systems such as 5G, 6G and beyond require ultra-low phase noise performance in the LO to satisfy the high data-rate requirement with higher modulation schemes. The sub-sampling PLL (SSPLL) was proposed by [91] to improve the in-band PLL phase noise and the block diagram is shown in Fig. 7.53. The main reason the SSPLL can achieve a

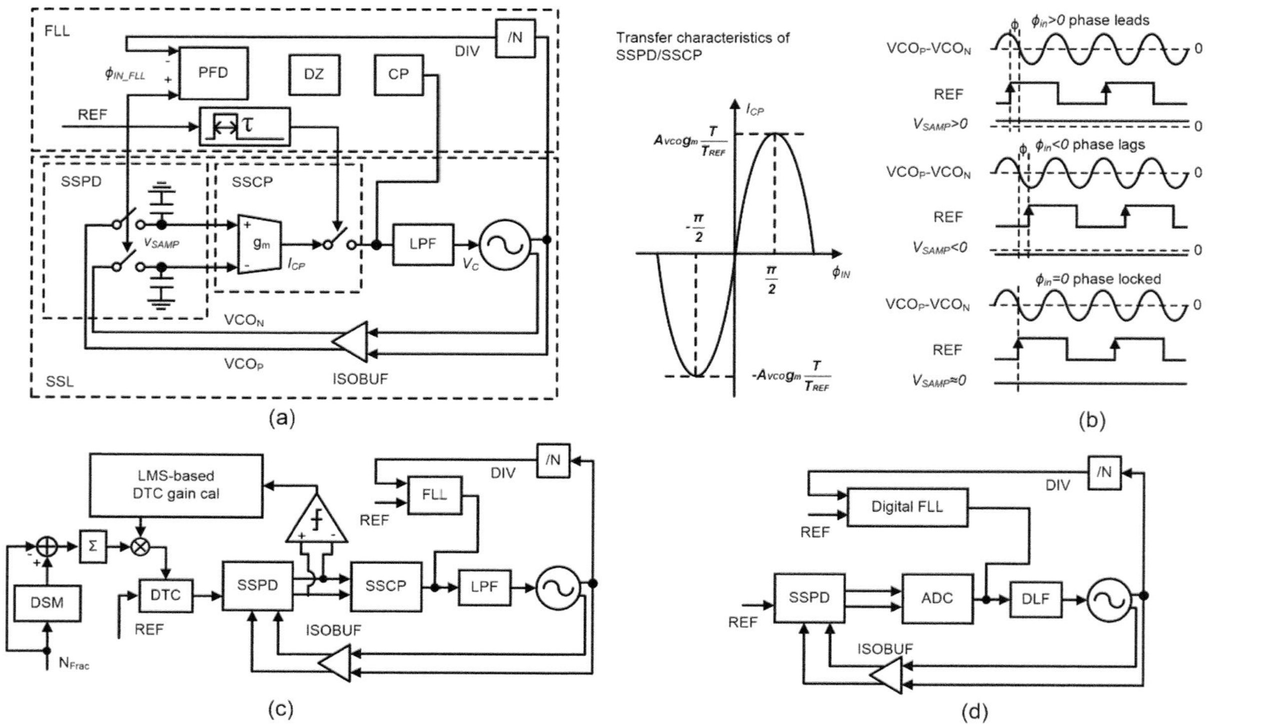

Figure 7.53 SSPLL (a) integer-*N* SSPLL, (b) transfer characteristics of SSPD/SSCP and timing diagram, (c) fractional-*N* SSPLL, and (d) digital SSPLL [100].

low in-band phase noise in comparison to CP-PLL is that the charge-pump noise is not multiplied by N^2 at the output of the PLL. As shown in Fig. 7.53, the SSPLL consists of a main sub-sampling loop and an auxiliary FLL. The FLL loop typically consists of a feedback divider, PFD, charge-pump and loop filter. In comparison to CP-PLLs, during the initial frequency acquisition phase, the FLL is active. As the VCO phase is closer to the reference phase and falls within the PFD dead-zone, the FLL is deactivated, and the phase acquisition starts. In the SSPD, the VCO outputs are sampled by the low-frequency reference. Based on the input phase error, the sampled output voltage is converted to current by sub-sampling charge-pump (SSCP) with a transconductance of gm to charge or discharge the loop filter.

The SSPD is an important block in the SSPLL. It only detects the phase. The conceptual timing diagram of the SSPD is shown in Fig. 7.54 [91]. The sampler samples the VCO output with the reference clock. When the VCO phase leads or lags the reference phase, the sampled voltage V_{sam} is different from V_{DC}. The VCO DC voltage is equivalent to the phase error between the VCO and reference phase. When the VCO and reference phase are aligned, V_{sam} is the same as V_{DC} and the frequency ratio between the VCO and reference is an integer equal to N.

The transfer characteristics and timing diagram of SSPD and SSCP are shown in Fig. 7.55 [91]. As shown in Fig. 7.55 [91], the SSPD directly samples the VCO frequency without any frequency down-scaling by feedback dividers in comparison to the CP-PLLs and maintains a high gain around the zero crossing of the VCO waveform. Also, since the VCO waveform is a sinusoidal signal, the acquisition range of SSPD is only $\pm\pi/2$ VCO phase. Due to the narrow acquisition range or capture range of the SSPD, only small disturbances in phase are tracked and eliminated by the SSPLL. The SSPLL may lose its lock due to any external interferences, sudden disturbances, and large frequency drift due to temperature variations, etc. To this end, as the phase error falls outside the range of the SSPLL, the FLL is automatically enabled to bring the VCO frequency close to the target frequency before phase locking by SSPLL.

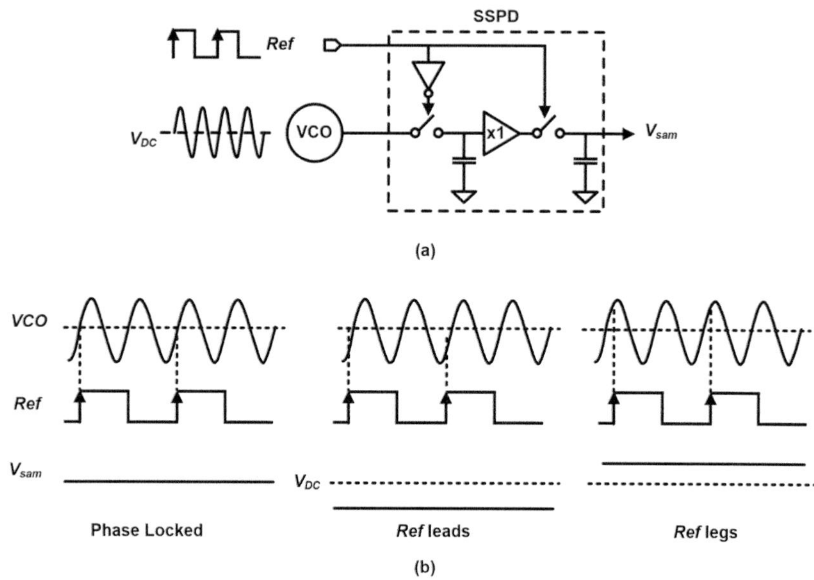

Figure 7.54 SSPD: (a) conceptual diagram and (b) timing diagram.

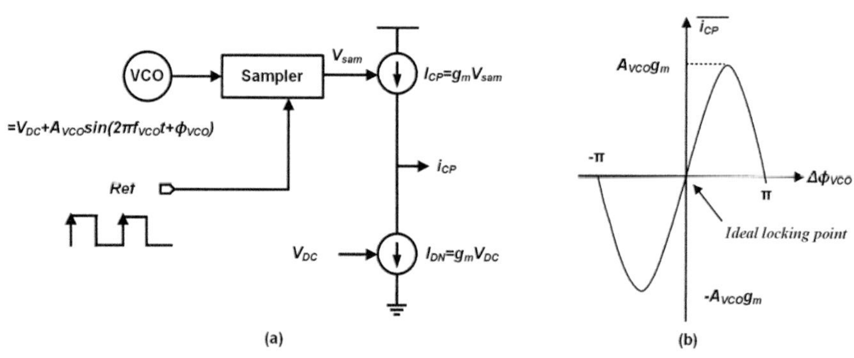

Figure 7.55 SSPD and SSCP: (a) conceptual diagram and (b) timing diagram.

The phase domain model of the SSPLL is shown in Fig. 7.56 [91]. Analogous to the CP-PLLs, the SSPLL can be modelled as a time continuous system because the PLL bandwidth is much less than the reference frequency [91]. In the phase domain model, the division ratio N is added after the reference frequency as the output frequency is virtually $N \times F_{\text{ref}}$ in the SSPLL. As can be seen in Fig. 7.56, due to the absence of the feedback divider in the SSPLL, the PFD and charge-pump noise is not multiplied by N^2. Therefore, the in-band phase noise of the SSPD to the PLL output can be expressed as [91]

$$L_{\text{SSPD,in-band}} = \frac{kT}{C_s A_{\text{VCO}}^2 f_{\text{ref}}}, \tag{7.43}$$

where C_s is the sampling capacitor and A_{VCO} is the amplitude of the VCO sine wave.

Though the SSPLL can achieve a very low in-band phase noise and low reference spur [101], the SSPLL requires an auxiliary FLL for frequency acquisition. Also, one of the important challenges for the SSPLL is to directly generate mmWave frequencies for the next-generation wireless communications. Due to the higher VCO frequencies employed at mmWave frequencies, the capture range of the SSPLL decreases considerably as the VCO frequency is directly sampled in SSPLL [102]. One solution is to employ a prescaler following the VCO as proposed in [103]. However, the in-band phase noise increases similar to CP-PLLs due to the use of prescaler in the feedback path.

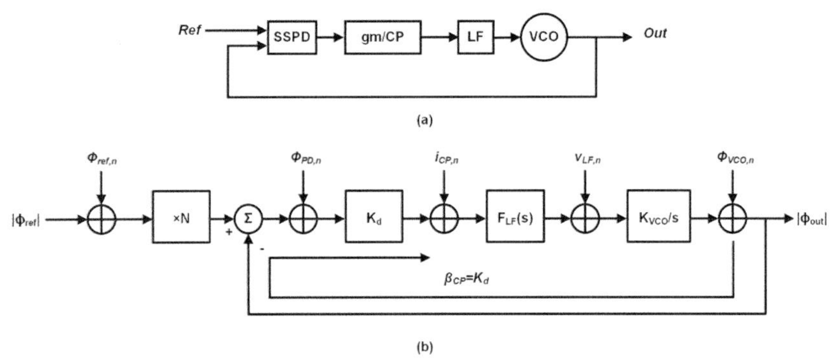

Figure 7.56 SSPLL: (a) architecture block diagram and (b) phase domain model.

7.5.1.1 Fractional-*N* SSPLL

More often, in the frequency synthesizer, it is necessary to generate output frequencies which are fractional multiples of reference frequency. Since SSPLL can only lock to the integer multiples of the reference frequency, [104] adopted a digital-to-time converter (DTC) to realize a fractional division in SSPLL by modulating the phase of the reference clock at each cycle and the block diagram is shown in Fig. 7.57.

The operation of fractional division in SSPLL is illustrated in Fig. 7.58. As shown in Fig. 7.58, we start by assuming that the VCO and reference are aligned in the first cycle. In the second cycle, when the timing error is 0.25 × T_{VCO}, the reference clock must be delayed by this time equivalent to the timing error. In the third and fourth cycles, this timing error increases to 0.5 × T_{VCO} and 0.75 × T_{VCO}, respectively, and the sampling edge needs to be delayed accordingly. However, in the fifth cycle, the sampling edge and VCO are aligned, and this cycle repeats periodically. Therefore, the delay required in each reference period is fixed and can be precisely calculated based on the reference frequency and the fractional frequency.

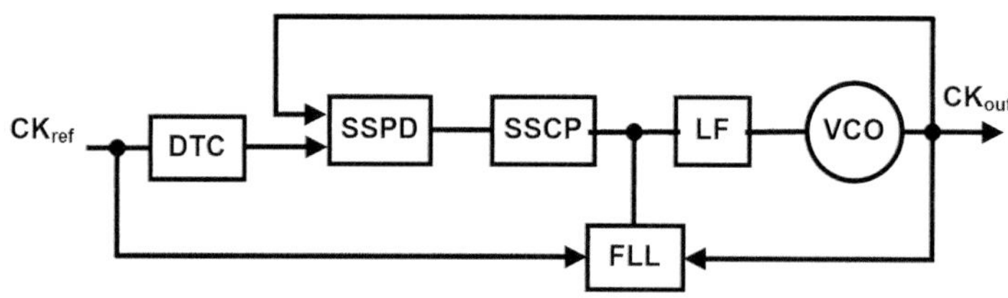

Figure 7.57 Block diagram of DTC-based fractional-N SSPLL.

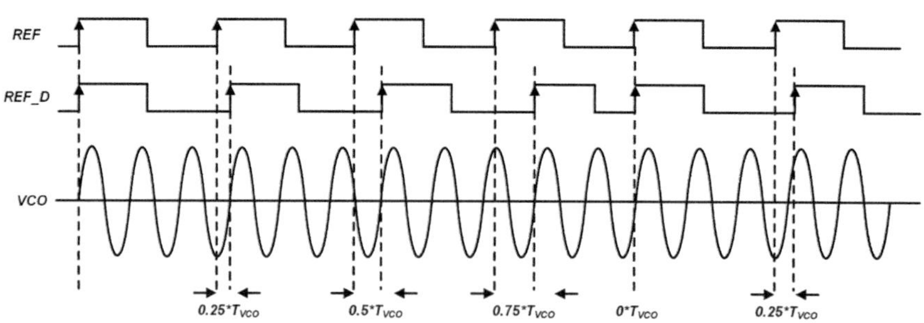

Figure 7.58 Principle of fractional-N SSPLL.

In the fractional SSPLL as proposed in [104], the DTC shifts the sampling edge of the reference clock to align it with the VCO zero crossing for phase locking. However, the required DTC tuning range has to be more than one VCO cycle with a fine resolution as it plays a major role in determining the in-band phase noise [104–106]. The jitter caused by DTC can also degrade the PLL in-band phase noise [107] and is proportional to the total delay required by the DTC [104]. In addition, DTC's non-idealities such as non-linearity, gain error and finite DTC quantization can generate fractional spurs in the PLL output [108]. Hence, several calibration schemes are proposed to overcome the DTC non-idealities [109–111].

In order to further increase the performance of fractional spurs in fractional-*N* SSPLL, a phase interpolator is proposed in [112, 113] and the block diagram is shown in Fig. 7.59. The conceptual fractional-*N* SSPLL based on DTC and the phase interpolator-assisted DTC are shown in Fig. 7.60. In the DTC-based fractional SSPLL, the DTC needs to have a minimum dynamic range of at least one VCO cycle for a specified resolution of ΔT. However, with the phase interpolator, the VCO cycle is split into 2^M equal phases. This greatly reduces the DTC dynamic range requirements by a factor of $1/2^M$ since the DTC needs to generate phases between two consecutive phases generated by the phase interpolator as shown in Fig. 7.60 [112]. The phase interpolator chooses the VCO phase that is closest to the reference phase and any residual phase error is resolved by DTC. The addition of a phase interpolator in the VCO feedback path reduces the DTC design effort.

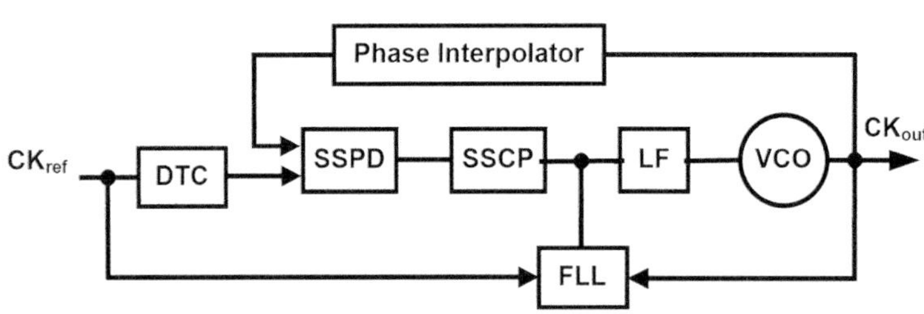

Figure 7.59 Block diagram of phase interpolator-based fractional-*N* SSPLL.

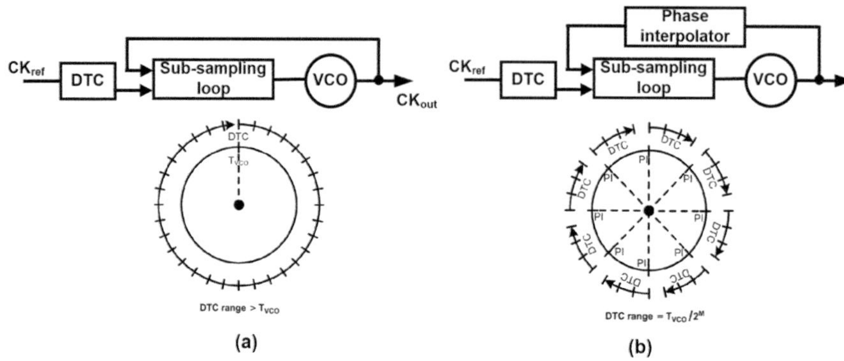

Figure 7.60 Fractional-*N* SSPLL: (a) conventional with DTC dynamic range, and (b) phase interpolator assisted with DTC dynamic range.

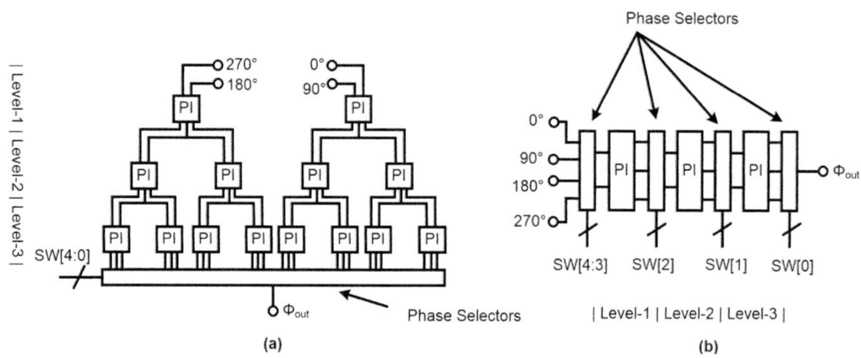

Figure 7.61 Phase interpolator architecture (a) tournament-type, and (b) pipelined.

Among the existing phase interpolator (PI) architectures, the tournament-type phase interpolator and pipelined phase interpolator are more popular and is shown in Fig. 7.61. As shown in Fig. 7.61(a), in the tournament PI, the individual slices are connected as a tree structure and finally a multiplexer is used to choose the required phase. The tournament-type PI exhibits very high linearity as each PI unit needs to synthesize the output phase exactly in the middle of the two input phases [107, 112]. However, since only one phase synthesized out of all the output phase is required in the tournament-type PI at any instant, the power consumption is large. This is addressed in the pipeline PI architecture as shown in Fig. 7.61(b). The output phase is chosen based on the control word and the number of individual PI units are reduced considerably to reduce the power consumption.

7.6 Conclusion

In this chapter, a brief overview of the frequency synthesizer is introduced. An understanding of the performance parameters of the frequency synthesizer is important to achieve the target specifications of the local oscillator and the transceiver. Next, the PLL dynamics are presented along with the analysis of noise and PLL loop parameters based on the open- and closed-loop transfer functions.

A detailed description of the various PLL building blocks, namely the divider, PFD, and charge-pump, is presented from the design standpoint. The chapter focuses on the various circuit level implementations with a brief description of the advantages and the disadvantages with respect to their topologies. Also, practical considerations for the PFD and the charge-pump, which are the major source of reference spur in the PLL, are discussed. Finally, sub-sampling PLLs, which achieve a less in-band noise contribution in comparison to charge-pump PLLs, are presented.

References

1. F. M. Gardner, *Phaselock Techniques*, 2nd ed., New York, Wiley and Sons, 1979.
2. D. H. Wolaver, *Phase-Locked Loop Circuit Design*, New Hersey, Prentice Hall, 1991.
3. W. F. Egan, *Frequency Synthesis by Phase Lock*, New York, Wiley and Sons, 2000.
4. V. Manassewitsch, *Frequency Synthesizers – Theory and Design*, Wiley-Interscience, pp. 283–284, 1987.
5. R. E. Best, *Phase Locked Loops: Design, Simulation, and Applications (Professional Engineering)*, 5th ed., McGraw-Hill Professional, 2003.
6. J. Craninckx and M. S. J. Steyaert, A fully integrated CMOS DCS-1800 frequency synthesizer, *IEEE Journal of Solid-State Circuits*, vol. 33, no. 12, pp. 2054-2065, Dec. 1998.

7. Y. Koo et al., A fully integrated CMOS frequency synthesizer with charge-averaging charge pump and dual-path loop filter for PCS- and cellular-CDMA wireless systems, *IEEE Journal of Solid-State Circuits*, vol. 37, no. 5, pp. 536–542, May 2002.
8. D. G. Wilson, W. Rhee, and B. S. Song, Integrated RF receiver front ends and frequency synthesizers for wireless, *Emerging Technologies: Designing Low Power Digital Systems*, pp. 369–396, Dec. 1996.
9. C. Kelly and S. Gupta, The digital phase-locked loop as a near-optimum FM demodulator, *IEEE Transactions on Communications*, vol. 20, no. 3, pp. 406–411, June 1972.
10. A. Sempel and H. van Nieuwenburg, A fully integrated HiFi PLL FM-demodulator, *IEEE International Solid-State Circuits Conference*, pp. 102–103, Feb. 1990.
11. B. Schlecker, M. Ortmanns, J. Anders, and G. Fantner, PLL-based high-speed demodulation of FM signals for real-time AFM applications, *IEEE International Symposium on Circuits and Systems*, pp. 197–200, Aug. 2013.
12. M. Rau, T. Oberst, R. Lares, A. Rothermel, R. Schweer, and N. Menoux, Clock/data recovery PLL using half-frequency clock, *IEEE Journal of Solid-State Circuits*, vol. 32, no. 7, pp. 1156–1159, Jul. 1997.
13. S. Y. Sun, An analog PLL-based clock and data recovery circuit with high input jitter tolerance, *IEEE Journal of Solid-State Circuits*, vol. 24, no. 2, pp. 325–330, Apr. 1989.
14. M. Wurzer, J. Bock, H. Knapp, W. Zirwas, F. Schumann, and A. Felder, A 40-Gb/s integrated clock and data recovery circuit in a 50 GHz f/sub T/silicon bipolar technology, *IEEE Journal of Solid-State Circuits*, vol. 34, no. 9, pp. 1320–1324, Sep. 1999.
15. A. Marques, M. Steyaert, and W. Sansen, Theory of PLL fractional-N frequency synthesizers, Wireless Networks, vol. 4, no. 1, pp. 79–85, Jan. 1998.
16. S. Goldman, *Phase-Locked Loop Engineering Handbook for Integrated Circuits*, Artech House, 2007.
17. B. Razavi, K. F. Lee, and R. H. Yan, Design of high-speed, low-power frequency dividers and phase-locked loops in deep submicron CMOS, *IEEE Journal of Solid-State Circuits*, vol. 30, no. 2, pp. 101–109, Feb. 1995.
18. F. Gardner, Charge-pump phase-lock loops, *IEEE Transactions on Communications*, vol. 28, no. 11, pp. 1849–1858, Nov. 1980.

19. K. Ma and K. S. Yeo, Low-Power Wireless Communication Circuits and Systems – 60 GHz and Beyond, 1st edition, Pan Stanford Publishing Pte. Ltd, 2018.

20. K. Shu and E. Sánchez-Sinencio, CMOS PLL Synthesizers: Analysis and Design, vol. 783: Springer Science and Business Media, 2006.

21. M. Nagarajan, Design of Low Power PLL Synthesizer for 60 GHz Applications, Ph.D Thesis, Nanyang Technological University, Singapore, 2016.

22. D. Abramovitch, Phase-locked loops: A control centric tutorial, *IEEE American Control Conference*, pp. 1–15, May 2002.

23. B. Razavi, *Monolithic Phase-Locked Loops and Clock Recovery Circuits: Theory and Designs*, Wiley, IEEE Press, 1996.

24. C. Lau and M. Perrott, Fractional-N frequency synthesizer design at the transfer function level using a direct closed loop realization algorithm, *IEEE Design Automation Conference*, pp. 526–531, Jun. 2003.

25. U. Singh and M. M. Green, High-frequency CML clock dividers in 0.13 μm CMOS operating up to 38 GHz, *IEEE Journal of Solid-State Circuits*, vol. 40, no. 8, pp. 1658–1661, Aug. 2005.

26. J. O. Plouchart, Jonghae Kim, V. Karam, R. Trzcinski, and J. Gross, Performance variations of a 66 GHz static CML divider in 90 nm CMOS, *IEEE International Solid-State Circuits Conference*, pp. 2142–2151, Feb. 2006.

27. A. Ghilioni, A. Mazzanti, and F. Svelto, Analysis and design of mm-wave frequency dividers based on dynamic latches with load modulation, *IEEE Journal of Solid-State Circuits*, vol. 48, no. 8, pp. 1842–1850, Aug. 2013.

28. H. M. Cheema, R. Mahmoudi, M. A. T. Sanduleanu, and A. van Roermund, A Ka Band, static, MCML frequency divider, in standard 90 nm-CMOS LP for 60 GHz applications, *IEEE Radio Frequency Integrated Circuits Symposium*, pp. 541–544, Jun. 2007.

29. H. Knapp, H. D. Wohlmuth, M. Wurzer, and M. Rest, 25 GHz static frequency divider and 25 Gb/s multiplexer in 0.12 μm CMOS, *IEEE International Solid-State Circuits Conference*, pp. 302–303, Feb. 2002.

30. J. O. Plouchart et al., A power-efficient 33 GHz 2: 1 static frequency divider in 0.12 μm SOI CMOS, *IEEE Radio Frequency Integrated Circuits Symposium*, pp. 329–332, Jun. 2003.

31. G. von Buren, C. Kromer, F. Ellinger, A. Huber, M. Schmatz, and H. Jackel, A combined dynamic and static frequency divider for a 40 GHz

PLL in 80 nm CMOS, *IEEE International Solid-State Circuits Conference*, pp. 246–247, Feb. 2006.

32. L. Bangli, C. Dianyong, W. Bo, C. Dezhong, and T. Kwasniewski, A 43 GHz static frequency divider in 0.13 μm standard CMOS, *Canadian Conference on Electrical and Computer Engineering*, pp. 111–114, May 2008.
33. C. Jung-Yu and L. Shen-Iuan, A 4–54 GHz static frequency divider with back-gate coupling, *International Symposium on VLSI Design, Automation and Test*, pp. 1–4, Apr. 2007.
34. H. Knapp et al., A 79 GHz dynamic frequency divider in SiGe bipolar technology, in *IEEE International Solid-State Circuits Conference*, pp. 208–209, Feb. 2000.
35. M. Kurisu et al., A Si bipolar 28 GHz dynamic frequency divider, *IEEE Journal of Solid-State Circuits*, vol. 27, no. 12, pp. 1799–1804, Dec. 1992.
36. R. H. Derksen and H. M. Rein, 7.3 GHz dynamic frequency dividers monolithically integrated in a standard bipolar technology, *IEEE Transactions on Microwave Theory and Techniques*, vol. 36, no. 3, pp. 537–541, Mar. 1988.
37. J. Lee and B. Razavi, A 40 GHz frequency divider in 0.18 μm CMOS technology, *IEEE Journal of Solid-State Circuits*, vol. 39, no. 4, pp. 594–601, Apr. 2004.
38. H. Knapp et al., 86 GHz static and 110 GHz dynamic frequency dividers in SiGe bipolar technology, *IEEE International Microwave Symposium*, pp. 1067–1070, Jun. 2003.
39. L. Tang Nian, B. Shuen Yin, and Y. J. E. Chen, A 60 GHz, 0.13 μm CMOS divide-by-three frequency divider, *IEEE Transactions on Microwave Theory and Techniques*, vol. 56, no. 11, pp. 2409–2415, Nov. 2008.
40. B. Razavi, A study of injection locking and pulling in oscillators, *IEEE Journal of Solid-State Circuits*, vol. 39, no. 9, pp. 1415–1424, Sep. 2004.
41. N. Mahalingam, K. Ma, K. S. Yeo, and W. M. Lim, Coupled dual LC tanks based ILFD with low injection power and compact size, *IEEE Microwave and Wireless Components Letters*, vol. 24, no. 2, pp. 105–107, Feb. 2014.
42. N. Mahalingam, K. Ma, K. S. Yeo, and W. M. Lim, Area efficient inter-coupled differential injection enhancement wide locking range injection locked frequency divider, *IEEE URSI General Assembly and Scientific Symposium*, pp. 1–4, Aug. 2014.

43. N. Mahalingam, K. Ma, K. S. Yeo, and W. M. Lim, Modified inductive peaking direct injection ILFD with multi-coupled coils, *IEEE Microwave and Wireless Components Letters*, vol. 25, no. 6, pp. 379–381, Jun. 2015.

44. N. Mahalingam, K. Ma, K. S. Yeo, and W. L. Yang, Coupled inductive peaking injection locked frequency divide-by-2 with wide locking range and compact size, *IEEE International SOC Design Conference*, pp. 199–200, Nov. 2015.

45. N. Mahalingam, K. Ma, and K. S. Yeo, A multi-mode multi-coil coupled tuned inductive peaking ILFD for low injected power with compact size, *IEEE Access*, vol. 07, pp. 59059–59068, May 2019.

46. N. Mahalingam, K. Ma, and K. S. Yeo, A multi-mode compact size multi-coil tuned inductive peaking ILFD for low injected power level, *IEEE International Microwave Symposium*, pp. 603–605, Jun. 2018.

47. N. Mahalingam, K. S. Yeo, and B. K. Thangarasu, A wide locking range harmonic enhanced injection locked frequency divide-by-4 with low injected power level, *IEEE International Wireless Symposium*, pp. 1–3, May 2018.

48. S. L. Jang, Y. S. Chen, C. W. Chang, and C. C. Liu, A wide-locking range 3 injection-locked frequency divider using linear mixer, *IEEE Microwave and Wireless Components Letters*, vol. 20, no. 7, pp. 390–392, Jul. 2010.

49. S. L. Jang, T. C. Kung, and C. W. Hsue, Wide-locking range divide-by-4 injection-locked frequency divider using linear mixer approach, *IEEE Microwave and Wireless Components Letters*, vol. 27, no. 4, pp. 398–400, Apr. 2017.

50. S. L. Jang, H. W. Lai, and J. Y. Sung, Current-reused divide-by-16 injection-locked frequency divider, *IEEE Microwave and Wireless Components Letters*, vol. 32, no. 5, pp. 426–429, May 2022.

51. L. Wang, Y. Z. Xiong, S. M. Hu, and T. G. Lim, A 0.13 μm HBT divide-by-6 injection-locked frequency divider, *IEEE Asian Solid-State Circuits Conference*, pp. 97–100, Nov. 2011.

52. A. Buonomo, A. Lo Schiavo, M. A. Awan, M. S. Asghar, and M. P. Kennedy, A CMOS injection-locked frequency divider optimized for divide-by-two and divide-by-three operation, *IEEE Transactions on Circuits and Systems I: Regular Papers*, vol. 60, no. 12, pp. 3126–3135, Dec. 2013.

53. J. W. Wu, C. C. Chen, H. W. Kao, J. K. Chen, and M. C. Tu, Divide-by-three injection-locked frequency divider combined with divide-by-two

locking, *IEEE Microwave and Wireless Components Letters*, vol. 23, no. 11, pp. 590–592, Nov. 2013.

54. T. Siriburanon, W. Deng, A. Musa, K. Okada, and A. Matsuzawa, A 13.2% locking-range divide-by-6, 3.1 mW, ILFD using even-harmonic-enhanced direct injection technique for millimeter-wave PLLs, *IEEE European Solid-State Circuits Conference*, pp. 403–406, Oct. 2013.
55. S. M. Kang and Y. Leblebici, *CMOS Digital Integrated Circuits: Analysis and Design*, McGraw-Hill Education, 2002.
56. J. Navarro Soares and W. A. M. Van Noije, A 1.6 GHz dual modulus prescaler using the extended true-single-phase-clock CMOS circuit technique (E-TSPC), *IEEE Journal of Solid-State Circuits*, vol. 34, no. 1, pp. 97–102, Jan. 1999.
57. J. Yuan and C. Svensson, High speed CMOS circuit technique, *IEEE Journal of Solid-State Circuits*, vol. 24, no. 2, pp. 62–70, February 1989.
58. K. S. Yeo, M. A. Do, and C. C. Boon, *Design of CMOS RF Integrated Circuits and Systems*, World Scientific Publishing Co. Pte. Ltd., 2010.
59. W. Zhu et al., A 5.8 GHz wideband TSPC divide-by-16/17 dual modulus prescaler, *IEEE Transactions on Very Large Scale Integration Systems*, vol. 23, no. 1, pp. 194–197, Jan. 2015.
60. M. Jung, J. Fuhrmann, A. Ferizi, G. Fischer, R. Weigel, and T. Ussmueller, Design of a 12 GHz low-power extended true single-phase clock (E-TSPC) prescaler in 0.13 μm CMOS technology, *IEEE Asia-Pacific Microwave Conference*, pp. 1238–1241, Dec. 2011.
61. W. H. Chen and B. Jung, High-speed low-power true single-phase clock dual-modulus prescalers, *IEEE Transactions on Circuits and Systems II: Express Briefs*, vol. 58, no. 3, pp. 144–148, Mar. 2011.
62. S. Jia, S. Yan, Y. Wang, and G. Zhang, Design of low-power high-speed divide-by-2/3 prescalers with improved true single-phase clock scheme, *IEEE Asia-Pacific Microwave Conference*, pp. 241–243, Nov. 2014.
63. C. S. Vaucher, I. Ferencic, M. Locher, S. Sedvallson, U. Voegeli, and Z. Wang, A family of low-power truly modular programmable dividers in standard 0.35 μm CMOS technology, *IEEE Journal of Solid-State Circuits*, vol. 35, no. 7, pp. 1039–1045, Jul. 2000.
64. J. C. Candy and G. C. Temes, *Oversampling Delta-Sigma Data Converters: Theory, Design, and Simulation*, IEEE Press, New York, 1992.
65. H. Sorensen, P. Aziz, and J. V. del Spiegel, An overview of sigma-delta converter. How a 1-bit ADC achieves more than 16-bit resolution, *IEEE Signal Processing Magazine*, pp. 61–84, Jan. 1996.

66. R. Schreier and Gabor C. Temes, *Understanding Delta-Sigma Data Converters*, 1st ed., IEEE Press, New York, 2004.

67. Y. Matsuya, K. Uchimura, A. Iwata, T. Kobayashi, M. Ishikawa, and T. Yoshitome, A 16-bit oversampling A-to-D conversion technology using triple-integration noise shaping, *IEEE Journal of Solid-State Circuits*, vol. 22, no. 6, pp. 921–929, Dec. 1987.

68. B. D. Muer and M. S. J. Steyaert, A CMOS monolithic ΣΔ-controlled fractional-N frequency synthesizer for DCS-1800, *IEEE Journal of Solid-State Circuits*, vol. 37, no. 7, pp. 835–844, Jul. 2002.

69. B. Miller and R. J. Conley, A multiple modulator fractional divider, *IEEE Transactions on Instrumentation and Measurement*, vol. 40, no. 3, pp. 578–583, Jun. 1991.

70. S. Kim, M. Keel, K. Lee, and S. Kim, CMOS delta-sigma frequency synthesizer with a new frequency divider and a simplified MASH structure, *Journal of the Korean Physical Society*, vol. 41, no. 6, pp. 967–973, Dec. 2002.

71. N. Mahalingam, Y. Wang, K. Ma, B. K. Thangarasu, and K. S. Yeo, A 30 GHz power efficient phase-locked loop frequency synthesizer for 60 GHz applications, *IEEE Transactions on Microwave Theory and Techniques*, vol. 65, no. 11, pp. 4165–4175, Nov. 2017.

72. Y. Wang, K. Ma, M. Nagarajan, and K. Seng Yeo, Delta-sigma fractional-N frequency divider for phase lock loop in 60 GHz, *IEEE International Conference on Electron Devices and Solid-State Circuits*, pp 1–2, Nov. 2011.

73. K. Shu and E. Sánchez-Sinencio, *CMOS PLL Synthesizers: Analysis and Design*, Springer Science and Business Media, 2006.

74. C. F. Liang, S. H. Chen, and S. I. Liu, A digital calibration technique for charge pumps in phase-locked systems, *IEEE Journal of Solid-State Circuits*, vol. 43, no. 2, pp. 390–398, Feb. 2008.

75. R. Woogeun, Design of high-performance CMOS charge pumps in phase-locked loops, *IEEE International Symposium on Circuits and Systems*, pp. 545–548, Jul. 1999.

76. T. H. Lee, *The Design of CMOS Radio-Frequency Integrated Circuits*, 2nd ed., Cambridge: Cambridge University Press, 2003.

77. B. Razavi, *Design of Analog CMOS Integrated Circuits*, 2nd ed., McGraw-Hill, 2003.

78. C. T. Charles and D. J. Allstot, A buffered charge pump with zero charge sharing, 2008 *IEEE International Symposium on Circuits and Systems*, pp. 2633–2636, Jun. 2008.

79. C. Zhang, T. Au, and M. Syrzycki, A high performance NMOS-switch high swing cascode charge pump for phase-locked loops, *IEEE International Midwest Symposium on Circuits and Systems*, pp. 554–557, Aug. 2012.

80. V. Manassewitsch, *Frequency Synthesizers: Theory and Design*, 3rd ed., Wiley-Interscience, 2005.

81. S. W. Hsiao, N. Tzou, and A. Chatterjee, A programmable BIST design for PLL static phase offset estimation and clock duty cycle detection, *IEEE VLSI Test Symposium*, pp. 1–6, Apr. 2013.

82. J. Choi, W. Kim, and K. Lim, A spur suppression technique using an edge-interpolator for a charge-pump PLL, *IEEE Transactions on Very Large Scale Integration Systems*, vol. 20, no. 5, pp. 969–973, May 2012.

83. T. C. Lee and W. L. Lee, A spur suppression technique for phase-locked frequency synthesizers, *IEEE International Solid State Circuits Conference*, pp. 2432–2441, Feb. 2006.

84. M. Elsayed, M. Abdul-Latif, and E. Sánchez-Sinencio, A spur-frequency-boosting PLL with a −74 dBc reference-spur rejection in 90 nm digital CMOS, *2011 IEEE Radio Frequency Integrated Circuits Symposium*, pp. 1–4, July 2011.

85. C. F. Liang, H. H. Chen, and S. I. Liu, Spur-suppression techniques for frequency synthesizers, *IEEE Transactions on Circuits and Systems II: Express Briefs*, vol. 54, no. 8, pp. 653–657, Aug. 2007.

86. P. Y. Hsieh, S. Y. Shu, and C. Y. Yang, A spur-suppression technique for frequency synthesizer with pulse-width to current conversion, *International Symposium on Intelligent Signal Processing and Communication Systems*, pp. 1–2, Dec. 2019.

87. G. J. Huang, C. S. Chen, W. S. Wuen, and K. A. Wen, A fast settling and low reference spur PLL with double sampling phase detector, *IEEE International Conference on Electronics, Circuits and Systems*, pp. 316–319, Nov. 2008.

88. C. T. Charles and D. J. Allstot, A calibrated phase/frequency detector for reference spur reduction in charge-pump PLLs, *IEEE Transactions on Circuits and Systems II: Express Briefs*, vol. 53, no. 9, pp. 822–826, Sep. 2006.

89. C. F. Liang, S. H. Chen, and S. I. Liu, A digital calibration technique for charge pumps in phase-locked systems, *IEEE Journal of Solid-State Circuits*, vol. 43, no. 1, pp. 390–398, Jan. 2008.

90. C. H. Chan, et al., Trending IC design directions in 2022, *Journal of Semiconductors*, vo. 43, no. 7, pp. 1674–1684, 2022. online: https://doi.org/10.1088/1674-4926/43/7/071401.

91. X. Gao, E. A. M. Klumperink, M. Bohsali, and B. Nauta, A low noise sub-sampling PLL in which divider noise is eliminated and PD/CP noise is not multiplied by N^2, *IEEE Journal of Solid-State Circuits*, vol. 44, no. 12, pp. 3253–3263, Dec. 2009.

92. J. Gong, E. Charbon, F. Sebastiano, and M. Babaie, A low-jitter and low-spur charge-sampling PLL, *IEEE Journal of Solid-State Circuits*, vol. 57, no. 2, pp. 492–504, Feb. 2022.

93. Z. Yang, Y. Chen, S. Yang, P. I. Mak, and R. P. Martins, 16.8 A 25.4-to-29.5 GHz 10.2 mW isolated sub-sampling PLL achieving –252.9 dB jitter-power FoM and –63 dBc reference spur, *IEEE International Solid-State Circuits Conference*, pp. 270–272, Feb. 2019.

94. Z. Zhang, G. Zhu, and C. Patrick Yue, A 0.65 V 12-16 GHz sub-sampling PLL with 56.4 fsrms integrated jitter and –256.4 dB FoM, *IEEE Journal of Solid-State Circuits*, vol. 55, no. 6, pp. 1665–1683, Jun. 2020.

95. D. G. Lee and P. P. Mercier, A sub-mW 2.4 GHz active-mixer-adopted sub-sampling PLL achieving an FoM of –256 dB, *IEEE Journal of Solid-State Circuits*, vol. 55, no. 6, pp. 1542–1552, Jun. 2020.

96. J. H. Seol, K. Choo, D. Blaauw, D. Sylvester, and T. Jang, Reference oversampling PLL achieving –256 dB FoM and –78 dBc reference spur, *IEEE Journal of Solid-State Circuits*, vol. 56, no. 10, pp. 2993–3007, Oct. 2021.

97. X. Yang, C. H. Chan, Y. Zhu, and R. P. Martins, A calibration-free ring-oscillator PLL with gain tracking achieving 9% jitter variation over PVT, *IEEE Transactions on Circuits and Systems I: Regular Papers*, vol. 67, no. 11, pp. 3753–3763, Nov. 2020.

98. W. Wu et al., A 14 nm ultra-low jitter fractional-N PLL using a DTC range reduction technique and a reconfigurable dual-core VCO, *IEEE Journal of Solid-State Circuits*, vol. 56, no. 12, pp. 3756–3767, Dec. 2021.

99. M. Mercandelli et al., A 12.5 GHz fractional-N type-I sampling PLL achieving 58 fs integrated jitter, *IEEE Journal of Solid-State Circuits*, vol. 57, no. 2, pp. 505–517, Feb. 2022.

100. Z. Zhang, CMOS analog and mixed-signal phase-locked loops: An overview, *Journal of Semiconductors*, vol. 41, no. 11, 2020. online: http://doi.org/10.1088/1674-4926/41/11/111402.

101. X. Gao, E. A. M. Klumperink, G. Socci, M. Bohsali, and B. Nauta, Spur reduction techniques for phase-locked loops exploiting a sub-sampling phase detector, *IEEE Journal of Solid-State Circuits*, vol. 45, no. 9, pp. 1809–1821, Sep. 2010.

102. X. Gao, Sub-sampling PLL for millimeter-wave applications: An overview, *IEEE MTT-S International Microwave Conference on Hardware and Systems for 5G and Beyond (IMC-5G)*, Aug. 2019, pp. 1–5.

103. V. Szortyka, Q. Shi, K. Raczkowski, B. Parvais, M. Kuijk, and P. Wambacq, 21.4 A 42 mW 230 fs-jitter sub-sampling 60 GHz PLL in 40 nm CMOS, *IEEE International Solid-State Circuits Conference*, Feb. 2014, pp. 366–367.

104. K. Raczkowski, N. Markulic, B. Hershberg, and J. Craninckx, A 9.2–12.7 GHz wideband fractional-N subsampling PLL in 28 nm CMOS with 280 fs RMS jitter, *IEEE Journal of Solid-State Circuits*, vol. 50, no. 5, pp. 1203–1213, May 2015.

105. W. S. Chang, P. C. Huang, and T. C. Lee, A fractional-N divider-less phase-locked loop with a subsampling phase detector, *IEEE Journal of Solid-State Circuits*, vol. 49, no. 12, pp. 2964–2975, Dec. 2014.

106. S. Levantino, G. Marucci, G. Marzin, A. Fenaroli, C. Samori, and A. L. Lacaita, A 1.7 GHz fractional-N frequency synthesizer based on a multiplying delay-locked loop, *IEEE Journal of Solid-State Circuits*, vol. 50, no. 11, pp. 2678–2691, Nov. 2015.

107. Y. T. Chen, P. J. Peng, and H. W. Lin, A 12–14.5 GHz 10.2 mW –249 dB FoM fractional-N subsampling PLL with a high-linearity phase interpolator in 40 nm CMOS, *IEEE Transactions on Very Large Scale Integration Systems*, vol. 30, no. 5, pp. 634–643, May 2022.

108. N. Markulic, K. Raczkowski, P. Wambacq, and J. Craninckx, A 10-bit, 550 fs step digital-to-time converter in 28 nm CMOS, *IEEE European Solid State Circuits Conference*, pp. 79–82, Sep. 2014.

109. S. Levantino, G. Marzin, and C. Samori, An adaptive pre-distortion technique to mitigate the DTC nonlinearity in digital PLLs, *IEEE Journal of Solid-State Circuits*, vol. 49, no. 8, pp. 1762–1772, Aug. 2014.

110. N. Pavlovic and J. Bergervoet, A 5.3 GHz digital-to-time-converter based fractional-N all-digital PLL, *IEEE International Solid-State Circuits Conference*, pp. 54–56, Feb. 2011.

111. R. B. Staszewski and P. T. Balsara, *All-Digital Frequency Synthesizer in Deep-Submicron CMOS*, New York, NY, USA: Wiley-Interscience, 2006.

112. A. T. Narayanan et al., A fractional-N sub-sampling PLL using a pipelined phase-interpolator with an FoM of −250 dB, *IEEE Journal of Solid-State Circuits*, vol. 51, no. 7, pp. 1630–1640, Jul. 2016.

113. D. Liao, F. F. Dai, B. Nauta, and E. A. M. Klumperink, A 2.4 GHz 16-phase sub-sampling fractional-N PLL with robust soft loop switching, *IEEE Journal of Solid-State Circuits*, vol. 53, no. 3, pp. 715–727, Mar. 2018.

Water without touching, cannot know its depth.
The road not taken, cannot determine the outcome.
A battle not fought, cannot decide the winner.
A person who does not socialize, cannot read people's true character.
A heart without virtue, cannot understand morality.

水不试，难分深浅。
路不走，难分远近。
战不打，难分强弱。
人不交，难分好坏。
心不修，难分善恶。

—Kiat Seng Yeo

Index

For Product Safety Concerns and Information please contact our EU representative GPSR@taylorandfrancis.com Taylor & Francis Verlag GmbH, Kaufingerstraße 24, 80331 München, Germany